Applied
Finite
Mathematics

Applied
Finite
Mathematics

Robert F. Brown
University of California, Los Angeles

Brenda W. Brown

Wadsworth
Publishing Company, Inc.
Belmont, California

To Geoff and Matthew

Mathematics Editor: Don Dellen

Production: Phyllis Niklas

Design: Dare Porter

Technical Illustrations: AYXA Art

Printed in the United States of America

1 2 3 4 5 6 7 8 9 10—81 80 79 78 77

Library of Congress Cataloging in Publication Data
Brown, Robert F   Date
    Applied finite mathematics.

    Includes bibliographies and index.
    1.   Mathematics—1961—
I.   Brown, Brenda W., joint author.
II.   Title.
QA39.2.B78   519   76—27695
ISBN 0—534—00499—7

Contents

Preface

Students taking a finite mathematics course bring with them a variety of backgrounds and interests. In our approach to the subject we seek to respond to this diversity. Our book is an attempt to communicate substantial mathematical content to finite mathematics students in a way that will hold their interest and convince them that finite mathematics is pertinent to their other academic and professional concerns.

To accomplish this goal, we have given each chapter a social science or business theme, an application, that we introduce in essays. The introductory essay in each chapter sets out a problem. After presenting the mathematical content of the chapter, we use some of the mathematics to obtain information that contributes to a solution to the problem. In Chapters 2, 3, and 7, we carry out the application to the

problem in a mathematical section. In Chapters 4–6, we describe the application to the problem of the introductory essay in a technical essay—expository writing that assumes the reader is familiar with the mathematical content of the chapter. Every chapter ends with a concluding essay that briefly mentions some other areas to which the mathematics of the chapter can be applied.

All the mathematical sections are completely independent of the essays. Consequently, the role the essays play can range from that of optional background reading to a more prominent position in the course.

We have tried to make our presentation of the mathematics relatively informal. The narrative style introduces and explains mathematical abstractions through examples. We believe that it is not sufficient in presenting mathematics just to state facts and explain techniques. On the other hand, we do not think that mathematical statements must necessarily be supported by formal proofs. Rather, we consider it our responsibility to make the mathematics in this book seem reasonable to the student, and we have employed a variety of means to accomplish this goal. However, in this type of presentation, the mathematical material does not always stand out as clearly as in more formal mathematical writing. Therefore, we have included in each section a brief summary of its mathematical content.

The book has been made as self-contained as possible by including a substantial amount of review material. Few students will need all of it, but when one lacks some information, we think the instructor would prefer to send the student to the text rather than to find a reference elsewhere or give a review lesson.

The contents of the book include the topics that are customary for finite mathematics. We have, however, integrated the material on sets into the chapter on probability instead of devoting a separate chapter to the subject. There is a short chapter on computers at the end of the book, because we believe that any book that is about applied mathematics should recognize the importance of computers.

So many people were generous with their help and advice while this book was being written, it is impractical to list them here. We must, nevertheless, express special thanks to David Cusick, whose detailed reading of the manuscript and constructive suggestions were of great value in its preparation. We appreciate the encouragement and support we received from Don Dellen and his thoroughly professional staff at Wadsworth Publishing Company.

Chapter 1
Applied
Mathematics

**Essay on
Applied
Mathematics**

By its title, this book claims to be about applied mathematics, which means mathematics put to practical use. Such a definition is vague enough to be acceptable to most mathematicians, but it is not precise enough to be of much use in discussing the subject. However, trying to arrive at a more specific definition of applied mathematics presents difficulties. For, in making the attempt we find ourselves in the middle of a long-lived, wide-ranging, and, at times, acrimonious debate along philosophical as well as technical lines on the delicate distinction between "pure" and "applied" mathematics.

Because philosophical debates are not the concern of this essay, we have adopted the definitions proposed

1

by a group of distinguished mathematicians in 1968. This group, the Committee on Support of Research in the Mathematical Sciences (COSRIMS) of the National Academy of Sciences, classified as applied mathematics the four major areas in the mathematical sciences that have particular direct and important relationships with other sciences and technologies: computer science, operations research, statistics, and physical mathematics (classical applied mathematics). Rather than call the rest of mathematics "pure," the committee called the traditional disciplines of logic, number theory, algebra, geometry, and analysis the "central core of mathematics." The separation between applied mathematics and the core of mathematics must not, however, be thought of as fixed. Applications may draw upon any part of core mathematics from the simple arithmetic that is used in everyday life to the sophisticated modern mathematics employed in physics and astronomy. In the other direction, many of the essential ideas and concepts of the core area can be traced to problems arising outside mathematics.

Throughout history the growth and development of mathematics has been stimulated by human needs to understand, describe, and explain the physical universe. Even geometry, developed as an abstract discipline by the Greeks, originated in practical problems of land measurement. However, with the exception of the Greek study of geometry, for centuries mathematics remained one of the practical arts connected with the necessities of trade, surveying, navigation, and astronomy. From the Renaissance to the present time the growth of mathematics has been intimately connected with the natural sciences, and particularly with the physical sciences.

In the seventeenth century, Newton and Leibniz, working independently, discovered the calculus. This discovery is regarded by many mathematicians as the beginning of modern mathematics. One reason the calculus is so important is that it is a mathematical method for describing physical motion. Consequently, throughout the eighteenth century many mathematicians concentrated on the calculus and its application to classical physics (mechanics). However, as Dirk J. Struik says in *A Concise History of Mathematics* (see References), in the nineteenth century new mathematical research gradually emancipated itself from the ancient tendency to see in mechanics and astronomy the final goal of the exact sciences. In fact, by 1870 mathematics had grown into an enormous and unwieldy structure divided into a large number of fields in which only specialists knew the way.

This proliferation of mathematics, which was noticeable in the nineteenth century, became even more pronounced in the twentieth. As mathematical knowledge and technique broadened in scope, it is not surprising that the range of subjects to which mathematics could be applied also increased far beyond the traditional concentration on the physical sciences. In particular, since World War II one of the major trends in mathematics has been the increase of mathematical applications in the biological, behavioral, and social sciences. For example, Richard Stone, writing in 1964, said, "Seventy-five years ago the American economist Irving Fisher stated that the entire world's literature contained scarcely 50 worthwhile books and articles on mathematical economics. Today the situation is different, not only in economics but also in all the other social sciences: each year sees thousands of additions to a mountain of mathematical literature."*

John G. Kemeny said, in his essay for COSRIMS (see References), that the difficulty in applying mathematics to the social sciences

lies in the fact that the social sciences are more complex than the physical sciences. The behavior of a committee of human beings is vastly more complex than the orbits of a planetary system. This accounts both for the late development of the social sciences and the difficulty in employing mathematical methods. . . . Perhaps the most interesting question for the

mathematician is whether present-day mathematics is adequate for coping with so complex a subject matter. If not, then the social sciences may serve as a source of inspiration for new branches of mathematics as the physical sciences have served in the past.

But what does this imply for the social scientist? How can the economist, the sociologist, the psychologist, the political scientist, and even the historian take advantage of this new direction in mathematics? The major problem is communication. To improve the possibilities for communication the social scientist must learn more about mathematics and mathematicians must become more aware of the kinds of problems the social scientists are likely to bring them.

General comments on the usefulness of mathematics are all very well, but the question now is, how is mathematics applied to a problem in the social sciences? The basic method is known as *building a mathematical model.* Daniel P. Maki and Maynard Thompson, in *Mathematical Models and Applications* (see References), describe the process in five steps.

Mathematical modeling begins with a problem in the real world. The problem may be how to devise a sampling method that will accurately predict an election, how to plan the use of farm acreage for maximum profit, how to determine whether a product should

*R. Stone, "Mathematics in the Social Sciences," in M. Kline, ed., *Mathematics in the Modern World* (San Francisco: W. H. Freeman, 1968), p. 284.

be marketed or discontinued, how to understand the economy of a country for effective economic planning, or any of thousands of others from real-life situations. The first step, the recognition and description of the problem, must be done by a person who knows the discipline from which the problem comes.

The second step is the effort to make the problem precise. Here the aim is to get rid of unessential information and to simplify what is left. For example, if a psychologist has designed a learning experiment using monkeys and an obstacle course, the number of obstacles to be climbed over compared to the number crawled through may make no difference to the final results, while the lighting of the course or the age and sex differences of the monkeys may be significant. After making a careful analysis of the problem, called by Maki and Thompson *forming the real model,* the researcher proceeds to the third step.

The goal of this next stage is to express the entire situation in symbolic terms. One must look at the real model and attempt to identify the operative processes. It is here that the real model becomes a mathematical model. This stage is the most important and most difficult, because if the correspondence set up between the real world and the mathematical world is incorrect, the results will not be very useful. Furthermore, there is not always one best mathematical model for a certain problem; in fact, often there are several different models that may be helpful in different ways.

The mathematical system that emerges from the third step is then studied using appropriate mathematical ideas and techniques. The results of this study are generally expressed in mathematical statements that are the basis for conclusions by the originator of the problem. Sometimes new mathematics results from this stage, but this is viewed only as a beneficial side effect. The main goal is to produce new information about the problem. From this stage of the study, the important contribution is often the recognition of the relationship between known mathematical results and the situation being analyzed.

The final step is the comparison of predicted results with the actual real-world event. The best of all possible solutions would be if everything actually observed were accounted for by the mathematical model. But writing about the process of mathematical modeling is much, much easier than doing it. The description of the process makes it seem more clear-cut than it actually is. In practice, problems in the social sciences, the behavioral sciences, and even the biological sciences often involve factors that are extremely difficult to quantify. In dealing with these problems, essential elements may be lost in the transition from the real to the mathematical model.

In one sense we have all had some experience with mathematical modeling. Solving story problems involves part of the process. So, in elementary school, when we had to decide whether to add or subtract to get the answer to such questions as: "The second-grade class made 25 paper flowers for the fiesta and the third-grade class made 37. How many more did the third-grade class make than the second-grade class?" we were, in a very limited way, translating a real problem into a mathematical one.

However, successful model building at a professional level requires a wider range of talents from its practitioners. An understanding of the real-life problem area and an appropriate background in mathematics are both necessary, but there is even more to it. The ability to see similar patterns in different problems and to recognize the basic structure is as important in model building as it is in pure mathematics. It seems quite reasonable that the problem of figuring out the share of the market a new product will attract and the problem of determining the proportion of stores that will place orders with a manufacturer's representative in a certain time period use basically the same mathematics for their solution. It is less apparent that the problem of determining the proportion of a population who will inherit a certain genetic defect is of the same basic type.

This book is not a textbook on mathematical modeling. However, each chapter makes contact with mathematical modeling by beginning with a problem from the real world that can be at least partially solved through the use of mathematics. For example, Chapter 2 begins with a discussion of underground nuclear testing and the difficulties involved in its detection. Most of the detection problem involves technological innovation in seismology, but the problem of determining the acceptable level of effectiveness of the detection devices requires the mathematical theory of probability. For the rest of Chapter 2 we talk about probability using story problems that show other examples of the application of probability. And, by the end of the chapter, we will have shown you how probability theory can make a contribution to the problem of detecting nuclear tests.

In each of the succeeding chapters we will briefly discuss an actual problem to which mathematical modeling techniques have been applied. We emphasize that we will not attempt to reproduce the entire modeling process. Rather, we will present a mathematical topic that is used in the modeling of the introductory problem. Then we will demonstrate that even the relatively modest amount of mathematical material covered can be used to give new, although necessarily limited, information about the problem.

References

A general discussion of mathematics, which can be read with no technical background, is found in the Life Science Library series book *Mathematics* by David Bergamini and the Editors of Life (New York: Time Inc., 1963). James R. Newman's four volume *The World of Mathematics* (New York: Simon and Schuster, 1956) contains a wide variety of articles dealing with all phases of mathematics. A very useful and interesting collection of articles touching on many aspects of mathematics is *Mathematics in the Modern World.* This is an anthology of articles originally published in *Scientific American* from 1948 to 1968 with introductions by Morris Kline (San Francisco: W. H. Freeman, 1968). An extremely informative volume that features essays on different areas of mathematics written by experts in those areas is *The Mathematical Sciences: A Collection of Essays,* edited by the Committee on Support of Research in the Mathematical Sciences (COSRIMS) for the National Academy of Sciences (Cambridge, Mass.: MIT Press, 1969). These essays were written to explain modern mathematics to nonmathematicians, but because they try to explain the current research problems in their fields, some of them are rather difficult. *The Mathematical Sciences: A Report,* also by COSRIMS (Washington, D.C.: National Academy of Sciences, 1968), is a statement on the progress and the possible future goals of mathematical research as of 1968.

Probably the most readable book on the history of mathematics is Dirk J. Struik's *A Concise. History of Mathematics* (New York: Dover, 1948), which covers mathematics through the nineteenth century. E. T. Bell's books *Men of Mathematics* (New York: Simon and Schuster, 1937), which concentrates on the lives and work of particular mathematicians, and *The Development of Mathematics* (New York: McGraw-Hill, 1945), which emphasizes modern mathematics, are both interesting. *Mathematics in Western Culture,* by Morris Kline (New York: Oxford University Press, 1953), is a history of the interaction between mathematics and culture rather than a history of pure mathematics.

If more advanced work in mathematical modeling is of interest, see the text *Mathematical Models and Applications* by Daniel P. Maki and Maynard Thompson (Englewood Cliffs, N.J.: Prentice-Hall, 1973). This book assumes a year or two of college-level mathematics including calculus, but some of its chapters may be read with less background.

Chapter 2
Underground
Nuclear Testing
and the Principles
of Probability

**Introductory
Essay**

On August 5, 1963, in Moscow, the Soviet Union, Great Britain, and the United States signed a treaty banning nuclear tests in the atmosphere, in outer space, and underwater. Underground nuclear tests were not included in the treaty, because they are much more difficult to detect than those above ground. The signers of the treaty promised to try to extend the ban to include underground nuclear tests and have continued to negotiate through a permanent Committee on Disarmament under the auspices of the United Nations.

These extended negotiations have made little progress toward a treaty banning underground nuclear tests, because neither the United States nor the Soviet Union has been willing to compromise on the issue of on-site inspection. In 1963, the Soviet Union stated (see References):

We are starting here from the premise that the ban on all tests, including underground tests, may be controlled by instruments which are already at the disposal of the national governments. The Soviet Union is prepared to continue efforts in order to complete the Moscow agreement by suitable provisions to eliminate completely all possibilities of further dangerous nuclear experiments. . . . The Soviet delegation feels that the question of an underground test ban will be solved and answered whenever the Western Powers give up their demand to institute inspections. The Soviet Union will not be prepared to accept any inspections inasmuch as they are not necessary.

At the same time, the United States representative declared (see References):

The United States also believes, like everyone else, that a ban on underground testing is the important next forward step in stopping the nuclear arms race. I am bound to say, however, that the U.S. remains steadfastly opposed to a voluntary unverified moratorium on underground testing. The scientific facts point to the amount and type of verification necessary to reach agreement. . . . The core of our effort will be to improve our knowledge of seismology and of nuclear test detection. . . . We only ask for as much verification as is necessary to give adequate assurance of compliance with the provisions of a comprehensive treaty. And, in the case of underground tests, adequate assurance requires on-site inspection in order to make it possible to dispel doubts as to the nature of certain seismic events.

To reach some solution to the problem of detecting underground nuclear testing without on-site inspection, to which the Soviet Union remains unalterably opposed, the United States and other countries have concentrated their efforts on improving available seismological detection capabilities. In 1967, the Swedish representative to the Committee on Disarmament of the United Nations stated that "identification methods . . . are indeed so effective that it now seems to have become meaningful to discuss verification without on-site inspection." But the United States remained unconvinced.

These political problems have not yet been solved. As of 1970, the positions of the United States and the Soviet Union remained much the same as they had been in 1963. The United States continued to support seismic research and the international exchange of seismic data, as suggested by Canada and Sweden. The Soviet Union did agree to exchange seismic data, but would not reveal where their stations were located.

In July 1974, the American and Soviet leaders negotiated and signed a treaty placing some limitation on underground nuclear weapon tests. Underground tests were not to exceed 150 kilotons. Tests were to be kept to a minimum, and negotiations with the goal of solving the problem of all underground nuclear tests were to continue. However, since an explosion of 150 kilotons (equal to 150,000 tons of TNT) would generate a force equal to a sizeable earthquake, this limitation is not worth much. Taking a more hopeful view, this rather limited treaty indicates that the United States and the Soviet Union are willing to consider banning tests which can easily be detected without on-site inspection.

The United States and the Soviet Union signed another treaty in April 1976 which will permit on-site inspection of some underground nuclear explosions for peaceful purposes. However, the Soviet Union is still firmly opposed to on-site inspection of tests of nuclear weapons. The treaties of 1974 and 1976 are not in effect because they have not yet been ratified by the United States Senate.

The question remains, how much improvement in test-ban monitoring will be necessary before the United States will feel it has adequate assurance of compliance to permit it to sign a treaty banning all underground nuclear weapon tests? That decision is ultimately a political one, but it is possible for mathematics to be of assistance in making the decision. In this chapter, we will discuss some of the mathematical tools that can be used for this purpose.

2-1   Probability

Most of the diplomats and scientists who are concerned with the problem agree that a seismic detection system which is to be effective in monitoring an underground nuclear test ban treaty must offer at least a 90% probability of detecting an explosion of significant size. The meaning of this statement is that if a country sets off an explosion, there are nine chances out of ten that it will show up on the instruments of a seismic observatory in another country.

The probability estimate is arrived at on the basis of past experience in detecting "seismic events," i.e., nuclear explosions and earthquakes. A record is kept of the events whose existence is known because of information from seismographic stations located near the event or from intelligence information. The records of stations outside the country of the occurrence are examined to determine whether the event would have been detected from that evidence alone.

Recent studies indicate that 90% of seismic events of magnitude at least 4.75 would have been detected by teleseismic methods, that is, by sensing stations far from the event. (The force of an earthquake is expressed by a number called its "magnitude.") On the authority of these studies, we say that there is a 90% probability of detecting an earthquake of magnitude at least 4.75 or, equivalently, a nuclear explosion with a force equal to at least 10 kilotons (10,000 tons) of conventional explosives. The 10 kiloton figure assumes that the explosion is set off in a hard rock such as granite. Otherwise, magnitude 4.75 corresponds to a much larger explosion.

The manner of estimating probability described above is subject to challenge on the grounds that past experience in detecting earthquakes and explosions is not sufficient information on which to predict what will happen in the future, i.e., that 90% of seismic events above magnitude 4.75 will be detected. There is, however, a mathematical law, the law of large numbers, which states, in essence, that if probability is computed on the authority of a large enough amount of experience, then the prediction will be very accurate. Thus, if information is collected on a continuing basis, then it can be expected that, in the long run, the predicted probability will be essentially correct. Consequently, continuing studies of seismographic detection are being carried out by groups of scientists in a number of countries.

When trying to calculate probability in other situations, the same approach can be used. Data is collected and made the basis for an **estimate,** that is, a guess or prediction of the value of the probability. The law of large numbers assures us that an estimate based on sufficient evidence will produce a number close to the true probability.

Of course, the actual situation with regard to the detection of seismic events is quite complicated. The law of large numbers is predicted on the assumption that the probability being measured does not change over time. However, as improved seismographic equipment is developed and more recording stations are built, the probability of detection can be expected to increase. Therefore, data gathered over a long period of time might be too conservative in predicting the probability of detection.

Past experience can also be used to estimate the number of seismic events that will occur, as Example 2-1 illustrates.

2-1 **Example**   The number of shallow earthquakes of magnitude at least 6.0 in the Mongolian region (34° to 42° N and 75° to 120° E) from 1930 to 1939 is given in the following table:

Year	1930	1931	1932	1933	1934	1935	1936	1937	1938	1939
Number of events	1	4	3	2	1	2	2	1	1	3

On the basis of this information, estimate the probability that next year there will be exactly two shallow earthquakes of at least this magnitude in the Mongolian region.

Solution   During the 10 years from 1930 to 1939, there were 3 years in which exactly two shallow earthquakes of magnitude at least 6.0 were observed (1933, 1935, and 1936). Therefore, we estimate the probability to be 3 out of 10 or 30% or, as we prefer to write it, .30.

There are many circumstances other than the detection of seismic events in which one may estimate probability from experience.

2-2   **Example**   A country permits a limited amount of hunting of a fairly scarce kind of antelope in order to control the size of the herd. One thousand hunting licences are issued and each hunter is permitted to kill only one antelope. If, last year, 160 antelope were killed and the conditions such as the size of the herd have not changed, estimate the probability that a hunter who obtains a licence will kill an antelope.

Solution   Since 160 of the 1000 hunters with licences killed antelopes, we estimate the probability to be 160 out of 1000 or $160/1000 = .16$.

In contrast to the seismic detection problem, there are situations in which probabilities can be predicted accurately just from the nature of the question being asked—without collecting any data at all. For example, if a coin is flipped, what is the probability that it will come up heads? We reason that since there are two possibilities, heads and tails, and we assume that the coin is evenly balanced, then each face is as likely to turn up as the other. Thus, heads should come up half the time. In other words, the probability that heads will turn up is 50%. It is customary in mathematics to express probability as a fraction or decimal rather than as a percentage. Since heads should come up half the time, the probability of heads will be said to be 1/2 or .50.

If a die is rolled, the probability of rolling a three is 1/6, because there are six numbers on a die and we assume that all are equally likely to turn up. As in the coin-toss example, no experience is needed to determine this probability. As a final example, suppose a card is drawn at random from a standard deck of cards. The probability of drawing an ace is four chances out of 52, that is, the probability equals 4/52 or, reducing the fraction, 1/13. Our reasoning is that there are 52 cards in all, each with the same chance of being selected in a random draw, and since there are four aces in the deck, there are four chances to draw an ace.

Observe that all the reasoned out probabilities were computed in the same way: The number of ways the desired event can occur (one way for heads in the coin toss; four ways for the four aces in a deck of cards) was divided by the number of possible outcomes (two when a coin is tossed; 52 when a card is drawn). When equally likely events are studied, this ratio of favorable outcomes to possible outcomes is precisely what is meant by **probability**.

2-3   **Example**   A die is rolled. Find the probability that it will come up a number greater than four.

Solution   Of the six numbers on the die, two of them, five and six, are greater than four. Since there are six equally likely outcomes and two of them satisfy the condition of the question, the probability is 2/6 or 1/3.

2-4   **Example** Toss two coins at the same time. What is the probability that they will both come up heads?

Solution Even though they are tossed at the same time, the two coins are physically distinct, so we distinguish them by calling one the "first coin" and the other the "second coin." Notice that the problem is the same whether the coins are tossed at the same time or one after the other. The possible outcomes are listed in Table 2-1. There are four outcomes in all. One of the four outcomes is that both the coins turn up heads, so the probability is 1/4.

Table 2-1	
First coin	Second coin
Heads	Heads
Heads	Tails
Tails	Heads
Tails	Tails

2-5   **Example** If one letter is chosen at random from the eleven letters making up the word *probability*, what is the probability that it will be an *i*?

Solution There are eleven letters in the word *probability*. Since the letter *i* appears twice, the probability is 2/11 that if a letter is chosen at random, it will be an *i*.

We can see the connection between this reasoned out sort of probability and the probability estimated on the basis of experience in the following way: Suppose a coin is flipped many times and the number of heads and tails are recorded. If we estimate the probability of heads in a single flip of the coin by dividing the number of times heads came up by the total number of flips, then the quotient we calculate will be very close to .50 provided we flipped the coin enough times. Thus, in the long run the probability estimated by actually flipping the coin is the same as the probability calculated for equally likely events. This relationship holds in general. That is, for equally likely outcomes, if the probability of an event is estimated by means of an actual experiment, the estimate will predict the ratio of favorable outcomes to possible outcomes—provided the experiment is repeated sufficiently many times.

Thus, what appears to be two concepts of probability, that estimated on the basis of experience and that calculated abstractly, are really the same thing. In this chapter, we will be concerned with certain basic principles of probability. These principles are most clearly seen in the context of equally likely outcomes where the calculation of probability is just a matter of arithmetic. Since probability in the broader sense, the kind of probability that in practice can only be estimated from experience, is really the same concept, then the principles which are seen to apply in the clear-cut situations must hold as well in the more realistic cases. Consequently, we will discuss these principles in the context of flipped coins, rolled dice, and the like—and then apply these principles to the more elusive seismic detection problem.

Summary

If in the past a certain type of outcome occurred k times out of n times in which it was possible for it to happen, then one **estimates** that the **probability** is k/n (or the same number in decimal form) that this type of outcome will occur in the future, provided the circumstances do not change. If all possible outcomes of an act are equally likely to happen, it is not necessary to estimate on the basis of experience the probability that a certain class of outcomes will occur. Rather, that **probability** can be calculated directly: Divide the number of favorable outcomes by the total number of possible outcomes.

Exercises

● *In Exercises 2-1 through 2-3, a die is rolled.*

2-1 What is the probability that the die will come up an odd number?

2-2 What is the probability that the number showing on the die is less than five?

2-3 What is the probability that the die will come up with a number at least as large as two?

2-4 In a marketing survey, a family is given four unmarked tubes of toothpaste, one of which is the product of the company paying for the survey. If the family cannot tell the four brands apart and so chooses one at random, what is the probability that it will choose the sponsoring company's product?

2-5 A study of the traffic at tollbooths on a bridge finds that 3721 cars used the right-hand tollbooth on a certain day, while during the same period of time, only 2210 cars used the left-hand booth. If a car chosen at random crosses the bridge, what should we estimate for the probability that it will pay at the right-hand tollbooth?

2-6 The trains in a certain mythical kingdom are seldom very late. In fact, in a survey of trains arriving at the capital city, only 15 out of 75 were more than 24 hours late. If you travel to this capital on a train chosen at random, what probability should you assign to the outcome of arriving no more than 24 hours later than scheduled?

- *In Exercises 2-7 through 2-9, we suppose that a man has one pair of socks for each day of the week. Two pairs are dark blue, two pairs are dark brown, and the rest are black. At the end of the week, he washes all his socks.*

2-7   When he reaches into the clothes dryer without looking, what is the probability that he will take out a dark blue sock?

2-8   What is the probability that he will take out a black sock?

2-9   If nine of the socks have holes in them, what is the probability that he will take out a sock without a hole?

2-10 A study of newspaper records establishes that snow has been on the ground on Christmas day in a certain northern city in 60 of the past 75 years. What probability should be assigned to the likelihood that the city will have a white Christmas this year?

- *In Exercises 2-11 through 2-14, we suppose a letter is chosen at random from the 28 letters making up the word* antidisestablishmentarianism.

2-11 What is the probability that the letter is a *t*?

2-12 What is the probability that the letter is a vowel?

2-13 What is the probability that the letter is one of the first four letters of the alphabet?

2-14 What is the probability that the letter is not one of the last four letters of the alphabet?

- *Exercises 2-15 through 2-17 are based on a study of a sporting goods store in which it was found that out of 348 customers who parked in the store's parking lot, 287 made at least one purchase, while of 518 customers who did not use the parking lot, 305 made purchases.*

2-15 Estimate the probability that a customer who uses the parking lot will make a purchase.

2-16 Estimate the probability that a customer who does not use the parking lot will make a purchase.

2-17 Estimate the probability that the next customer who comes into the sporting goods store will make at least one purchase.

- *In Exercises 2-18 through 2-20, a garden supply shop has a bin of mixed tulip bulbs containing 22 bulbs which will produce red tulips, 14 bulbs which will grow yellow tulips, and 7 bulbs for white tulips. Since all the bulbs look alike, a customer chooses a bulb at random.*

2-18 What is the probability that the bulb chosen is one that will grow a white flower?

2-19 What is the probability that the bulb will grow either a white or a yellow flower?

2-20 What is the probability that the flower produced will not be yellow?

2-21 Edgar Allan Poe's story *The Gold Bug* contains the solution to a code. A coded message in the story consists of 159 symbols in all, including 33 repetitions of the symbol "8", 26 of ";", and so on. Poe wrote "Now, in English, the letter which most frequently occurs is e . . . [it] predominates so remarkably that an individual sentence of any length is rarely seen, in which it is not the prevailing character." Thus, the narrator of the story assumes, correctly as it turns out, that the symbol "8" represents the letter e. If the message in Poe's story is typical of English, what is the probability that a letter chosen at random from an English sentence will be an e?

2-22 Three sleeping pills are, by accident, added to a bottle containing 100 aspirin. If a person takes out a pill at random without noticing its appearance, what is the probability that the pill chosen is a sleeping pill?

• *Exercises 2-23 through 2-26 are concerned with the gambling game of American roulette. The American roulette wheel has 38 sections numbered 0, 00, and 1 through 36. The game amounts to choosing one of these numbers at random.*

2-23 What is the probability that the number 12 will be chosen?

2-24 The numbers 0 and 00 are neither odd nor even. What is the probability that the outcome will be an even number?

2-25 What is the probability that the game will choose a number greater than 24?

2-26 What is the probability that the game will choose an odd number greater than 24?

• *For Exercises 2-27 through 2-30 we imagine that lottery tickets numbered 000 to 999 are all sold. One such number is chosen at random and the holder of that ticket wins $100. All other ticket holders whose first two digits are the same as the winner's receive $10. Thus, if number 124 is the winning ticket, then the holder of ticket number 128 will get $10.*

2-27 How many tickets were sold?

2-28 What is the probability of winning $100?

2-29 What is the probability of winning $10?

2-30 What is the probability of not winning anything?

• *In Exercises 2-31 through 2-36, a card is chosen from a standard deck. (See the Appendix, p. 122, if you are not a card player.)*

2-31 What is the probability that the card is a club?

2-32 What is the probability that the card is a black jack?

2-33 What is the probability that the card is either a jack or a queen?

2-34   What is the probability that the card is neither a jack nor a queen?

2-35   If aces are considered to be higher than kings, what is the probability that the card drawn is higher than a nine?

2-36   What is the probability that the card drawn will be lower than a nine when aces are considered higher than kings?

●   *In Exercises 2-37 through 2-39, a mouse is placed in the maze pictured below. Suppose the mouse never reverses direction and continues until it can go no further.*

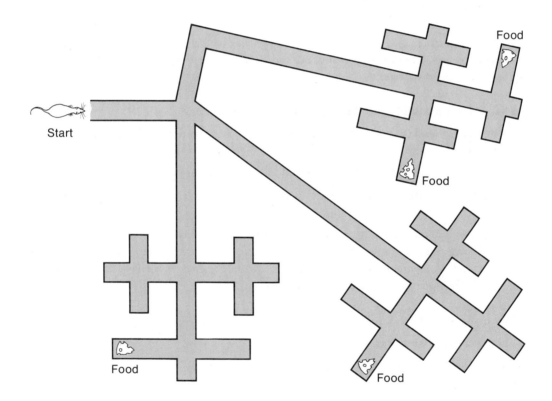

2-37   How many distinct destinations may the mouse reach?

2-38   How many such destinations contain food?

2-39   If the mouse is as likely to reach one destination as another, what is the probability that it will reach food?

2-2   Counting

As a consequence of the definition, computing the probability that a certain class of outcomes among equally likely events will occur involves two counting problems: counting the total number of outcomes possible and counting the number of favorable outcomes. In the examples and exercises of Section 2-1, these two counting problems were not very difficult. In many important types of probability problems, however, the counting problems which arise are not so straightforward. Thus, in order to be sure that the counts are correct, it is desirable to have enough counting theory available to carry out the job. Such a theory eliminates the need for reasoning out each new type of probability problem from first principles. Instead, appeal to the counting procedures which we will discuss can turn the calculation of probability in even quite intricate problems into a relatively routine exercise.

Automobile dealers and manufacturers sometimes complain that their business is substantially complicated by the fact that the buyer has a wide choice of styles, colors, and optional extras. The number of possible combinations is large enough so that one could almost say that no two cars purchased are exactly alike; that each new car is virtually a custom product. Let us see how this problem arises by first considering a simple example. Suppose that the purchaser of a certain make of automobile has only two types of decisions concerning the new car: the style and the color. Further assume that there are just three styles available: two-door, four-door, and wagon. And there are just five colors: black, gray, blue, green, and red. We wish to count the number of distinct combinations of style and color the manufacturer must be prepared to supply.

The purchaser of the automobile must make two choices, style and color. We will represent the choices by means of a simple but useful mathematical concept—an **ordered pair.** An ordered pair is simply two symbols with a specified order, usually written in the form (first symbol, second symbol). The purchaser's choices will be an ordered pair where the first symbol (word in this case) determines the automobile's style and the second its color. Schematically, the buyer must choose an ordered pair (style, color), for example, (two-door, red) or (wagon, gray). Thus, the question is: How many ordered pairs (style, color) are there when there are three choices of style and five choices of color?

One convenient way to represent all the possible ordered pairs is by means of a rectangular diagram like the one shown in Table 2-2. Each dot represents the ordered pair determined by its row and column headings, e.g., the circled dot represents the pair (two-door, blue) and the dot in the square corresponds to (wagon, green). It is evident that each possible ordered pair is represented by a dot, and the number of ordered pairs equals the number of dots. In this case, since there are three rows of five dots each, we have fifteen ordered pairs. Thus, even with these restricted choices, the manufacturer must produce fifteen different automobiles.

Table 2-2

Style \ Color	Black	Gray	Blue	Green	Red
Two-door	·	·	⊙	·	·
Four-door	·	·	·	·	·
Wagon	·	·	·	⊡	·

2-6 **Example**   Two dice are rolled. How many different outcomes are possible?

Solution   Even though the two dice are rolled at the same time, they are physically distinct, and so the possible outcomes will be the same as if they were rolled one at a time. Thus, we refer to them as the "first die" and the "second die." (Compare Example 2-4.) A possible outcome is an ordered pair of the form (first die, second die), for example, (2, 4) represents a two on the first die and a four on the second. The possible outcomes are described by Table 2-3. Since there are six rows of six dots each, there are 36 possible outcomes. Notice the significance of the fact that we consider *ordered* pairs, e.g., (2, 4) and (4, 2) represent different outcomes even though the same numbers are involved. The distinction represents the fact that the two dice are physically distinct.

Table 2-3

First \ Second	1	2	3	4	5	6
1	·	·	·	·	·	·
2	·	·	·	·	·	·
3	·	·	·	·	·	·
4	·	·	·	·	·	·
5	·	·	·	·	·	·
6	·	·	·	·	·	·

Now let us consider a different, but as it turns out closely related, type of counting problem. Suppose the four aces are separated from the rest of a standard deck of playing cards (see the Appendix, p. 122). An ace is chosen and then a second ace is selected from among the three that remain. In how many ways can this be done? We emphasize that we will want to preserve the idea of an ordered pair. The outcome of the ace of spades on the first draw and the ace of hearts on the second is not considered to be the same as the ace of hearts on the first draw and the ace of spades on the second. We are interested in ordered pairs of the form (first ace, second ace). But this situation is rather different from the cases above, because the choices available for the second ace depend on which ace was chosen first. Thus, if the first ace was the ace of spades, the choices for the second ace are restricted to hearts, clubs, and diamonds. Consequently, the method we used before for representing ordered pairs is not convenient, and instead we list the possible outcomes as shown in Table 2-4. Now the ordered pairs are represented by symbols, e.g., the circled symbol represents the pair $(\heartsuit, \clubsuit)$ and the symbol in the square represents the pair $(\diamondsuit, \spadesuit)$. Each possible outcome is represented by a symbol in the rectangular array and, since there are four rows of three symbols each, we conclude that there are twelve outcomes in all. The important thing to notice about this example is that although the choices available for the second ace are dependent on the choice of the first ace, the *number* of such choices—three—is the same no matter what the first choice is. This is the reason we were able to use a rectangular display with three symbols in each row.

Table 2-4

First ace	Second ace		
♠	♡	◇	♣
♡	♠	◇	⊛
◇	⊡	♡	♣
♣	♠	♡	◇

Of course, the number of dots or symbols in a rectangular array will equal the number of rows of the array times the number of columns. As long as the number of choices for the second member of an ordered pair does not depend on the choice of the first member, the possible ordered pairs can always be described by a rectangular array with, say, each row corresponding to a choice of first element in the pair and the number of dots or symbols in each row equal to the number of choices for the second element. We combine these observations into the following counting rule:

> If there are n_1 choices for the first member of an ordered pair and n_2 choices for the second no matter what the first member is, then there are $n_1 n_2$ (n_1 times n_2) ordered pairs in all.

2-7   **Example**   How many two digit numbers (including 00, 01, etc.) are there and how many of them involve two distinct digits?

Solution   By "digits" we mean the numbers 0, 1, 2, 3, 4, 5, 6, 7, 8, and 9. The two digit numbers may be thought of as ordered pairs (first digit, second digit) where, for example, (2, 5) is the ordered pair corresponding to the number 25. Since there are ten different digits,

$$n_1 = 10 \quad \text{and} \quad n_2 = 10$$

so there are

$$(10)(10) = 100$$

two digit numbers in all. However, if the digits are to be distinct, for example (1, 4) and (1, 8) are permitted, but (1, 1) is not. Then, after the first digit is selected, the second digit must be chosen from the remaining nine digits. In this case,

$$n_1 = 10 \quad \text{but} \quad n_2 = 9$$

so there are

$$(10)(9) = 90$$

two digit numbers involving two distinct digits.

Now let us return to the car buyer and allow some choice of optional extras on the new car. In addition to choosing the style from among two-door, four-door, and wagon, and the color from among black, gray, blue, green, and red, we will assume that the manufacturer also offers two options: high-quality steel-belted radial tires in place of standard tires, and an air-conditioning unit. Thus, in addition to choosing style and color, the purchaser must decide whether to choose both, neither, or one of the options.

In order to describe more complicated choices such as this, we will need the concept of an **ordered k-tuple**. An ordered k-tuple is a collection of k symbols in a fixed order. Consequently, an ordered pair could also be called an ordered 2-tuple.

We can represent the buyer's decision by means of an ordered 3-tuple (also called an ordered triple) of the form (style, color, options), where the alternatives in the options position are both, neither, tires only, and air only. In a more realistic situation where the buyer has choices of engine size and type, upholstery, external decoration, and so on, the choices could be described by an ordered k-tuple (, , ... ,) if there were k decisions to be made.

Extending the rule for ordered pairs above, the general counting rule for ordered k-tuples is:

> If there are n_1 choices for the first member of an ordered k-tuple, n_2 for the second no matter what the first is, n_3 for the third no matter what the first two choices are, and so on to n_k for the kth (independent of what the previous choices are), then there are $n_1 n_2 n_3 \cdots n_k$ such k-tuples in all.

2-8   **Example**   (*a*) How many three digit numbers (including 000, 001, and so on) are there? (*b*) How many in which all the digits are distinct? (*c*) How many involve exactly two digits?

Solution   (*a*) A three digit number may be thought of as an ordered triple, e.g., 124 corresponds to (1, 2, 4). Since there are ten digits, $n_1 = n_2 = n_3 = 10$ and the rule tells us that there are $n_1 n_2 n_3 = 1000$ such numbers.

(*b*) If the three digits are to be distinct, then there are $n_1 = 10$ choices for the first digit in the number, but only $n_2 = 9$ choices for the second since it must be different from the first. No matter what the first two digits are, the last digit, which must be different from the others, will be chosen from the eight digits remaining, so $n_3 = 8$. Then there are $(10)(9)(8) = 720$ such three digit numbers.

(*c*) Note that there are 1000 numbers with the digits by part (*a*). By part (*b*), 720 of them involve three different digits. On the other hand, ten of the numbers, namely 000, 111, and so on, use only one digit. Thus, there are

$$1000 - (720 + 10) = 270$$

three digit numbers involving exactly two different digits. Another way to solve this part of the example is to note that a three digit number made up of two different digits, for example, 545, can be described by listing three choices: which digit will appear twice in the number (here 5), which other digit will appear once (here 4), and where in the number the single digit will appear (since 455, with the 4 in the first position, is not the same number as 545, in which the 4 occupies the second position). Thus, a three digit number involving just two digits can be described by an ordered triple of the form

(digit to appear twice, digit to appear once, position of single digit)

There are ten digits to choose from for the first member of the triple and nine for the second no matter what the first is. The single digit can occupy any of three possible positions in the number so, by the counting rule, there are again $(10)(9)(3) = 270$ such numbers.

Summary

An **ordered pair** consists of two symbols in a fixed order, written in the form (first symbol, second symbol). An **ordered k-tuple** is a collection of k symbols in a fixed order.

Counting Rule

If there are n_1 choices for the first member of an ordered k-tuple, n_2 for the second no matter what the first is, n_3 for the third no matter what the first two are, and so on to n_k for the kth (independent of what the previous choices are), then there are $n_1 n_2 n_3 \cdots n_k$ such k-tuples in all.

Exercises

2-40   A game is played by rolling a die and then drawing a card from a standard deck. How many different outcomes can this game have?

2-41   A sport shirt of a certain style can be purchased in a choice of sizes: small, medium, large, and extra large. There is also a choice of color from among white, blue, green, gray, and brown. Suppose a store wishes to keep in stock exactly one shirt of this style in each combination of size and color. How many shirts must the store keep in stock?

- *In Exercises 2-42 and 2-43, a "three letter word" means a sequence of three letters such as bat, tab, qcc, ccq, and xxx.*

2-42   How many three letter words are there?

2-43   How many three letter words are there that use three different letters?

- *In Exercises 2-44 through 2-47, three dice are rolled.*

2-44   How many different outcomes are there?

2-45   If the first die is even, how many different outcomes are there?

2-46   In how many different ways can the dice be rolled so that all three dice show odd numbers?

2-47   In how many different ways can the dice be rolled so the numbers that come up are all different?

2-48   A polling organization divides the population into groups, called "strata," according to sex, race (white or nonwhite), age (18–35, 36–50, over 50), and education (college-educated or not). For example, college-educated white male between the ages of 18 and 35 is one stratum. How many different strata has the polling organization set up?

2-49   In how many ways can three boys and two girls line up so that boys and girls alternate?

2-50   How many different ways are there to fill in an answer sheet if an examination consists of five true–false questions?

2-51   How many different ways are there to fill in an answer sheet if an examination consists of five multiple-choice questions with four possible answers to each question?

2-52   A psychologist attempts to teach a monkey to distinguish "words" made up of the symbols $\bigcirc$, $\square$, $\triangle$, and $\diamond$. If words are exactly five symbols long and repetitions of symbols are allowed, how large can the monkey's vocabulary be?

2-53   A family plans to drive from Los Angeles to New York, stopping in Chicago along the way. The routes they are considering are indicated in the diagram. How many different choices of routes are being considered?

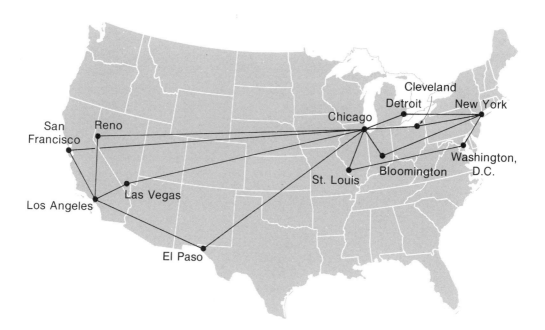

2-54   A botanist wishes to determine the effects of temperature, humidity, and light on the growth rate of a certain type of plant. The botanist wishes to test growth at constant temperatures of 10, 15, 20, and 25°C; at 20, 50, and 80% constant humidity, and under the following lighting schedules: constant light, 12 hours of light alternating with 12 hours of darkness, and 16 hours of light alternating with 8 hours of darkness. If each choice of temperature, humidity, and light schedule constitutes a distinct experiment, how many different experiments must the botanist set up?

2-55   Suppose the members of the parliament of some country belong to three political parties: Liberal, Moderate, and Conservative. The parliament has 20 Liberal members, 50 Moderates, and 30 Conservatives. If a three member committee of the parliament is to be made up of one member from each party, how many different such committees can there be?

•   *In Exercises 2-56 and 2-57, we suppose that shoppers in a supermarket with three checkout counters choose the counter they will use at random. Unknown to each other, each of three customers decides on a counter.*

2-56   How many different choices of counter can be made by the three shoppers?

2-57   In how many different ways can the three shoppers choose three different counters?

2-58   A woman has a choice of six different routes between her home and her work. How many different choices does she have for a round trip if she never uses the same route both for going to work and for returning on the same day?

2-59   A college requires its students to take a basic course in each of three areas: natural sciences, social sciences, and humanities. It offers six natural science courses, five social science courses, and eight humanities courses that satisfy the requirement. In how many different ways can a student fulfill the basic requirements?

2-60   A large corporation uses four digit numbers for telephone extensions. Dialing 8 or 9 will connect the phone with an outside line, and 0 calls the operator, so the first digit of the extension cannot be any of these. How many telephone extension numbers does the company have available?

•   *In Exercises 2-61 and 2-62, a bakery wishes to find the best recipe for chocolate cake, so it decides to vary the amounts of salt, cocoa, sugar, and vanilla extract in a basic recipe to determine which combination produces the tastiest cake. The amount of salt will be either 1/8 teaspoon or 1/4 teaspoon; the amount of cocoa is 1/2 cup, 1/2 cup plus 1 tablespoon, or 1/2 cup plus 2 tablespoons; there is either 1¼ cups or 1½ cups of sugar; and the amount of vanilla extract is either 3/4 teaspoon or 1 teaspoon. Suppose the company bakes a chocolate cake for each possible combination of amounts of these ingredients.*

2-61   How many cakes must the company bake?

2-62   How much sugar will the company use in baking these cakes?

•   *In Exercises 2-63 through 2-66, a thief attempts to find a lock combination which consists of three digits, each between 0 and 6. Thus, 062 is a possible combination, as is 511, and so on. The thief decides to try every possible combination starting with 000, then 001, and on to 066, 100, 101, and finally to 666. Suppose it takes the thief 1 second to try each combination.*

2-63   If the correct combination is 666, how long will it take the thief to find it?

2-64   How long will it take the thief to reach 366?

2-65   How long will it take the thief to reach 446? [*Hint:* Use Exercise 2-64.]

2-66   If the correct combination is 453, how long will it take the thief to find it?

2-3   Permutations and Combinations

We begin this section by applying the counting rule of Section 2-2 to a problem which is a simplified version of a type that frequently arises in probability theory.

2-9   **Example**   In how many ways can the letters of the words *red, brown,* and *dangerously* be ordered?

Solution   An ordering of the letters of the word *red,* such as red, dre, erd, and so on, can be thought of as an ordered triple (, ,) involving all three letters r, e, and d. Choosing any one of the $n_1 = 3$ available letters for the first position, there is a choice of $n_2 = 2$ left for the second, and when we get to the third position there is only a single letter remaining, so $n_3 = 1$. Thus, there are $(3)(2)(1) = 6$ orderings of the word *red.*

Similarly, an ordering of the letters of *brown* is an ordered 5-tuple involving each of the letters b, r, o, w, and n exactly once. Clearly, $n_1 = 5$ since any letter can be used in the first position, while $n_2 = 4$ since there are four letters left, $n_3 = 3$, $n_4 = 2$, and in the final position there is only a single letter remaining so $n_5 = 1$. By the counting rule, there are $(5)(4)(3)(2)(1) = 120$ different ways to order the letters of *brown.*

Finally, *dangerously* consists of 11 different letters, so $n_1 = 11$, $n_2 = 10$, $n_3 = 9$, and so on, until we get to the last position of the 11-tuple, and $n_{11} = 1$. By the counting rule, there are

$$(11)(10)(9)(8)(7)(6)(5)(4)(3)(2)(1) = 39{,}916{,}800$$

ways to order the letters of *dangerously.*

Example 2-9 illustrates an important consequence of the counting rule. If we are given a set of k distinct symbols and asked in how many ways they can be ordered, we are being asked to count the number of ordered k-tuples with different symbols in each position from a list of k symbols. Since all k symbols are available for the first position, $n_1 = k$ always. At each succeeding position, the choice is reduced by one from the list of choices in the previous position so $n_2 = k - 1$, $n_3 = k - 2$, and so on. When we reach the last, kth, position in the ordered k-tuple, all but one symbol has been used previously, so there is just a single choice and $n_k = 1$. By the counting rule, the number of different orderings must be

$$k(k - 1)(k - 2) \cdots (3)(2)(1)$$

A product of this form occurs so frequently in mathematics that it has a special name, **k factorial**, and symbol, **k!**. Thus, by definition

$$2! = (2)(1) \qquad = 2$$
$$3! = (3)(2)(1) \qquad = 6$$
$$4! = (4)(3)(2)(1) = 24$$

and so on. We extend the definition by letting $1! = 1$. In addition, it is convenient to give a meaning to the symbol 0! (zero factorial). The value $0! = 1$ has been chosen, because it works properly within the definition of factorial, as we shall see below. In this language then, the number of orderings of k distinct symbols is equal to $k!$.

2-10   **Example** (*a*) How many slates of officers, president, vice-president, secretary, and treasurer, are possible out of an organization consisting of 15 members? (*b*) Express the answer to part (*a*) in terms of factorials.

Solution (*a*) We must count 4-tuples of the form

(president, vice-president, secretary, treasurer)

Since any member could be president, $n_1 = 15$. This leaves $n_2 = 14$ choices for vice-president, and then $n_3 = 13$ for secretary, and $n_4 = 12$ for treasurer, so by the counting rule there are

$$(15)(14)(13)(12) = 32{,}760$$

different slates of officers possible in the organization.

(*b*) The product $(15)(14)(13)(12)$ looks like the beginning of the definition of 15!; the product was terminated after 12. To turn $(15)(14)(13)(12)$ into 15!, we would multiply by $(11)(10) \cdots (2)(1)$, which we observe is the definition of 11!. Thus,

$$15! = [(15)(14)(13)(12)](11)(10) \cdots (2)(1)$$

or

$$15! = [(15)(14)(13)(12)](11!)$$

Dividing both sides of the latter equation by 11!, we find that

$$\frac{15!}{11!} = (15)(14)(13)(12)$$

which expresses the answer to part (*a*) as a quotient of factorials.

We might wonder what relation 11! has to the statement of Example 2-10. Notice that there are fifteen members and four of them will occupy elected positions, so the remaining eleven will not. In other words, $11 = 15 - 4$. More generally, if there were n members of the organization and k different offices, then there would be $n_1 = n$ choices for the highest office, $n_2 = n - 1$ choices for the next, and so forth. If we wish to compute the number of different slates of officers, we will come up with a product beginning $n(n - 1)(n - 2) \cdots$. The first k numbers, starting at n and reducing by 1 each time, are

$$n, \quad n - 1, \quad n - 2, \quad \ldots, \quad n - k + 1$$

so, by the counting rules, there are

$$n(n - 1)(n - 2) \cdots (n - k + 1)$$

slates of k officers out of an organization with n members. As in Example 2-10,

$$n! = [n(n - 1)(n - 2) \cdots (n - k + 1)](n - k)(n - k - 1) \cdots (2)(1)$$
$$= [n(n - 1)(n - 2) \cdots (n - k + 1)](n - k)!$$

so, in factorial notation, there are $n!/(n - k)!$ such slates of officers. If all n members are to be officers, each with a different title, then a slate of officers is just an ordering of all the members. As we observed above, there are $n!$ such orderings. Since $n = k$ in this case and we have the definition $0! = 1$, we can express the number of slates as

$$\frac{n!}{(n - k)!} = \frac{n!}{0!} = n!$$

just as before.

An ordering of k elements out of a set of n elements in all, which is just what a slate of k officers for an organization with n members represents, is called a **permutation** of k elements out of n. The number of such permutations is denoted by the symbol P_k^n. The argument above shows us that

$$P_k^n = n(n - 1)(n - 2) \cdots (n - k + 1) = \frac{n!}{(n - k)!}$$

2-11 **Example** How many different five digit numbers are there which involve five different digits?

Solution There are $n = 10$ different digits, and a five digit number involving five different digits is just a permutation of the five digits, so the answer is

$$P_5^{10} = \frac{10!}{5!} = (10)(9)(8)(7)(6) = 30{,}240$$

There are many counting problems for which the concept of ordered k-tuple is inappropriate because there is no sense of order. For example, how many five card poker hands are there out of the standard 52 card deck? To a card player, it makes no difference whether the cards are picked up in the order ($3\heartsuit$, $5\clubsuit$, $3\spadesuit$, $3\diamondsuit$, $5\diamondsuit$) or the order ($5\diamondsuit$, $3\spadesuit$, $5\clubsuit$, $3\heartsuit$, $3\diamondsuit$) or in any other order; the player will be delighted in any case. So if we consider a poker hand to be an ordered 5-tuple of cards, i.e., a permutation of five cards out of the deck of 52, then we are viewing it in an unrealistic way. Similarly, if we wish to determine the number of different six to three majorities possible in a United States Supreme Court decision, there is no sense of ordering among the six justices who make up the majority. Or, if a television dealer wants to select two sets to test out of a shipment of ten in order to decide whether to accept the shipment, the dealer may need to know the number of such samples (as we shall see below), but again there is no reason to distinguish between the first set chosen and the second.

Thus, a very common kind of counting problem requires that we determine the number of distinct sets of k objects out of a collection of n, without imposing an ordering on the k objects. An unordered set of this kind is called a **combination** of k elements out of n, and the number of combinations is represented by the symbol C_k^n.

Combination problems occur when samples are taken. Suppose that an instructor in a psychology class of 40 students needs three of these students to take part in a demonstration experiment for the class. If the group of three students is chosen all at once, without regard for any order of selection, then the three students form a combination of three out of the 40 students in the class. We would like to calculate C_3^{40}, the number of different (unordered) samples of three students possible out of a class of 40 students.

Since we already have a formula for P_3^{40}, the number of permutations (ordered triples) of three members out of 40, it seems reasonable to try to relate C_3^{40} to P_3^{40} in order to take advantage of what we already know. We would be interested in counting permutations if the three students in the sample were to be distinguished from each other in some way. For instance, the first student might be told the correct purpose of the experiment, the second given misleading information as to the purpose of the experiment, and the third told nothing at all. Suppose for the moment that the three students are given different information as indicated and let us see how a permutation might be selected. As a first step, three students could be selected as a group by, say, drawing three slips of paper from a box with, for example, this result:

We picture the names in this way to emphasize the lack of order among them. As a second step, the instructor decides what information will be given to each, e.g.,

JANE—correct   TOM—misleading   MARY—none

Thus, the choice of three students, each given different information, is a permutation obtained by a two step process. We may therefore represent the process as an ordered pair:

(choose combination of three students,   assign amount of information to each)

Now the number of ways we can assign different information, correct, misleading, none, to three students certainly has nothing to do with which students are chosen. From earlier in this section we recall that there is always $3! = 6$ ways to assign the information. By the counting rule of Section 2-2, the number of permutations, that is, the number of ordered pairs

(combination, ordering)

is just the number of combinations times the number of orderings. In symbols, keeping in mind that we are discussing permutations and combinations of three students in a class of 40, the counting rule tells us that

$$P_3^{40} = (C_3^{40})(3!)$$

Another way to write this equation is

$$P_3^{40} = (3!)(C_3^{40}) \quad \text{or} \quad (3!)(C_3^{40}) = P_3^{40}$$

We already have the computational formula

$$P_3^{40} = \frac{40!}{(40-3)!} = \frac{40!}{37!}$$

so we can write

$$(3!)(C_3^{40}) = \frac{40!}{37!}$$

Dividing through the equation by 3!, we find that

$$C_3^{40} = \frac{40!}{(3!)(37!)} = \frac{(40)(39)(38)(37!)}{(3!)(37!)} = \frac{(40)(39)(38)}{(3)(2)(1)} = 9880$$

Therefore, there are 9880 different unordered samples (combinations) of three students in a class of 40. In contrast, the number of samples of three that are ordered with respect to which information is given is

$$P_3^{40} = \frac{40!}{37!} = 59,280$$

To obtain a general formula for C_k^n, the number of combinations of k members from a set of size n, we just repeat the preceding argument for any choice of n and k. A permutation of k elements out of n consists of k elements in some fixed order. Thus, the selection of a permutation requires two choices: choose k elements out of n (a combination of k objects out of a set of n objects) and then choose an ordering for the k elements. We may view a permutation, then, as an ordered pair of choices:

(combination of k objects out of n, ordering of the k objects)

Observe that by our discussion above, no matter which combination of k objects was selected, there are always precisely $k!$ different ways to order them. Since P_k^n and C_k^n denote the number of permutations and combinations, respectively, of k elements out of a set of n elements and $k!$ is the number of orderings of k elements, the form of the counting rule for ordered pairs (of choices) implies that

$$P_k^n = (C_k^n)(k!)$$

because P_k^n was described as the number of ordered pairs where $n_1 = C_k^n$ and $n_2 = k!$. Another way to write the equation is

$$P_k^n = (k!)(C_k^n) \qquad \text{or} \qquad (k!)(C_k^n) = P_k^n$$

We have a computational formula for P_k^n, namely $P_k^n = n!/(n-k)!$, so we can write

$$(k!)(C_k^n) = \frac{n!}{(n-k)!}$$

Dividing through the equation by $k!$ produces the required computational formula for C_k^n:

$$C_k^n = \frac{n!}{k!(n-k)!}$$

2-12 **Example** (*a*) How many different six to three majorities are possible in the United States Supreme Court? (*b*) How many different five card poker hands are there?

Solution (*a*) There are nine justices in all and a six person majority is a combination of any six, so there are

$$C_6^9 = \frac{9!}{(6!)(3!)} = \frac{(9)(8)(7)(6)(5)(4)(3)(2)(1)}{(6)(5)(4)(3)(2)(1)(3)(2)(1)} = \frac{(9)(8)(7)}{(3)(2)(1)} = 84$$

different majorities.

(*b*) There are 52 cards in the deck and we choose five at a time, so

$$C_5^{52} = \frac{52!}{(5!)(47!)} = \frac{(52)(51)(50)(49)(48)(47!)}{(5)(4)(3)(2)(1)(47!)} = \frac{(52)(51)(50)(49)(48)}{(5)(4)(3)(2)(1)} = 2,598,960$$

is the total number of poker hands.

We conclude this section by applying our counting theory to some probability problems. We will use the notation $p(\ )$ to represent the probability that an outcome of the type described by the words or symbols within the parentheses will occur. For example, if we toss a coin, we conclude that $p(\text{Heads}) = .50$; if we roll two dice, then $p(\text{Sum} = 3) = 2/36$; if we draw a card at random from a standard deck, then $p(5\diamondsuit) = 1/52$.

2-13   **Example**   What is the probability of being dealt a royal flush (ace, king, queen, jack, and ten all of the same suit) in poker?

Solution   If the cards are dealt fairly, any hand is as likely to appear as any other. We know from Example 2-12 that there are 2,598,960 poker hands in all. Since there are just four royal flush hands, one for each suit, then by the definition of probability in Section 2-1,

$$p(\text{Royal flush}) = \frac{4}{2,598,960} = .0000015 \qquad \text{approximately}$$

2-14   **Example**   A three digit number is chosen at random. What is the probability that all three digits will be different?

Solution   If the number is chosen at random, that means any number is as likely to be selected as any other. Thus, the probability that the three digits will be distinct is the number of three digit numbers with this property divided by the total number of three digit numbers. From Example 2-8 we know there are 1000 numbers with three digits, and 720 with three distinct digits. Therefore,

$$p(\text{All digits distinct}) = \frac{720}{1000} = .72$$

2-15   **Example**   A television dealer receives ten sets from the manufacturer; two are defective. If the dealer selects two sets at random to test, what is the probability that neither of them will be defective?

Solution   Since the sample of two is chosen at random,

$$p(\text{Selecting two good sets}) = \frac{\text{Number of samples consisting of two good sets}}{\text{Number of samples of two sets}}$$

There are eight good sets, so there are C_2^8 samples consisting of two good sets. There are ten sets in all, so there are C_2^{10} samples of two sets altogether. Therefore,

$$p(\text{Selecting two good sets}) = \frac{C_2^8}{C_2^{10}} = \frac{28}{45} = .62 \qquad (\text{approximately})$$

which means that there is a probability of about .62 that the testing procedure will fail to give the dealer any hint that the shipment is not perfect.

Summary

The term **k factorial** refers to the product

$$k! = k(k - 1)(k - 2) \cdots (3)(2)(1)$$

where, by definition, $0! = 1$ and $1! = 1$.

A **permutation** of k elements out of a set containing n elements is an ordered k-tuple of k different elements from the set of n. The number of permutations of k elements out of a set of n is denoted by P_k^n and computed by the formula

$$P_k^n = \frac{n!}{(n - k)!}$$

A **combination** of k elements out of a set containing n elements is an unordered subset of k elements out of the set of n. The number of combinations of k elements out of a set of n is denoted by C_k^n and computed by the formula

$$C_k^n = \frac{n!}{k!(n - k)!}$$

Exercises

2-67 Calculate C_2^7, P_6^8, P_5^{10}, and C_6^9.

2-68 Calculate C_1^3, C_1^4, and C_1^5. What is the value of C_1^n for any n?

2-69 A student who applies to the University of California is asked to list three of the eight general campuses of the university in order of preference. In how many ways can the student answer this question?

2-70 The members of a kindergarten class were asked to list their order of preference in ice cream flavors from among chocolate, vanilla, strawberry, and pumpkin. Each student was asked to list all four flavors and no ties were permitted. If each child chose a different ordering, how large was the class at most?

2-71 How many different triangles can you draw by connecting three of the dots below?

2-72 The worst possible bridge hand consists of thirteen cards in which none is higher than a five (aces are the highest cards in bridge). How many different such hands are there?

2-73 Eight swimmers are in the final of an Olympic event. The first three finishers win gold, silver, and bronze medals, respectively. How many different outcomes—awards of medals—can there be?

• *In Exercises 2-74 and 2-75, a bookstore decides to feature four current best-selling novels in its display windows. There are ten novels on the best-seller list.*

2-74 How many choices does the bookstore have if the order in which the books appear in the windows makes no difference?

2-75 How many choices does the bookstore have if there is a most desirable location in the windows, next most desirable, third most desirable, and least desirable location?

• *In Exercises 2-76 through 2-78, a handsome prince is informed by his fairy godmother that he is to select the three wishes from a list of ten that he wants most. However, one of the wishes has the undesirable side effect that if he chooses it, he will be instantly, and permanently, turned into a frog. The prince has no way of knowing which of the ten wishes on the list has this side effect, so he selects three wishes at random.*

2-76   In how many different ways can the prince choose his wishes?

2-77   In how many different ways can the prince choose three which avoid the undesirable wish?

2-78   What is the probability that the prince will avoid being turned into a frog?

 ●   *In Exercises 2-79 through 2-82, two sleeping tablets have been put in a box with eight aspirin tablets which look exactly the same.*

2-79   In how many different ways can two tablets be removed from the box?

2-80   In how many ways can two aspirin tablets be removed from the box?

2-81   If two tablets are removed at random from the box, what is the probability that both of them are aspirin tablets?

2-82   What is the probability of removing the two sleeping tablets if two tablets are taken at random from the box?

 ●   *In Exercises 2-83 and 2-84, we suppose that in order to complete the requirements for a major in mathematics, a student must take four required courses which are offered during each of the three quarters of the year. Each quarter, the four courses are scheduled so that they meet at four different hours. This makes it possible to take all of them at once.*

2-83   If the student decides to take two of the required courses this quarter, in how many ways can this be done?

2-84   If the student decides to take one of the required courses in each of the three quarters this year, in how many ways can this be done?

2-85   A committee selecting teams of one professional golfer and one amateur for a pro-am golf tournament finds itself left with four professionals who need partners and nine amateurs who would like to play in the tournament. In how many different ways can the committee select the teams so that each professional has a partner?

 ●   *In Exercises 2-86 through 2-91, imagine a card game in which each player receives three cards.*

2-86   How many different three card hands are possible, using a standard deck?

2-87   How many different hands consist of three cards of the same suit (flush)?

2-88   How many different hands consist of three cards of the same value, such as three kings (three of a kind)?

2-89   What is the probability of drawing a flush?

2-90   What is the probability of drawing three cards of the same value?

2-91   If you were writing the rules for three card poker, you would make a hand more valuable if the probability of obtaining it is lower. Which, then, should be the more valuable hand in three card poker, flush or three of a kind?

- *In Exercises 2-92 through 2-94, a family consisting of a father, mother, and child is chosen at random, and each member is asked on what day of the week they were born.*

2-92   How many possible answers are there?

2-93   In how many ways can the three be born on three different days?

2-94   What is the probability that the three were born on three different days?

2-95   The Dow-Jones stock averages are based on the prices of the stocks of 30 industrial companies, 20 transportation companies, 15 utilities, and 65 other companies. A stockbroker wishes to recommend a portfolio of stocks, that is, an unordered list of companies, to a customer. All of the companies in the portfolio are to be taken from those used by Dow-Jones. If the portfolio is to consist of one industrial company, one transportation company, and three utilities, how many different portfolios could the stockbroker make up?

2-96   How many different five card poker hands contain exactly four spades?

2-97   How many different five card poker hands contain exactly four cards of the same suit?

- *In Exercises 2-98 and 2-99, seven mountain climbers must choose three of their number to make the final attempt at the summit, while the other four remain behind at the base camp.*

2-98   In how many ways can the three be chosen?

2-99   If the expedition leader must be included in the three chosen, how many choices are there?

2-4   Principles of Probability: "or" Statements

We will think of an occurrence the probability of which we wish to measure as the outcome of an experiment (flipping a coin, dealing a poker hand, monitoring a seismic event) which will be a **success** if there is a certain outcome (heads, a royal flush, detecting the event) and a **failure** otherwise. This language of success and failure of experiments is traditional in probability theory because it helps to isolate the mathematical facts being discussed from the context of the particular situation under consideration.

Table 2-5 shows the possible outcomes in the game of American roulette. An honest roulette game amounts to the selection of one of the possible numbers at random. The player bets on the outcome of the random choice. While it is possible to bet that a particular number will be chosen, a more common sort of bet involves blocks of numbers. For example, the player may bet that the number that comes up will be in the first third (1–12), the second third (13–24), or the last third (25–36). In another type of bet, the gambler chooses a column of numbers from among the first column (1, 4, 7, and so on to 34), the

---

Table 2-5

	0	00	
1	2	3
4	5	6
7	8	9
10	11	12
13	14	15
16	17	18
19	20	21
22	23	24
25	26	27
28	29	30
31	32	33
34	35	36

---

second column (2, 5, 8, and so on to 35), and the third column (3, 6, 9, and so on to 36). Roulette players often bet on more than one block of numbers in a single play of the game.

As indicated in the table, we are to suppose that a gambler has placed a bet on the first third of the numbers and also on the last column. The next play of the roulette game will be a success from the gambler's point of view if the number chosen turns out to be in the first third or the last column, since in either case the gambler will win some money. We wish to compute the probability that the gambler will be successful, that is, to compute $p(\text{First third or last column})$.

Since we assume that the roulette game is honest so that any number is as likely to come up as any other, the definition in Section 2-1 tells us that the computation of the probability amounts to a counting problem. We will use the notation $n(\ \ )$ from now on to indicate the number of outcomes of the type described by the words or symbols within the parentheses. For example, $n(\text{Outcomes possible from one die}) = 6$ and $n(\text{Kings in a standard card deck}) = 4$, while $n(\text{Poker hands}) = C_5^{52} = 2{,}598{,}960$. The definition of probability in Section 2-1 then becomes

$$p(\text{Success}) = \frac{n(\text{Successful outcomes})}{n(\text{Possible outcomes})}$$

In particular for the roulette problem, we have

$$p(\text{First third or last column}) = \frac{n(\text{First third or last column})}{n(\text{Outcomes in roulette})}$$

The numbers 0, 00, 1, 2, 3, . . . , 36 give 38 different outcomes, so the denominator of the fraction is 38. The significant part of the problem then is the calculation of the numerator.

We can compute n(First third or last column) by listing the outcomes and counting up how many there are. We want to determine how many numbers between 1 and 36 are either less than or equal to 12 (first third) or evenly divisible by 3 (since the last column consists of 3, 6, 9, . . . , 36). The numbers which satisfy these requirements are

$$1, 2, 3, 4, 5, 6, 7, 8, 9, 10, 11, 12, 15, 18, 21, 24, 27, 30, 33, 36$$

If we count these numbers, we find that n(First third or last column) = 20. Therefore,

$$p(\text{First third or last column}) = \frac{20}{38}$$

However, in problems involving large numbers of outcomes, it may not be practical to make such a list, so we need a more sophisticated counting procedure.

It often happens that when there are two classes of favorable outcomes, call them A and B, the calculation of $n(A)$ and $n(B)$, the number of outcomes of each class, is simpler than the calculation of $n(A$ or $B)$. Thus, n(First third) is certainly 12 since first third means the numbers 1–12. Similarly, n(Last column) is 12 because the numbers 3, 6, 9, . . . , 36 that make up the last column are just the numbers 1–12 each multiplied by 3. In general, it is not possible to calculate $n(A$ or $B)$ only knowing $n(A)$ and $n(B)$. If we try adding $n(A)$ to $n(B)$ in the example, guessing that the number of outcomes in either A or B is just the number of outcomes in A plus the number in B, we get

$$n(\text{First third}) + n(\text{Last column}) = 12 + 12 = 24$$

Recall, however, that n(First third or last column) was 20. The problem is that a number like 6, which is both between 1 and 12 and a multiple of 3 is counted twice in the sum n(First third) + n(Last column).

The correct way to relate n(First third) + n(Last column) to n(First third or last column) is to adjust the sum so that no single number is counted twice. The numbers that are counted twice in n(First third) + n(Last column) are precisely those numbers that are both between 1 and 12 and also multiples of 3, namely 3, 6, 9, and 12. If we subtract the number of these, which should be written n(First third *and* last column), from the sum, then we are counting each of the numbers 3, 6, 9, and 12 just once. Thus, we have

$$n(\text{First third}) + n(\text{Last column}) - n(\text{First third and last column})$$
$$= 12 + 12 - 4 = 20 = n(\text{First third or last column})$$

The roulette example illustrates a general counting rule. Let us visualize all possible outcomes of some experiment as the points shown in Figure 2-1. If the experiment consists

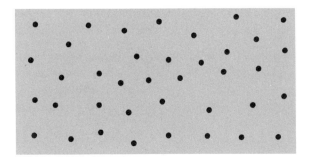

Figure 2-1

of a play of roulette, each dot represents one of the numbers listed in Table 2-5. The set of all outcomes is called the **sample space** of the experiment.

Suppose there are two classes of favorable outcomes for this experiment, class A and class B. The outcomes in these classes are represented in Figure 2-2 by the points within the indicated curves. The computation of the number of favorable outcomes either in class A or in class B amounts to counting the number of points within at least one of the curves. If we count up the number $n(A)$ of points in class A and add to it the number $n(B)$ of points in class B, we are counting those points which are in A and not in B once each and those points which are in B and not in A once each. But points in both A and B are counted twice, once as members of A and once as members of B. Thus, if we subtract from the sum $n(A) + n(B)$ the number $n(A \text{ and } B)$ of points which are both in A and B, then we eliminate the double counting of these elements and we obtain a correct count of the number of favorable outcomes. We now have the counting rule:

$$n(A \text{ or } B) = n(A) + n(B) - n(A \text{ and } B)$$

The visualization of outcomes by means of the sample space suggests a convenient language for stating the counting rule. The sample space is just a set of points. The outcomes

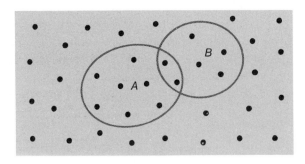

Figure 2-2

of class A are points in a subset of the sample space. We will call the subset A. Similarly, the outcomes of class B form a subset. The set of points that are either in the subset A or the subset B or both form a set which is the **union** of A and B; this set is written $A \cup B$. The set of points that are both in A and in B form another subset of the sample space, the **intersection** of A and B; this set is written $A \cap B$. Now taking $n(\ \ )$ to mean the number of points in the set named in the parentheses, the counting rule can be stated as

$$n(A \cup B) = n(A) + n(B) - n(A \cap B)$$

Thus, the counting rule which helps us calculate the probability of success by means of either of two types of outcomes is really a general counting rule for counting the number of points in the union of two sets.

2-16   **Example**   How many numbers from 1 through 1000 are either even or greater than 800?

Solution   It would certainly take a long time to list all the numbers required, so let A be the set of even numbers from 1 through 1000 and let B be the set of numbers from 801 through 1000. We wish to compute $n(A \cup B)$. Since there are 1000 numbers in all and half are even, $n(A) = 500$. There are 200 numbers from 801 through 1000, so $n(B) = 200$. The set $A \cap B$ consists of the even numbers from 801 through 1000, which is half the numbers in B. So $n(A \cap B) = 100$. By the counting rule, the answer we seek is

$$n(A \cup B) = n(A) + n(B) - n(A \cap B) = 500 + 200 - 100 = 600$$

The counting rule leads us to our first important principle of probability. If U is the symbol for the set of all points in the sample space (where each point represents a possible outcome to an experiment), A and B are two classes of favorable outcomes, and all outcomes are, by the nature of the experiment, equally likely, then

$$p(\text{Success}) = p(A \text{ or } B) = \frac{n(A \text{ or } B)}{n(U)}$$

In this equation, $n(U)$ is the number of all possible outcomes.

Just as, in arithmetic, we can rewrite fractions like

$$\frac{4 + 3 - 2}{7} \qquad \text{as} \qquad \frac{4}{7} + \frac{3}{7} - \frac{2}{7}$$

so the counting rule tells us that

$$p(A \text{ or } B) \doteq \frac{n(A \text{ or } B)}{n(U)} = \frac{n(A) + n(B) - n(A \text{ and } B)}{n(U)}$$

$$= \frac{n(A)}{n(U)} + \frac{n(B)}{n(U)} - \frac{n(A \text{ and } B)}{n(U)}$$

By definition, $n(A)/n(U)$ is the probability $p(A)$ of success from class A, and $n(B)/n(U)$ is $p(B)$. Also, the probability $p(A \text{ and } B)$ that both classes of success will be achieved simultaneously is the fraction $n(A \text{ and } B)/n(U)$. We will examine $p(A \text{ and } B)$ more closely in Section 2-7. For now, we restate the equations above entirely in terms of probability to obtain our first principle of probability:

$$p(A \text{ or } B) = p(A) + p(B) - p(A \text{ and } B)$$

Another way of expressing this principle of probability comes from the set theory description of probability we introduced above. In that setup, A and B are sets of points in the sample space, $(A \text{ or } B)$ is replaced by the set $(A \cup B)$ and $(A \text{ and } B)$ by $(A \cap B)$. Thus, in set theory language, the principle of probability is

$$p(A \cup B) = p(A) + p(B) - p(A \cap B)$$

In Section 2-1, we stated that probability is the same no matter whether it is calculated by counting equally likely outcomes or by accumulating experience in some complicated situation such as the detection of seismic events. Consequently, we will use the principle

$$p(A \text{ or } B) = p(A) + p(B) - p(A \text{ and } B)$$

whenever it is convenient in a probability problem, even though we reasoned it out only in the counting probability setting.

Now let us return to the gambler who bet that the roulette game would select a number either in the first third or the last column. To compute the probability of success, we recall that there are 38 possible outcomes represented by the numbers 0, 00, 1, 2, . . . , 36. Therefore,

$$p(\text{First third}) = p(\text{Last column}) = \frac{12}{38}$$

because there are 12 numbers of each type, and

$$p(\text{First third and last column}) = \frac{4}{38}$$

because 3, 6, 9, and 12 are the numbers of 1–12 that are divisible by 3. By the probability principle,

$$p(\text{First third or last column}) = p(\text{First third}) + p(\text{Last column})$$

$$- p(\text{First third and last column})$$

$$= \frac{12}{38} + \frac{12}{38} - \frac{4}{38} = \frac{20}{38} = .53 \quad (\text{approximately})$$

2-17   **Example**   A three digit number (including 000, 001, and so on) is chosen at random. What is the probability that it will either begin with a 9 or that all three digits will be different?

Solution   Let A represent all three digit numbers in which the first digit is 9 and let B represent all three digit numbers with three different digits. We wish to compute $p(A \text{ or } B)$. A number such as 922 in A may be viewed as an ordered triple of the form

(9, second digit, third digit)

There is only one choice for the first entry in the triple—it must be 9—but the other two members of the triple can each be any of the ten digits 0, 1, 2, . . . , 9. By the counting rule of Section 2-2, we know that $n(A) = (1)(10)(10) = 100$. From Example 2-8, we know there are 1000 three digit numbers in all, so $p(A) = 100/1000 = .10$. In Example 2-8 we also calculated that $n(B) = 720$, so $p(B) = 720/1000 = .72$. Now, $(A \text{ and } B)$ consists of those three digit numbers, or ordered triples of digits, of the form

(9, second digit, third digit)

where all three digits are different. Thus, the second digit must be a digit other than 9 and the third must be different from both 9 *and* the digit in the second position. The counting rule of Section 2-2 implies that $n(A \text{ and } B) = (1)(9)(8) = 72$, so that $p(A \text{ and } B) = 72/1000 = .072$. The principle of probability then produces the answer.

$$p(A \text{ or } B) = p(A) + p(B) - p(A \text{ and } B) = .10 + .72 - .072 = .748$$

Summary

If there are two classes of success, A and B, for an experiment, then the probability of success in the experiment is

$$p(A \text{ or } B) = p(A) + p(B) - p(A \text{ and } B)$$

The set of all outcomes is the **sample space** of the experiment. If the classes of successes A and B are thought of as subsets of the sample space, then their **union,** written $A \cup B$, is the set of outcomes which is of at least one such class. Their **intersection,** written $A \cap B$, is the set of outcomes which are both in class A and in class B. And

$$p(A \cup B) = p(A) + p(B) - p(A \cap B)$$

Exercises

- *In Exercises 2-100 and 2-101, one card is chosen from a standard deck.*

2-100   What is the probability of drawing either a face card or a club?

2-101   What is the probability of drawing either a red card or an ace?

2-102   A student needs money to stay in school. He can do so either if he finds a part-time job or if he is awarded a scholarship. He estimates the probability that he will get a job at .75, that he will be awarded a scholarship as .20, and that both will happen as .15. What is the probability that he will be able to stay in school?

2-103   In a faculty of 30 people, 22 have received doctorates and 10 have published books. Of the ten who have published books, seven have doctorates. How many members of the faculty have either a doctorate or a published book to their credit?

2-104   The European roulette wheel consists of the number 0, which is neither even nor odd, and the numbers 1–36. The game consists of choosing one of these numbers at random. What is the probability of winning if you bet on the odd numbers and the numbers greater than 24?

2-105   A publisher is assured of financial success from a novel if it is chosen as a book-club selection or if the reprinting rights are purchased by a paperback publisher. Suppose that there is a probability of .05 that any book will be selected by a book club, .15 that the reprint rights will be sold, and .045 that both will happen. What is the probability that a novel will be an assured financial success in at least one of these ways?

2-106   A professor asked her students how many had taken a course in trigonometry, and 32 raised their hands. She then asked how many had studied statistics, and 61 responded. When she asked how many students had taken both trigonometry and statistics, 16 said they had. If, in order to take the professor's course, a student must have taken at least one of trigonometry or statistics, how many students were in the class?

2-107   Two people walk into a department store. One wishes to buy a hammer and the other a tennis raquet. Suppose the probability is .40 of being served within 5 minutes in the hardware department and is .60 of being served within 5 minutes in the sporting goods department. If we are told that the probability is .76 that at least one of the two will be served within 5 minutes, what is the probability that they will both be served within 5 minutes?

2-108 Suppose that in a certain city 80% of all households have a television set, 50% subscribe to a newspaper, and 90% of the households that subscribe to a newspaper also have television sets. What is the probability that a household chosen at random will either have a television set or subscribe to a paper?

2-109 There is a country with a telephone system that is so bad there is a probability of .55 that a caller will be connected with the wrong number, .35 that the call will be cut off before the answering party can be identified, and .20 that the caller will both be connected with the wrong number and also cut off before discovering the number reached was the wrong one. What is the probability that at least one of these two catastrophes, wrong number or uncompleted call, will befall a person who makes a call?

2-110 An oil company drills two wells. It estimates that there is a probability of .20 that it will strike oil at the location of the first well and .30 that it will strike oil at the second. The company reports to its stockholders that there is a probability of .44 that it will strike oil in at least one of these wells. What is the estimated probability of striking oil at both wells?

2-111 Suppose that in a study of blue-eyed people it was found that 45% had a blue-eyed father, 45% had a blue-eyed mother (for each figure the other parent may be blue-eyed or not), and 35% had both blue-eyed parents. What percentage of the blue-eyed population had no blue-eyed parent?

2-112 The price of the stock of a shipbuilding company will rise if the company receives a large number of orders for its ships or if it is granted a low-interest government loan. The president of the company estimates that there is a probability of .40 that the stock will rise in price, because there is a probability of .30 that the company will receive a large number of orders for its ships and of .05 that it will both receive the orders and get the government loan. What does the president estimate to be the probability that his company will be granted the government loan?

2-113 When two dice are rolled, what is the probability that either the sum will be eight or at least one die will be even?

2-5 Disjoint Sets of Outcomes

An important special case of the probability principle

$$p(A \text{ or } B) = p(A) + p(B) - p(A \text{ and } B)$$

arises when there is no outcome which is both in class A and in class B. A gambler wins on the first roll of two dice if the sum of the numbers on the dice adds up to either 7 or 11. Recall that the possible outcomes when two dice are rolled may be viewed as ordered pairs (first die, second die) and that there are thus 36 outcomes in all. If A represents outcomes in which the numbers add up to 7 and B those where the sum is 11, then the probability that the gambler will win on the first roll is $p(A \text{ or } B)$. There is no way that the numbers on the dice can, on the same roll, add up to both 7 and 11, so there is no outcome which is in both classes. That is, $n(A \text{ and } B) = 0$. Consequently, $p(A \text{ and } B) = 0/36 = 0$ and the principle of probability reduces in this case to the simple formula

$$p(A \text{ or } B) = p(A) + p(B)$$

The outcomes of type A (sum of 7) are (1, 6), (2, 5), (3, 4), (4, 3), (5, 2), and (6, 1), so $p(A) = 6/36$. The only rolls with a sum of 11 are (5, 6) and (6, 5), so $p(B) = 2/36$. Therefore,

$$p(\text{Win on first roll}) = p(A \text{ or } B) = p(A) + p(B)$$

$$= \frac{6}{36} + \frac{2}{36} = \frac{8}{36} = .22 \quad (\text{approximately})$$

The special case we have just illustrated arises so often, it is worth stating by itself:

> If there is no outcome which is both in class A and in class B, then
>
> $$p(A \text{ or } B) = p(A) + p(B)$$

In the language of sets, two sets are said to be **disjoint** if no point is a member of both. In other words, the set $A \cap B$ contains no points. Thus, if the sets A and B of outcomes are disjoint, then

$$p(A \cup B) = p(A) + p(B)$$

Suppose a gambler at American roulette bets that the number will be either in the first third (1–12) or the last third (25–36). Then since no number can be simultaneously less than 13 and greater than 24, the special case applies and the probability that the gambler will now win some money is

$$p(\text{First third or last third}) = p(\text{First third}) + p(\text{Last third})$$

$$= \frac{12}{38} + \frac{12}{38} = \frac{24}{38} = .63 \quad (\text{approximately})$$

in contrast to the probability $p(\text{First third or last column})$ of about .53 we calculated in Section 2-4. This is not to say that first third or last third is really a better bet than is first third or last column. It can be shown that the gambler should expect to lose money at exactly the same rate on either bet (in the long run), as we will see in Section 2-11.

The special case of the probability principle permits us to relate the probability that an experiment will succeed with the probability that it will fail. An experiment is performed and we ask for the probability $p(\text{Success or failure})$. If all outcomes are equally likely, then

$$p(\text{Success or failure}) = \frac{n(\text{Success or failure})}{n(\text{Outcomes})}$$

Every outcome is either a success or a failure, so certainly $n(\text{Success or failure}) = n(\text{Out-comes})$. We conclude that

$$p(\text{Success or failure}) = \frac{n(\text{Outcomes})}{n(\text{Outcomes})} = 1$$

In fact in any kind of probability situation, every outcome is either a success or a failure, so there is a 100% probability that the outcome will be one or the other. In decimal terms, $p(\text{Success or failure}) = 1$ again. On the other hand, no outcome is both a success and a failure at the same time, so the principle of probability applies in its special case to give

$$1 = p(\text{Success or failure}) = p(\text{Success}) + p(\text{Failure})$$

We can relate $p(\text{Success})$ to $p(\text{Failure})$ by writing either

$$p(\text{Success}) = 1 - p(\text{Failure})$$

or, if it is more appropriate to our problem,

$$p(\text{Failure}) = 1 - p(\text{Success})$$

2-18   **Example**   A television dealer receives ten sets from the manufacturer. The dealer decides to select two of the sets at random to test and send the entire shipment back to the manufacturer if anything is wrong. If there are in fact two faulty sets among the ten, what is the probability that the dealer will reject the shipment?

Solution   The probability that the dealer will succeed—find at least one of the defective sets in the sample—is most easily computed by first determining the probability of failure to find anything wrong. In Example 2-15, we calculated that there was a probability of about .62 that the dealer would select two good sets for the sample. Therefore,

$$p(\text{Reject the shipment}) = p(\text{Succeed}) = 1 - p(\text{Failure})$$

$$= 1 - .62 = .38 \qquad (\text{approximately})$$

If there are three or more ways to succeed in an experiment, we would also like to know how to calculate the probability of success in the sense of coming up with an outcome which is in one of the favorable classes. In other words, we would like a probability principle for, say, $p(A$ or B or $C)$. There is such a formula, but it is pretty complicated. Fortunately, in the cases that will concern us, we will know that the various classes of success are all distinct; that the sets of favorable outcomes are all disjoint. When this is true, the probability principle is quite simple.

Figure 2-3 illustrates a sample space with three distinct types of successes, A, B, C. We can count up the total number of successful outcomes by merely adding up the number of outcomes of each type, that is,

$$n(A \text{ or } B \text{ or } C) = n(A) + n(B) + n(C)$$

There is nothing to subtract, because no outcome is in more than one class and so there is no way any outcome can be counted more than once. More generally, if there are k classes $A_1, A_2, \ldots, A_k$ of successful outcomes for an experiment and no outcome is in more than one class, then

$$n(A_1 \text{ or } A_2 \text{ or } \ldots \text{ or } A_k) = n(A_1) + n(A_2) + \cdots + n(A_k)$$

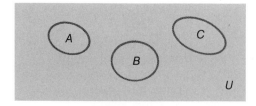

Figure 2-3

Once we can count the number of favorable outcomes, there is again no difficulty in calculating probability when all outcomes are equally likely. If A, B, and C denote types of outcomes, there is no outcome which is in more than one class, and U represents the set of all possible outcomes to the experiment, then

$$p(A \text{ or } B \text{ or } C) = \frac{n(A \text{ or } B \text{ or } C)}{n(U)} = \frac{n(A)}{n(U)} + \frac{n(B)}{n(U)} + \frac{n(C)}{n(U)}$$

$$= p(A) + p(B) + p(C)$$

The general principle of probability toward which we have been headed is now clear:

> If A_1, A_2, . . . , A_k are classes of outcomes for an experiment and no outcome is in more than one such class, then
>
> $$p(A_1 \text{ or } A_2 \text{ or} \ldots \text{or } A_k) = p(A_1) + p(A_2) + \cdots + p(A_k)$$

2-19   **Example**   Five coins are flipped. What is the probability that at least two of them will come up heads?

Solution   Since the five coins are physically distinct, we will think of them as first coin, second coin, and so on. Then the result of the experiment—flipping all five coins—may be represented by an ordered 5-tuple with either heads (H) or tails (T) in each of the five positions. Since there are two possible symbols in each position in the 5-tuple, the appropriate counting rule implies that there are $(2)(2)(2)(2)(2) = 32$ different outcomes in all. We will next list all 32 of the outcomes, grouped according to the number of heads in the outcome (Table 2-6). In Section 2-8, we will develop a much more efficient method for working with probability problems of this type. If there are exactly two heads in the five flips, then the experiment is successful in the sense that there were at least two heads. However, exactly three heads out of five flips is also an outcome with at least two heads, so this class is successful as well. The same may be said of outcomes with four heads out of five flips and of the single outcome in which all coins turned up heads. The various classes of outcomes represented by the columns of the table below are certainly disjoint; no outcome can, for example, contain both exactly two heads and exactly three. Let $n(\text{Two})$, $n(\text{Three})$, and so on stand for the number of outcomes with exactly two heads, exactly three, etc. From the table, we count $n(\text{Two})$, $= 10$, $n(\text{Three}) = 10$, $n(\text{Four}) = 5$, and $n(\text{Five}) = 1$. To illustrate the probability principle we have been discussing, we will use it in this example to conclude that

$$p(\text{At least two heads}) = p(\text{Two or three or four or five})$$

$$= p(\text{Two}) + p(\text{Three}) + p(\text{Four}) + p(\text{Five})$$

$$= \frac{10}{32} + \frac{10}{32} + \frac{5}{32} + \frac{1}{32} = \frac{26}{32} = .81 \qquad \text{(approximately)}$$

Table 2-6
Number of Heads

0	1	2	3	4	5
TTTTT	HTTTT	HHTTT	HHHTT	HHHHT	HHHHH
	THTTT	HTHTT	HHTHT	HHHTH	
	TTHTT	HTTHT	HHTTH	HHTHH	
	TTTHT	HTTTH	HTHHT	HTHHH	
	TTTTH	THHTT	HTHTH	THHHH	
		THTHT	HTTHH		
		THTTH	THHHT		
		TTHHT	THHTH		
		TTHTH	THTHH		
		TTTHH	TTHHH		

2-20   **Example**   If seven seismic events of a certain magnitude take place in some part of the world where there is a probability of .85 that each will be detected by teleseismic means, then the following table, based on methods which will be described in Section 2-9, states the approximate probability that exactly a given number of the events will be detected by such means:

Detections	0	1	2	3	4	5	6	7
Probability	.000	.000	.001	.011	.062	.210	.396	.320

(The probability .000 of no detections out of seven events does not mean that this outcome is impossible, i.e., that there is zero probability that it will happen. Rather, .000 indicates that the probability is too small to be measured to within three places after the decimal point. In fact, the probability of no detections out of seven is about .0000016.) What is the approximate probability that more than four of the seismic events will be so detected?

Solution   The classes of outcomes in which more than four events are detected are those in which exactly five, exactly six, and all seven are detected. Since no outcome can be of more than one such type, the principle of probability tells us that

$$p(\text{More than four detections}) = p(\text{Five or six or seven})$$

$$= p(\text{Five}) + p(\text{Six}) + p(\text{Seven})$$

$$= .210 + .396 + .320 = .926 \quad (\text{approximately})$$

Examples 2-19 and 2-20 point out some of the distinctions in language which arise in probability problems. In order to make clear the meanings of phrases which occur fre-

	0	1	2	3	4	5	6	7
				Table 2-7				
At least four					$\times$	$\times$	$\times$	$\times$
More than four						$\times$	$\times$	$\times$
At most four	$\times$	$\times$	$\times$	$\times$	$\times$			
Fewer than four	$\times$	$\times$	$\times$	$\times$				

quently in such problems, we have assumed that seven events have occurred and a probability problem asks for the probability that a certain number will be detected. A cross ($\times$) in Table 2-7 means that the probability of that exact number must be included in solving the problem. Remember also that *no more than* means *at most* and that *no fewer than* or *no less than* means *at least*.

2-21 **Example** What is the probability that a poker hand will contain at least one ace?

Solution On the model of the previous two examples, we would expect to compute the probability that the hand will contain exactly one ace, the probability that it will contain exactly two aces, and the same for three and four aces. We would then add up these probabilities to solve the problem. While this approach would work, it is very tedious and there is, fortunately, a much neater way to arrive at the solution. We use the principle

$$p(\text{Success}) = 1 - p(\text{Failure})$$

In this example, failure amounts to being dealt a five card poker hand in which no ace appears. A hand of that type is just the same as a poker hand drawn from an incomplete deck, namely, the standard 52 card deck with the four aces removed. A poker hand taken from such a 48 card deck is a combination (since order does not matter) of five cards out of the 48. According to Section 2-3, there are then

$$C_5^{48} = \frac{48!}{(5!)(43!)} = \frac{(48)(47)(46)(45)(44)}{(5)(4)(3)(2)(1)} = 1{,}712{,}304$$

different poker hands with no aces. We found in Example 2-12 that there are 2,598,960 poker hands in all, so

$$p(\text{Failure}) = p(\text{No aces}) = \frac{1{,}712{,}304}{2{,}598{,}960} = .659 \qquad (\text{approximately})$$

We conclude that

$$p(\text{At least one ace}) = 1 - p(\text{No aces}) = 1 - .659 = .341 \qquad (\text{approximately})$$

Summary

If there is no outcome which is both in class A and in class B, then

$$p(A \text{ or } B) = p(A) + p(B)$$

Consequently,

$$p(\text{Success}) = 1 - p(\text{Failure})$$

and

$$p(\text{Failure}) = 1 - p(\text{Success})$$

If $A_1, A_2, \ldots, A_k$ are classes of outcomes for an experiment and there is no outcome which is in more than one class, then

$$p(A_1 \text{ or } A_2 \text{ or } \ldots \text{ or } A_k) = p(A_1) + p(A_2) + \cdots + p(A_k)$$

Exercises

2-114   Two dice are rolled. What is the probability that both are odd? What is the probability that at least one is even?

2-115   A list of the rulers of England since the Norman Conquest indicates the age of each at death. Of these 42 rulers before Elizabeth II, two died before they turned 30, four died in their thirties (that is, from 30 through 39), nine died in their forties, ten in their fifties, ten in their sixties, four in their seventies, and two in their eighties. If a ruler is chosen at random, what is the probability that the ruler was at least 50 years of age when he or she died?

2-116   With respect to a certain gene, 15% of a population is of genotype AA, 80% of genotype Aa, and the remaining 5% of genotype aa. An individual either of genotype AA or of genotype aa is said to be *homozygous*. If an individual is chosen at random from this population, what is the probability that the individual is homozygous with respect to this gene?

●   *In Exercises 2-117 through 2-119 there is a country in which 30% of the population is color-blind. If ten individuals are chosen at random, the approximate probability p(k) that exactly k of them are color-blind is given by the following table:*

k	0	1	2	3	4	5	6	7	8	9	10
$p(k)$	.03	.12	.23	.27	.20	.10	.04	.01	.00	.00	.00

2-117   What is the approximate probability that at least five people in the sample will be color-blind?

2-118   What is the approximate probability that there will be at least one color-blind person in the sample?

2-119   What is the approximate probability that fewer than three people in the sample will be color-blind?

2-120   In a test of extrasensory perception, six cards, each from a different standard deck, are placed face down on a table and a "psychic" is asked to tell which suit each card is. He gets three correct out of the six. If a person guessed at random what suit each card was, the approximate probability $p(k)$ of getting exactly k correct out of the six is given by the following table:

k	0	1	2	3	4	5	6
$p(k)$	.18	.36	.30	.13	.03	.00	.00

What is the approximate probability that a person who guesses will do at least as well as the psychic did?

●   In Exercises 2-121 and 2-122, we suppose that in a certain city 30% of the families with children under 18 years of age have no car, 50% have one car, 15% have two cars, and 5% have more than two cars.

2-121   If a family is chosen at random from among all those with children under 18 years of age, what is the probability that it will have more than one car?

2-122   What is the probability that such a family chosen at random will have no more than two cars?

●   Exercises 2-123 through 2-125 use the following information: Two evenly matched teams are in a basketball playoff. That is, we assume that each team has exactly .50 probability of winning any given game. The winner of the playoff is the first team to win four games. Then the probability is 1/8 that the playoff will last only four games, 1/4 that it will last exactly five games, 5/16 that it will last exactly six games, and 5/16 that it will require all seven games.

2-123   What is the probability that the playoff will last at least five games?

2-124   What is the probability that the playoff will require fewer than seven games?

2-125   What is the probability that the playoff will require fewer than four games?

● *In Exercises 2-126 through 2-129, a city health department has received complaints that a supermarket is selling hamburger that is mostly fat. A health inspector takes three packages of hamburger at random from ten such packages in the meat counter to test whether any of them exceed the limit for fat content permitted by state laws.*

2-126   In how many ways can the inspector choose the sample of three packages?

2-127   If, in fact, four of the packages in the meat counter contain too much fat, in how many ways can the inspector choose three packages with legal fat content?

2-128   What is the probability that the inspector will fail to detect any violations?

2-129   What is the probability that the inspector will choose at least one package with too high a fat content?

2-130   A man owns six vacation cabins that he rents out by the week. He supplies television sets at extra cost. He observes that about 2/3 of the families who rent cabins from him want television sets, so he buys four of them. In a week in which all his cabins are rented, the following table gives the approximate probability $p(k)$ that exactly k of the families will want a television set:

k	0	1	2	3	4	5	6
$p(k)$	.00	.02	.08	.22	.33	.26	.09

What is the approximate probability that he will not have enough sets to satisfy all his customers?

2-131   Suppose that 55% of the voters in an election for mayor will vote for Smith while the rest will vote for Jones. The approximate probability $p(k)$ that out of a random sample of five voters exactly k will be for Smith is given by the following table:

k	0	1	2	3	4	5
$p(k)$	.02	.11	.27	.34	.21	.05

What is the approximate probability that the sample will predict correctly the outcome of the election (that is, it will contain a majority of voters who favor Smith)?

2-132   A candidate for governor concludes from the polls that the probability of election for each of her two opponents is .41 and .15, respectively. What does the candidate estimate her own probability of election to be?

2-133   In rolling three dice, what is the probability of obtaining a sum either at least as large as 17 or no greater than 4?

● *In Exercises 2-134 through 2-138, a garden supply store has a bin containing nine tulip bulbs. Four of the bulbs will produce red tulips, three will give yellow tulips, and two will give white tulips. Since the bulbs all look alike, a customer chooses two at random.*

2-134   In how many ways can the bulbs be chosen?

2-135   In how many ways can the customer choose two bulbs for red tulips? Two for yellow? Two for white?

2-136   What is the probability that the customer will end up with two red tulips? Two yellow tulips? Two white tulips?

2-137   What is the probability that the customer will buy two tulips of the same color?

2-138   What is the probability that the customer will get tulips of different colors?

2-6 Conditional Probability

Let us imagine that two gamblers agree to play the following simple card game: Each player bids $1 and is dealt one card from a standard deck. After looking at the card, one of the players has the opportunity to raise the value of the bet by $1. The players alternate each time the game is played as to which player has the right to raise the value of the bet. If one player raises the bet, the other player must meet it. The winning card of the two dealt is the higher one in the order

$$2\ 3\ 4\ 5\ 6\ 7\ 8\ 9\ 10\ J\ Q\ K\ A$$

and, if both cards are of the same value, the winning card is the higher suit in the order (with the higher suit to the right)

Thus, an ace beats a jack and the 6♠ beats the 6◊. The player holding the higher card wins the $2 bet, or, if the value of the bet has been raised, the entire $4 bet.

A player is dealt the 8◊ and has the opportunity to decide whether to raise the value of the bet. The strategy should be to raise the value of the bet if you think you hold the higher card, but keep the value of the bet at $1 otherwise. In this case, raise the value of the bet only if the probability is at least .50 that the 8◊ will win the game.

The cards that are higher than the 8◊ are all the aces, kings, queens, jacks, tens, and nines, together with the two higher eights: the 8♠ and the 8♡. Thus, there are 26 cards higher than the 8◊. Suppose the player reasons that since there are 52 cards in all the standard deck, the probability is 26/52 = .50 that the opponent holds a higher card. Then the player concludes that the probability of winning is also .50, and so there is no harm in raising the value of the bet. But a gambler who consistently uses this kind of reasoning is likely to have a gambling career that is both short and costly.

In order to compute correctly the probability that the opponent holds a higher card, the gambler must use all the information available. The error in the last paragraph was that the gambler failed to take account of the fact that the 8 ◇ in the gambler's own hand was from the same deck as the card dealt to the opponent. Thus, the opponent's card can be considered as dealt from a deck of 51 cards, the standard deck with the 8 ◇ removed. The 26 cards that can beat the 8 ◇ are still in the deck of 51, so the correct probability is 26/51, or about .51, that the opponent will hold a higher card. The gambler with the 8◇ should not raise the value of the bet, because the probability of winning is only about .49.

The reasoning in the last paragraph may seem rather difficult to carry over to other probability problems. Let us, therefore, try to visualize the analysis by means of the set theory language we introduced in Section 2-4. Let U be the set of all cards in the standard 52 card deck. Then U is the sample space for choosing cards at random from a deck. The subset A will consist of all aces, kings, queens, jacks, tens, nines, and the 8♠ and 8♡. When the gambler first asked the probability that the opponent would be dealt a card higher than the 8◇, the gambler was visualizing the problem as shown in Figure 2-4. With this in mind, the gambler computed the probability as

$$p(A) = \frac{n(A)}{n(U)} = \frac{26}{52} = .50$$

Now let B be the set of outcomes (cards) which are really possible for the opponent when we know that the gambler holds the 8◇. Then B consists of all the cards in the standard deck except the 8◇, and we should visualize the situation as shown in Figure 2-5.

When we took into account the additional information from the gambler's card, we computed the **conditional probability** that the outcomes in A will take place—under the conditions imposed by the set B. The conditional probability is represented by the symbol

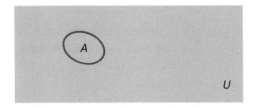

Figure 2-4

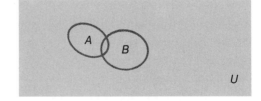

Figure 2-5

$p(A|B)$. The condition B tells us that the set of possible outcomes is not really the entire sample space U, but must actually be an outcome in the subset B. Thus, the denominator in the calculation of $p(A|B)$ is not $n(U)$, the number of all possible outcomes, but rather $n(B)$, the number of possibilities which really have any chance of occurring. One might expect that $p(A|B)$ would then be $n(A)$, the number of successful outcomes, divided by the number of outcomes that are really possible, $n(B)$. This is almost right, except that there may now be outcomes in A which we know cannot occur because they are not in B. While $p(A)$ is the number of successful outcomes divided by the total number of outcomes, $p(A|B)$ intuitively is the number of possible successful outcomes (possible after knowing the information in B) divided by the number of possible outcomes. Thus, the numerator of $p(A|B)$ is the number of outcomes which are both successful (in A) and possible (in B). Recall that the set of outcomes which belongs to both set A and set B is the intersection of A and B, denoted by $A \cap B$.

When considering equally likely outcomes, the definition of **conditional probability** is

$$p(A|B) = \frac{n(A \cap B)}{n(B)}$$

If we prefer to think of A and B as verbal descriptions of outcomes so that $(A$ and $B)$ means outcomes which are in both classes, then

$$p(A|B) = \frac{n(A \text{ and } B)}{n(B)}$$

In the gambling example above, B consists of all cards except for the $8\diamondsuit$, while A consists of the $8\spadesuit$, $8\heartsuit$, all nines, tens, jacks, queens, kings, and aces. Thus, in this case, $A \cap B$ is the same as A, because every card in A is a possible outcome for the gambler's opponent. We can now use the definition to compute the probability that the opponent will win when the gambler holds the $8\diamondsuit$:

$$p(A|B) = \frac{n(A \cap B)}{n(B)} = \frac{26}{51} \text{ (about .51)}$$

The definition

$$p(A|B) = \frac{n(A \cap B)}{n(B)}$$

that A will occur, knowing that the possibilities are limited to B, makes sense for the simple counting probability of equally likely events, provided only that B contain some outcome [since we cannot divide by $n(B) = 0$]. However, we would like a definition of conditional probability that avoids the idea of counting so that we can extend the definition to probability based on experience. We wish to replace the definition of conditional probability with an equivalent one expressed entirely in terms of probability itself. The trick is to divide both the numerator and the denominator of the fraction $n(A \cap B)/n(B)$ by the total number of outcomes if we ignore B, that is, by $n(U)$. Such division does not change the value of the fraction, but now we can use the definition of counting probability from Section 2-1 to write

$$p(A|B) = \frac{n(A \cap B)}{n(B)} = \frac{\dfrac{n(A \cap B)}{n(U)}}{\dfrac{n(B)}{n(U)}} = \frac{p(A \cap B)}{p(B)}$$

We take the equation we have just worked out for the simplest kind of probability and use it to define conditional probability in general.

> If an experiment is carried out, then the **conditional probability** that the outcome will come from a subset A, given that the possible outcomes are restricted to a subset B, is defined to be
>
> $$p(A|B) = \frac{p(A \cap B)}{p(B)}$$

If we prefer to think of A and B as words describing classes of outcomes, then we can define the conditional probability that an outcome will be in class A, given the additional information that the only possible outcomes are those in class B, by

$$p(A\,|B) = \frac{p(A \text{ and } B)}{p(B)}$$

2-22   **Example**   The faculty of a school consists of 20 men, of whom 8 have master's degrees, and 30 women, of whom 18 have master's degrees. (*a*) If a teacher is chosen at random to represent the school on a county education committee, what is the probability that the teacher selected will have a master's degree? (*b*) If before the announcement of the representative, the principal lets slip the fact that a woman was chosen by referring to the representative as "she," what now is the probability that a teacher with a master's degree will represent the school?

Solution (*a*) There are 50 teachers in all and 26 of them have master's degrees, so the probability $p(A)$ that a teacher chosen at random will have such a degree is

$$p(A) = \frac{n(A)}{n(U)} = \frac{26}{50} = .52$$

(*b*) Now we know that the teacher chosen comes from the set B of women teachers. According to the problem, there are 18 teachers who are both holders of master's degrees and also women. Therefore, the probability that the representative has a master's degree, given that she is a woman, is

$$p(A|B) = \frac{n(A \text{ and } B)}{n(B)} = \frac{18}{30} = .60$$

2-23 **Example** Suppose there is a card game in which each of the two players is dealt two cards. (*a*) What is the probability that a player will be dealt two face cards? (The face cards are the kings, queens, and jacks.) (*b*) If a player is dealt two face cards, what is the probability that the opponent will also have two face cards?

Solution (*a*) We wish to compute $p(A) = n(A)/n(U)$, where U represents all possible two card hands and A represents those in which both cards are face cards. In the terminology of Section 2-3, a hand of cards is a combination—in this case a combination of two cards out of some larger set of cards. The hands in U are all the two card combinations from the 52 card deck, so according to Section 2-3,

$$n(U) = C_2^{52} = \frac{52!}{(2!)(50!)} = \frac{(52)(51)}{(2)(1)} = 1326$$

The hands in A must consist just of face cards, so they may be considered to have been taken from the twelve card set of face cards. Consequently,

$$n(A) = C_2^{12} = \frac{(12)(11)}{(2)(1)} = 66$$

so the probability that a player will receive two face cards is

$$p(A) = \frac{66}{1326} = .05 \quad \text{(approximately)}$$

(*b*) We know the contents of one player's hand, so the opponent's possible hands, which we denote by B, are taken from the remaining 50 cards. Therefore,

$$n(B) = C_2^{50} = \frac{(50)(49)}{(2)(1)} = 1225$$

Since the cards already accounted for in the one player's hand are two face cards, the two card hands that both consist of face cards (are in class A) and come from the remaining 50 card deck (are in class B) are precisely those two card hands that can be drawn from the ten face cards that remain in the 50 card deck. We see that

$$n(A \text{ and } B) = C_2^{10} = \frac{(10)(9)}{(2)(1)} = 45$$

which tells us that

$$p(A|B) = \frac{n(A \text{ and } B)}{n(B)} = \frac{45}{1225} = .037 \qquad \text{(approximately)}$$

Thus, while the opponent is unlikely to have two face cards in any event, it will be even less likely given the first player's all-face-card hand.

In both Examples 2-22 and 2-23, the probability that an outcome from class A would occur changed when we were given additional information B. Thus, the probability that the school's representative would have a master's degree changed from .52 to .60 once it was known that she was a woman. The reader should not get the idea, however, that $p(A|B)$ is necessarily different from $p(A)$. Suppose a coin is tossed five times and we wish to calculate the probability that it will come up heads on the fifth flip. A sequence of five tosses may be thought of as an ordered 5-tuple with each position either heads (H) or tails (T). Then we recall that there are $(2)(2)(2)(2)(2) = 32$ possible outcomes. An outcome in which heads occurs on the last toss is represented by a 5-tuple of the form $(\ ,\ ,\ ,\ ,\text{H})$. There are $(2)(2)(2)(2)(1) = 16$ of these outcomes, because there is only one possibility for the last position. We conclude that

$$p(\text{Fifth toss heads}) = p(A) = \frac{16}{32} = .50$$

Now suppose we were told that the first four tosses came up heads and we were asked to find the conditional probability that the last flip would come up heads as well. As usual, we would let B denote the set of outcomes possible under the given condition; in this case that the first four tosses came up heads. Then B consists of ordered 5-tuples of the form (H, H, H, H,), of which there are just two: (H, H, H, H, H) and (H, H, H, H, T). The first of these outcomes is also in A, because the last toss in that outcome came out heads. Therefore,

$$p(A|B) = \frac{n(A \text{ and } B)}{n(B)} = \frac{1}{2} = .50$$

The calculation we have just concluded indicates that the outcome of the fifth toss of the coin is not affected by the fact that it has just come up heads four times in a row. This is hardly surprising, since the coin has no way of remembering what it did in previous flips. (We are assuming, of course, that it is an honest coin and that the first four heads were the result of chance and not an indication that the coin has heads on both sides.) The outcomes A, heads on the fifth toss, and B, heads on the first four tosses, are not related to each other. In technical language, we say that these classes of outcomes are **independent.**

Two classes of outcomes A and B to some experiment are **independent** if

$$p(A|B) = p(A)$$

As a consequence, we can test whether types of outcomes are independent by computing the two probabilities $p(A|B)$ and $p(A)$ to see whether they are the same.

2-24   **Example**   Two dice are rolled and the first die comes up a three. (a) What is the probability that the sum of the numbers on the two dice is 6? (b) Is the outcome (Sum equals 6) independent of the outcome (First die is three)? (c) What is the probability that the roll is a double, i.e., both dice show the same number? (d) Is the outcome (Double) independent of (First die is three)?

Solution   (a) Representing rolls of the dice as ordered pairs, the set B of pairs in which the first die is a three consists of the six outcomes $(3, 1)$, $(3, 2)$, $(3, 3)$, $(3, 4)$, $(3, 5)$, and $(3, 6)$. If A is the outcomes in which the sum is 6, namely the five outcomes $(1, 5)$, $(2, 4)$, $(3, 3)$, $(4, 2)$, and $(5, 1)$, then only one result, $(3, 3)$, is in both A and B. Since the question asks us to compute the probability that the sum is 6 when, as a condition, the first die is a three, we compute

$$p(A|B) = \frac{n(A \cap B)}{n(B)} = \frac{1}{6}$$

(b) To determine whether A (Sum equals 6) is independent of B (First die is three), we calculate $p(A) = n(A)/n(U)$, where U stands for the set of all possible outcomes of the experiment of rolling two dice. The answer, $p(A) = 5/36$, is not the same one we got for $p(A|B)$ in part (a), so the outcomes are not independent. Intuitively, we would expect the probability of getting a sum of 6 to be improved by knowing that the first die is a three, because this eliminates the possibility that the first die could be a six and therefore could not possibly give rise to a sum of 6 from two dice.

(c) Now let A stand for the six doubles outcomes $(1, 1)$, $(2, 2)$, $(3, 3)$, $(4, 4)$, $(5, 5)$, and $(6, 6)$. Again, only $(3, 3)$ belongs to B, the set of outcomes with a three on the first die, so

$$p(A|B) = \frac{n(A \cap B)}{n(B)} = \frac{1}{6}$$

(d) This time, since there are six doubles,

$$p(A) = \frac{n(A)}{n(U)} = \frac{6}{36} = \frac{1}{6} = p(A|B)$$

from part (c), so A and B are independent by definition. Our intuition tells us that the result makes sense, because in order to form doubles the second die must match the first, but it does not matter which of the six possible numbers the first die happens to be.

Summary

The **conditional probability** that A will occur, given that B happens, is denoted by $p(A|B)$ and defined in the language of set theory by

$$p(A|B) = \frac{p(A \cap B)}{p(B)}$$

or, taking A and B to be descriptions of classes of outcomes, by

$$p(A|B) = \frac{p(A \text{ and } B)}{p(B)}$$

When all outcomes are equally likely, so that probability depends on counting, then we can use the computational formula

$$p(A|B) = \frac{n(A \cap B)}{n(B)}$$

Outcomes of class A and class B are said to be **independent** if

$$p(A|B) = p(A)$$

Exercises

2-139   A publisher estimates that there is a probability of .05 that a new novel will be a book-club selection and .01 that it will both be a book-club selection and be made into a movie. What does he estimate to be the probability that a novel which is selected by a book club will be made into a movie?

•   *In Exercises 2-140 through 2-142, imagine that at a food-processing plant, an employee forgot to change the labels in a labeling machine, so the day's production of 2000 cans of cherries, 1000 cans of peaches, 4000 cans of peas, and 4000 cans of popping corn are all labeled identically as "beans" and cannot otherwise be distinguished because the cans are all the same size. The processor sells off the cans at a reduced price and the first customer chooses a can at random.*

2-140   What is the probability that the customer chooses a can of peaches?

2-141   What is the probability that the customer will choose a can of fruit?

2-142   The customer picks up a can and shakes it. Since it does not rattle, he knows it cannot contain popping corn. Now, what is the probability that the can contains fruit?

● *For Exercises 2-143 and 2-144, suppose that in a certain city 30% of the families with children under 18 years of age have no car, 50% have one car, 15% have two cars, and 5% have more than two cars.*

2-143   What is the probability that a family with children under 18 years of age has at least two cars?

2-144   If it is observed that there is a car belonging to the family parked in its garage, what now is the probability that the family has at least two cars?

● *In Exercises 2-145 through 2-147, a mouse is placed in the maze pictured below. The mouse runs until it can go no further, and it never reverses direction.*

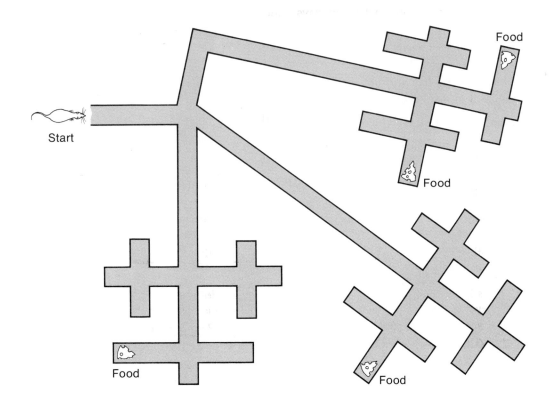

2-145   What is the probability that the mouse will reach food, assuming that it chooses directions at random so that all destinations are equally likely?

2-146   What is the probability that the mouse will reach food, given that it turned left at the first opportunity?

2-147   What is the probability that the mouse will reach food, given that if it has a choice of directions, it will never turn right?

2-148   In a certain city, 80% of all households have television sets, while 20% have both a television and a stereo. What is the probability that a household with a television selected at random also has a stereo?

2-149   An oil company plans to drill two wells. It estimates that there is a probability of .20 that it will strike oil at the first location and of .03 that it will strike oil at both locations. If the oil company assumes that the events of striking oil at the two locations are independent, what is the probability they estimate for striking oil at the second location?

2-150   A manufacturer of model airplane kits finds that 3% of its production is defective because the kit contains a broken part, 7% of the boxes in which the parts are packed are defective, and in 1% of its production the box is damaged and there is a broken part in the kit. Are the outcomes (Part broken) and (Box damaged) independent, or should the manufacturer suspect that there is a connection between its two problems?

●   *In Exercises 2-151 through 2-154, a family consisting of a father, mother, and child is chosen at random and asked on what day of the week each was born.*

2-151   What is the probability that all three were born on different days, given that the father was born on Monday?

2-152   Is the outcome (All born on different days) independent of (Father was born on Monday)?

2-153   What is the probability that all were born on different days, given that the father was born on Monday and the mother on Tuesday?

2-154   Is the outcome (All born on different days) independent of (Father was born on Monday and mother on Tuesday)?

2-155   An automobile manufacturer estimates the probability that the government will make emission standards for engines tougher at .60, that the cost of steel will go up at .40, and that both will happen at .24. Is the manufacturer assuming that the outcomes (Tougher emission standards) and (Higher steel prices) are independent?

●   *Exercises 2-156 through 2-158 concern the gambling game called "blackjack" or "21." In blackjack, each player is dealt two cards and, after looking at them, may ask for more cards. The player automatically loses if the total value of all cards exceeds 21. Aces count as 1 or 11 and face cards count as 10. A player is dealt a ten and a six.*

2-156   What is the probability that the player will exceed 21 if another card is taken?

2-157   In blackjack, the dealer gives himself (or herself) two cards; one is dealt face up so the other players can see it. If the dealer's visible card is a nine, what now is the player's probability of exceeding 21 if another card is taken?

2-158   If the dealer's visible card is a two, what is the probability that the player will exceed 21 if another card is taken?

● *In Exercises 2-159 and 2-160, suppose the price of the stock of a certain oil company will rise substantially this year if either its current oil explorations uncover a large amount of oil or the government permits a rise in domestic oil prices. The president of the company estimates there is a probability of .25 that the company will find a large amount of oil, of .60 that the government will permit oil prices to rise, and of .09 that both will happen.*

2-159   What does the president of the company estimate to be the probability that the government will permit oil prices to rise if the company does find a large amount of oil?

2-160   Is the president of the oil company assuming that a rise in the government ceiling on the price of oil is independent of whether the company finds large amounts of oil?

2-161   There is a country where the telephone service is so bad that there is a probability of .55 that a caller will be disconnected before the party on the other end can answer and of .15 that a caller will both be connected with a wrong number and then have the connection cut off before finding out the identity of the party on the other end of the line. If you make a telephone call in this country and the line is disconnected before you can find out who is on the other end, what is the probability that it was a wrong number?

2-162   A poker player has a pair of aces in a five card hand. What is the probability that an opponent also has a pair of aces?

● *In Exercises 2-163 through 2-165, two dice are rolled.*

2-163   What is the probability that the sum is greater than or equal to 4 and less than or equal to 8?

2-164   If the first die is a three, what is the probability that the sum is greater than or equal to 4 and less than or equal to 8?

2-165   Are the outcomes (First die a three) and (Sum is greater than or equal to 4 and less than or equal to 8) independent?

● *In Exercises 2-166 and 2-167, a bridge player picks up his cards one at a time as they are dealt. The first five cards are 2♡, 3♧, 2◇, 4♧, and 2♧. He begins to suspect that he may be headed for a worst-possible bridge hand: thirteen cards, all five or lower.*

2-166 Before he picked up any cards, what was the probability that he would receive a worst-possible bridge hand?

2-167 After receiving the five cards listed, what was the probability that he would receive a worst-possible bridge hand?

2-7 Principles of Probability: "and" Statements

One of the most popular card games in gambling casinos is the card game known as "blackjack" or "21." At the start of the game, each player is dealt two cards. A very desirable opening hand consists of two aces. One way to compute the probability of getting such a hand is to use the counting procedures of Section 2-3. The number of two card hands of two aces is

$$C_2^4 = \frac{4!}{(2!)(2!)} = 6$$

because these hands must come from just the four aces. The total number of two card hands is, as we have observed before,

$$C_2^{52} = \frac{52!}{(2!)(50!)} = \frac{(52)(51)}{(2)(1)} = 1326$$

We conclude that

$$p(\text{Two aces}) = \frac{6}{1326} = \frac{1}{221}$$

There is a quite different way to solve the same problem based on a principle which, as we shall see, is of great utility in the study of probability. Essentially, this principle is nothing more than a way of rewriting the definition of conditional probability which, we recall, is

$$p(A|B) = \frac{p(A \text{ and } B)}{p(B)}$$

Multiply both sides of the equation by $p(B)$, so that

$$p(B)p(A|B) = p(A \text{ and } B)$$

or, as we prefer to write it,

$$p(A \text{ and } B) = p(A|B)p(B)$$

Thus, the probability that both A and B will take place is the conditional probability that A will occur given that B has occurred, times the probability that B will happen. Although we will make use of this formula for $p(A \text{ and } B)$ below, there is a different formula based on the same idea which is the one we need for the blackjack problem.

The symbols A and B in the definition of conditional probability had no special significance. We could just as well have used

$$p(Z|W) = \frac{p(Z \text{ and } W)}{p(W)}$$

Thus, we can reverse the order of the letters A and B to define

$$p(B|A) = \frac{p(B \text{ and } A)}{p(A)}$$

The important thing about the definition is that the letter *after* the symbol $|$ appears in the denominator of the fraction. Now, $p(B \text{ and } A)$ means the probability that both B and A will occur, while $p(A \text{ and } B)$ means the probability that both A and B will take place. But then $p(B \text{ and } A)$ and $p(A \text{ and } B)$ mean precisely the same thing; there is no reason to use one order in preference to the other. Thus, we can replace $p(B \text{ and } A)$ by $p(A \text{ and } B)$ in the definition above:

$$p(B|A) = \frac{p(A \text{ and } B)}{p(A)}$$

This time we multiple both sides of the equation by the number $p(A)$ so that

$$p(A)p(B|A) = p(A \text{ and } B)$$

or, as we prefer to write this principle of probability,

$$p(A \text{ and } B) = p(B|A)p(A)$$

In words, the probability that both A and B will occur is the conditional probability that B will happen under the condition that A takes place, times the probability that A will happen. We will need both forms of the principle, namely,

$$p(A \text{ and } B) = p(A|B)p(B) \qquad \text{and} \qquad p(A \text{ and } B) = p(B|A)p(A)$$

In dealing two cards, let A represent the hands in which the first card dealt is an ace and B those in which the second card is an ace. We compute $p(A \text{ and } B)$ to solve the blackjack problem. The value of $p(A)$ is the probability of drawing an ace from a deck, and that, of course, is 4/52. The symbol $p(B|A)$ means the probability of drawing an ace from the deck that remains after an ace is withdrawn. Since there are three aces left in this 51 card deck, $p(B|A)$ is 3/51. Now we apply the principle

$$p(\text{Two aces}) = p(A \text{ and } B) = p(B|A)p(A) = \left(\frac{3}{51}\right)\left(\frac{4}{52}\right) = \frac{1}{221}$$

2-25   **Example**   Many automobiles have dual braking systems. That is, if the main system fails, there is a second, less effective, system which attempts to stop the car. Suppose the probability that the main brake system will fail is .001. The probability that if the main system fails, the backup system will be unable to stop the car is .02. What is the probability that when the brakes are applied, the car will not stop?

Solution   The car will be unable to stop if A, the main brake system fails, and B, the backup system also fails. Now, $p(A)$ is given as .001. The probability that the backup system will fail under the condition that the main system fails, $p(B|A)$, is stated as .02. (We do not know $p(B)$, the probability of failure of the backup system by itself. It is probably lower than .02, because that given figure takes into account the likelihood that whatever damaged the main braking system was major enough to take the backup system with it. In any event, we are not interested in how the backup system would behave if it were the only system on the car.) By the principle of probability,

$$p(\text{Car unable to stop}) = p(\text{Both brake systems fail})$$
$$= p(A \text{ and } B) = p(B|A)p(A) = (.02)(.001) = .00002$$

In other words, with the main braking system alone the chances are 999 in 1000 that when the driver steps on the brake pedal the car will stop, but the dual system brings the chances up to 99,998 out of 100,000.

At the end of Section 2-6, we discussed independent events, i.e., those in which the conditional probability was the same as the probability without any condition. More precisely, outcomes of type B are independent of the outcomes of a class A if $p(B|A) = p(B)$. If B is independent of A, then we may substitute $p(B)$ for the conditional probability in the principle $p(A \text{ and } B) = p(B|A)p(A)$ if A and B are independent. We then obtain the very useful formula $p(A \text{ and } B) = p(B)p(A)$ or, equivalently,

$$p(A \text{ and } B) = p(A)p(B) \qquad \text{if } A \text{ and } B \text{ are independent}$$

One immediate use of the formula is to answer a question that may have occurred to you in the last section. If B is independent of A, does it necessarily follow that A is independent of B? That is, does the fact that $p(B|A) = p(B)$ necessarily mean that $p(A|B) = p(A)$? The answer is "yes," because if B is independent of A, then we just saw that we get the formula

$$p(A \text{ and } B) = p(A)p(B)$$

while earlier in this section we obtained the form of the principle of probability

$$p(A \text{ and } B) = p(A|B)p(B)$$

Since $p(A)p(B)$ and $p(A|B)p(B)$ are both equal to $p(A \text{ and } B)$, they are equal to each other:

$$p(A)p(B) = p(A|B)p(B)$$

Dividing through both sides by the common term $p(B)$ gives

$$p(A) = p(A|B)$$

which, by definition, means that A is independent of B. We can run the same argument backwards to establish that if A is independent of B, then B is also independent of A. Thus, from now on we will just say "A and B are independent" and know that this statement makes sense.

If one engine of a two-engine airplane stops during a flight, the plane is designed so that it can fly at a reduced speed using just the remaining engine until it can land safely. However, if both engines fail, the situation is rather more unpleasant. For that reason, aircraft engines are made as independent of one another in the design of the plane as is possible. Let us assume that the probability that either engine fails is entirely independent of the other (which is not quite true since, for example, a fire started in one engine might spread to the other). If the probability that an engine will fail is .001, what is the probability that both will fail? Let A represent failure of the left engine and B failure of the right. Then, since we assumed the events were independent, the formula above gives

$$p(A \text{ and } B) = p(A)p(B) = (.001)(.001) = .000001$$

This computation tells us that two-engine planes are much safer than single-engine planes, assuming that all engines are equally dependable. We can expect a four-engine plane to be even more dependable, at least to the extent that it is very unlikely that all four engines will fail. In this case, the failure of each engine is represented by the letters A, B, C, and D, and we wish to know $p(A \text{ and } B \text{ and } C \text{ and } D)$. In general, the formula for computing that all of four outcomes will take place is pretty complicated, but when they are independent it is just what we might expect:

$$p(A \text{ and } B \text{ and } C \text{ and } D) = p(A)p(B)p(C)p(D)$$

Thus, if we assume that the failure of each engine is independent of what happens to the others and the probability of failure for each engine is .001, then

$$p(\text{All four fail}) = p(A \text{ and } B \text{ and } C \text{ and } D)$$
$$= p(A)p(B)p(C)p(D)$$
$$= (.001)(.001)(.001)(.001) = .000000000001$$

We give a general expression for the probability principle we have just used:

> Suppose that $A_1, A_2, \ldots A_k$ are all independent classes of outcomes to some experiment, which means that $p(A_i|A_j) = p(A_i)$ for all i and j. Then
>
> $$p(A_1 \text{ and } A_2 \text{ and} \ldots \text{and } A_k) = p(A_1)p(A_2)\cdots p(A_k)$$

In particular, if an experiment is performed several times in succession under identical circumstances (implying that each time the outcome is independent), then this principle may be applied. A classical example of repeated events under identical circumstances is provided by drawing a card at random from a deck, replacing it, drawing a card at random, and so on. If the drawn card is withheld and the next card is taken from the remaining deck, as it is in dealing a poker hand, then the experiment changes because the available deck changes, but if the card is always replaced, then the experiment is always the same.

2-26   **Example**   (a) What is the probability of being dealt a five card poker hand without any aces or face cards (kings, queens, jacks) in it? (b) What is the probability of drawing a card and replacing it in the deck, repeated five times in all, without drawing any aces or face cards?

Solution   (a) The five card poker hand is drawn from among the twos, threes, and so on to the tens; 36 cards in all. Therefore, by the sort of counting argument we have used several times before,

$$p(\text{No aces or face cards}) = \frac{C_5^{36}}{C_5^{52}} = \frac{376,992}{2,598,960} = .145 \qquad \text{(approximately)}$$

(b) The probability of drawing neither an ace nor a face card from a full deck is 36/52, so

$$p(\text{No aces or face cards}) = \left(\frac{36}{52}\right)\left(\frac{36}{52}\right)\left(\frac{36}{52}\right)\left(\frac{36}{52}\right)\left(\frac{36}{52}\right) = \frac{59,049}{371,293} = .159 \qquad \text{(approximately)}$$

Summary

$$p(A \text{ and } B) = p(A|B)p(B)$$

$$p(A \text{ and } B) = p(B|A)p(A)$$

If $A_1, A_2, \ldots, A_k$ are all independent classes of events $[p(A_i|A_j) = p(A_i)$ for all i and $j]$, then

$$p(A_1 \text{ and } A_2 \text{ and } \ldots \text{ and } A_k) = p(A_1)p(A_2) \cdots p(A_k)$$

Exercises

● *In Exercises 2-168 through 2-172, a spinning arrow is set up on a circular card. Half of the card is colored white and the other half is equally divided among the colors red, blue, and green. The arrow is spun twice.*

2-168   What is the probability of the arrow stopping at red both times?

2-169   What is the probability of getting red on the first spin and either white or green on the second?

2-170   What is the probability of spinning white on the first spin and green on the second?

2-171   What is the probability of spinning white on one spin and green on the other?

2-172   What is the probability of getting either red or blue on the first spin and a color other than green on the second?

2-173   A life insurance agent calls families who have just had babies. His experience is that 15% of those he calls agree to make an appointment with him to discuss insurance. Of those who agree to the appointment, one in four buys insurance from him. If he calls a family with a new baby, what is the probability that they will make an appointment and buy insurance from him?

● *Exercises 2-174 through 2-176 concern a children's party at which each child plays three games. In the first game, a child rolls two dice and wins if both numbers are the same. In the second, a child picks one card from a standard deck and wins if a face card appears. In the third, a child flips two coins and wins if they are both heads or both tails.*

2-174   What is the probability that a child will win all three games?

2-175   What is the probability that a child will win the dice game and lose the others?

2-176   What is the probability that a child will win exactly one game?

2-177   In an experiment conducted by Stanford Research Institute, the psychic Uri Geller on eight successive occasions correctly called the uppermost face of a die that had been shaken in a steel box. If Geller was guessing at random, what is the probability that this would have happened?

-    *In Exercises 2-178 and 2-179, a student will do well on an examination if either the student studies hard or the professor is an easy grader. The student estimates that there is a probability of .65 of studying hard for the examination and of .30 that the professor is an easy grader. Assume that the student's study habits and the professor's grading standards are independent.*

2-178   What is the probability that the student will study hard and that the professor is an easy grader as well?

2-179   What is the probability that the student will do well on the examination?

2-180   In an attempt to stay in business, a financially troubled company takes three actions. It markets a new model of its main product, it applies to its bank for a loan, and it attempts to sell a money-losing subsidiary. The company can stay in business if at least one of these actions is successful. Otherwise, the company will have to cease operation. It estimates the probability that the new model will sell successfully at .35, that it will get a loan at .20, and that it will sell the subsidiary at .45. If these actions are independent, what is the probability that the company will stay in business? [*Hint:* What is the probability that all three actions will be failures?]

-    *Exercises 2-181 and 2-182 are concerned with the following situation: After dinner, each of three people flips a coin to see who will pay the bill. If two come up the same (for example, two heads and one tails) then the person whose coin fails to match the others pays. If all three coins are the same, then the game is repeated until only two agree.*

2-181   What is the probability that it will require exactly two plays of the game to determine who will pay the bill?

2-182   What is the probability that it will require exactly three plays of the game to determine who will pay the bill?

-    *In Exercises 2-183 through 2-187, a high school student takes courses in English, history, mathematics, and typing. The grades he receives in the four courses are independent. He estimates the probability that he will get an A in each subject to be .30 in English, .40 in history, .20 in mathematics, and .90 in typing.*

2-183   What is the probability that he will not get any As?

2-184   What is the probability that he will get just one A—in typing?

2-185 What is the probability that he will get just one A—in mathematics?

2-186 What is the probability that he will get exactly one A?

2-187 What is the probability that he will get at least one A?

2-188 Suppose that 70% of the members of a mathematics class are women and that 20% of the women in the class are left-handed. If a person is chosen at random from the class, what is the probability that the person is a left-handed woman?

2-189 We will use a baseball player's batting average as an estimate of the probability that the player will get a hit next time at bat. Suppose the first two batters in a team's lineup have batting averages of .315 and .290, respectively. What is the probability that they will both get hits to start the game?

● *In Exercises 2-190 through 2-193, a card is drawn from a standard deck and then replaced, and a second card is drawn.*

2-190 What is the probability that both cards are spades?

2-191 What is the probability that the first card is a heart and the second card is either a face card or a club?

2-192 What is the probability that the second card is a spade?

2-193 What is the probability that exactly one of the cards is an ace?

2-194 A soft-drink bottler conducts a marketing survey in which families are given five unmarked bottles of cola drinks and asked to rank them for flavor in order of preference, with no ties permitted. The five different brands of cola include both the bottler's product and that of the leading competitor. If a family thinks that all five colas taste alike and so ranks them in a random order, what is the probability that they will rank the bottler's product as first choice and the leading competitor's as last?

2-195 A manufacturer of electronic parts must be sure that its product is very dependable, so it not only hires an inspector but also a second inspector who checks everything the first inspector approves. Suppose the first inspector finds 98% of all defective units and the second inspector catches 75% of all the defective units approved by the first inspector. What is the probability that a defective unit will get by both inspectors?

2-8 Repeated Experiments

Suppose that a baseball player has a probability of 3/10 of getting a hit when he comes up to bat. If he has four chances to bat in a single game, what is the probability that he will get exactly one hit in four times at bat? We will assume that the player has the same probability, 3/10, of getting a hit each time he comes to the plate, that is, that the quality of pitching he must face does not change during the course of the game and that he is enough of a professional not to be influenced by his previous performance that day.

If we think of the baseball player's four times at bat during the game as an experiment consisting of four parts, then we are assuming that each part of the experiment occurs under identical conditions and that there is the same probability, 3/10, of success in each part. Thus, the overall experiment consists of a single experiment repeated four times. If getting a hit is considered to be a success in a time at bat and any other outcome is a failure, then we are concerned with computing the probability of exactly one success out of four tries. More generally, in this section we will be concerned with the type of experiment that consists of the exact repetition of some action a number of times, say n, with a probability we denote by p of success in each repetition. We wish to obtain a formula for the probability of some number k of successes out of the n trials. This probability will be represented by the symbol $p(k|n)$ because it is a conditional probability—the probability of k successes under the condition that n repetitions of the action took place.

In order to get an idea of how to find the formula for $p(k|n)$, let us return to the example of the baseball player. In this case, we wish to compute $p(1|4)$, where p, the probability of success in each part of the experiment, is 3/10. One way in which the overall experiment can succeed, that is, in which the player can get one hit out of four times at bat, is for the player to get a hit his first time at bat and then fail to get a hit the other times. Let S denote success in a trial and F failure. Then the pattern just described is (S, F, F, F). Since the single hit can occur in any of the four times at bat, we see that each of the following patterns of success and failure in four tries also satisfies the requirement for one hit in four times at bat: (F, S, F, F), (F, F, S, F), (F, F, F, S). There is no other way the player can succeed in getting exactly one hit out of four, so we conclude from the principle of probability,

$$p(A_1 \text{ or } A_2 \text{ or } \ldots \text{ or } A_k) = p(A_1) + p(A_2) + \cdots + p(A_k)$$

when the A_i are disjoint, that

$$p(1|4) = p(S, F, F, F) + p(F, S, F, F) + p(F, F, S, F) + p(F, F, F, S)$$

Let us now determine the probability that the pattern will be (S, F, F, F). Since each time at bat is independent of the others, a principle of probability implies that

$$p(S, F, F, F) = p(S)p(F)p(F)p(F)$$

We were told that $p(S)$, the probability of getting a hit, is 3/10 = .30. The probability $p(F)$ of failing to get a hit is then

$$p(F) = 1 - p(S) = 1 - .30 = .70$$

and we conclude that

$$p(S, F, F, F) = (.30)(.70)(.70)(.70) = .1029$$

To save space, let us use **exponential notation** in writing the answer we just obtained. That is, for a number x, write xx (by which we mean "x times x") as x^2, write xxx as x^3, write $xxxx$ as x^4, and so on. Thus, we write

$$p(S, F, F, F) = (.30)(.70)^3$$

We also recall the notation $x^1 = x$ and, provided that x is not zero, the definition $x^0 = 1$. By the reasoning above,

$$p(F, S, F, F) = p(F)p(S)p(F)p(F) = (.70)(.30)(.70)(.70)$$

But since the order in which we write the numbers does not matter, again,

$$p(F, S, F, F) = (.30)(.70)^3$$

The same argument produces

$$p(F, F, S, F) = p(F, F, F, S) = (.30)(.70)^3$$

We have computed that

$$p(1|4) = 4(.30)(.70)^3 = .4116$$

If we change the problem slightly to allow the player five times at bat and ask for the probability that he will get exactly two hits, the basic method remains the same, but the calculations become rather tedious. The possible patterns are

(S, S, F, F, F)	(S, F, S, F, F)	(S, F, F, S, F)	(S, F, F, F, S)
(F, S, S, F, F)	(F, S, F, S, F)	(F, S, F, F, S)	(F, F, S, S, F)
(F, F, S, F, S)	(F, F, F, S, S)		

Next we compute the probability that one of these patterns can occur, say,

$$p(S, S, F, F, F) = p(S)p(S)p(F)p(F)p(F)$$
$$= (.30)(.30)(.70)(.70)(.70)$$

Since the probability of each of the ten patterns of two successes and three failures will be the product of two .30s and three .70s in some order, we conclude that

$$p(2|5) = 10(.30)^2(.70)^3 = .3087$$

More generally, suppose that the baseball player has a probability p of getting a hit at each time at bat. In the notation above, this means that $p(S) = p$. By the principles of probability, we know that $p(F) = 1 - p$. Since the argument used no special property of the number $3/10$, we may replace .30 by p and .70 by $1 - p$ in the formulas above to conclude that, for an arbitrary probability p,

$$p(1|4) = 4p(1 - p)^3$$

and

$$p(2|5) = 10p^2(1 - p)^3$$

2-27 **Example** A die is rolled three times. What is the probability that a three will come up exactly twice?

Solution We consider success to be rolling a three, so we wish to compute $p(2|3)$, where $p = p(S)$ is the probability of rolling a three, namely $1/6$. There are three patterns of two successes and one failure which are (S, S, F), (S, F, S), and (F, S, S). Since, for example,

$$p(S, S, F) = p(S)p(S)p(F) = \left(\frac{1}{6}\right)\left(\frac{1}{6}\right)\left(\frac{5}{6}\right)$$

it follows that

$$p(2|3) = 3\left(\frac{1}{6}\right)^2\left(\frac{5}{6}\right) = \frac{15}{216} = .07 \qquad \text{(approximately)}$$

We should now feel confident in stating the general case. In computing $p(k|n)$, the probability of exactly k successes in n tries, the probability of any particular pattern of k successes out of n tries is the probability p of success multiplied by itself k times, times the probability $1 - p$ of failure multiplied by itself the number of times the experiment is to fail. There are n trials in all and k of them are successes, so the remaining $n - k$ will fail. In symbols, the probability of any particular pattern of k successes in n trials is $p^k(1 - p)^{n-k}$. Since the probability $p(k|n)$ is, by the principles of probability, just the sum of the probabilities of all possible such patterns and the probability of each pattern is $p^k(1 - p)^{n-k}$, then

$$p(k|n) = (\text{Number of patterns})p^k(1 - p)^{n-k}$$

In order to calculate $p(k|n)$ in practice, we still need to determine the number of possible patterns of k successes in n tries. The trick is to think of the n repetitions of the action as a deck consisting of n cards with a different number 1, 2, and so on to n, printed on each card. If we are dealt a "card hand" containing k cards, then that is the same thing as choosing k numbers from out of the numbers from 1 to n. If we are dealt the card with the number 3 on it, for example, that means that the third time the act takes place it will be a success. As in poker, it does not matter whether our "card hand" received the 3 as the first card dealt, the second card, or whatever. Recall from Section 2-3 that there are

$$C_5^{52} = \frac{52!}{(5!)(47!)}$$

5-card poker hands in the standard 52 card deck. For exactly the same reason, the number of k card "hands" from a "deck" of n cards is

$$C_k^n = \frac{n!}{k!(n - k)!}$$

On a somewhat more abstract level, a pattern of k successes out of n tries is a subset of k distinct numbers from the set of numbers $1, 2, \ldots, n$. Recall that a combination of k elements out of a set of n was by definition just such a subset. We observed in Section 2-3 that there were exactly

$$C_k^n = \frac{n!}{k!(n - k)!}$$

such combinations, so again there are C_k^n different patterns.

We have solved the problem of computing $p(k|n)$. The formula is

$$p(k|n) = C_k^n p^k(1 - p)^{n-k}$$

where

$$C_k^n = \frac{n!}{k!(n - k)!}$$

2-28   **Example**   If a coin is weighted so that the probability of it coming up heads is 2/3, what is the probability that it will come up heads exactly twice out of four tosses?

Solution   There are $n = 4$ tosses of the weighted coin and we wish to know the probability of $k = 2$ successes where the probability p of success (heads) on each toss is 2/3. By the formula

$$p(2|4) = C_2^4 \left(\frac{2}{3}\right)^2 \left(\frac{1}{3}\right)^2$$

where

$$C_2^4 = \frac{4!}{2!2!} = 6$$

Thus,

$$p(2|4) = 6 \left(\frac{2}{3}\right)^2 \left(\frac{1}{3}\right)^2 = \frac{24}{81} = .30 \qquad \text{(approximately)}$$

2-29   **Example**   Each time a sharpshooter fires the probability that she will hit a target is 9/10. What is the probability that she will score at least four hits out of five tries?

Solution   If the sharpshooter hits the target exactly four times, the experiment will be a success under the stated conditions, and if she hits it all five times, that is also a success. Since four hits out of five is a success that is distinct from five hits, the appropriate principle of probability implies that the probability of at least four hits is the sum of the probability of four hits out of five, $p(4|5)$, plus the probability $p(5|5)$ of five hits in a row. Applying the formula with $p = 9/10 = .90$, gives

$$p(4|5) = C_4^5 (.90)^4 (.10)^1 = 5(.90)^4 (.10) = .33 \qquad \text{(approximately)}$$

and

$$p(5|5) = C_5^5 (.90)^5 (.10)^0 = .59 \qquad \text{(approximately)}$$

Therefore,

$$p(\text{At least four hits}) = p(4|5) + p(5|5)$$

$$= .33 + .59 = .92 \qquad \text{(approximately)}$$

which means that there is about a 92% probability that the sharpshooter will get at least four hits out of five.

2-30 **Example** A die is rolled five times. What is the probability that a number less than three appears at least once?

Solution We have $n = 5$ repeated trials with $p = 2/6 = 1/3$, because a one or two on the die indicates a success in the trial. Since $k = 1$ success out of five satisfies the condition of at least one success, as does $k = 2, 3, 4,$ and 5, we could compute

$$p(\text{Number less than three at least once}) = p(1|5) + p(2|5) + p(3|5) + p(4|5) + p(5|5)$$

However, we may use the principles of probability to obtain a much easier solution to the problem. If success means rolling a number less than three at least once, then failure occurs when there is no occasion among the five rolls when a number less than three appears. Therefore, $p(\text{Failure})$ means $p(0|5)$. Consequently,

$$p(\text{Number less than three at least once}) = p(\text{Success})$$
$$= 1 - p(\text{Failure})$$
$$= 1 - p(0|5)$$

If we calculate

$$p(0|5) = C_0^5 \left(\frac{1}{3}\right)^0 \left(\frac{2}{3}\right)^5 = 1\left(\frac{2}{3}\right)^5 = \frac{32}{243}$$

Then we have our answer:

$$p(\text{Number less than three at least once}) = 1 - \frac{32}{243} = \frac{211}{243} = .87 \quad \text{(approximately)}$$

Summary

Exponential Notation

$$x^0 = 1 \quad \text{(if } x \text{ is not zero)}$$

$$x^1 = x \qquad x^2 = xx \qquad x^3 = xxx \qquad x^4 = xxxx \quad \ldots$$

The probability of exactly k successes out of n independent, identical repetitions of the same experiment with the same probability p of success each time is denoted by $p(k|n)$. The computational formula is

$$p(k|n) = C_k^n p^k (1 - p)^{n-k}$$

where

$$C_k^n = \frac{n!}{k!(n - k)!}$$

Exercises

2-196   An insurance agent sells a policy at 15% of the appointments he makes. If, one day, the agent has five appointments, what is the probability that he will make exactly two sales?

2-197   In an election for mayor in which 60% of the voters prefer White and 40% prefer Green, the individuals in a random sample of seven voters are asked their preference. What is the probability that the sample will predict the election correctly, that is, contain a majority of voters (at least four) who prefer White?

2-198   Suppose that 80% of all drivers on a certain highway use their seat belts. What is the probability that if there were accidents involving six cars on that highway, more than four of the drivers would be wearing seat belts?

2-199   What is the probability that a family with six children will have three boys and three girls, assuming the probability that a child will be a boy is .50.

2-200   In a certain country, the probability that an infant will live to the age of 60 is .65. If six infants are chosen at random in that country, what is the probability that at least two of them will reach the age of 60?

●   *Exercises* 2-201 *and* 2-202 *concern a test of extrasensory perception in which five cards, each from a different deck, are placed face down on a table and a psychic is asked what value (such as six, queen, and so on) each card has. The psychic gets two right out of the five.*

2-201   What is the probability that a person guessing at random would choose incorrect values for all five cards?

2-202   What is the probability that a person guessing at random would do at least as well as the psychic did?

●   *In Exercises* 2-203 *through* 2-205, *determine the probability of getting at least eight answers correct on an examination with ten questions.*

2-203   It is a true–false examination and you are guessing at random.

2-204   It is a true–false examination and you know enough about the subject to have a probability of .80 of getting the correct answer to each question.

2-205   It is a multiple-choice examination with four alternative answers to each question and you are guessing at random.

•   *For Exercises 2-206 and 2-207, suppose that there is a country in which 15% of the population is color-blind.*

2-206   If seven individuals from that country are chosen at random, what is the probability that exactly two of them are color-blind?

2-207   If seven individuals from that country are chosen at random, what is the probability that no more than two of them are color-blind?

2-208   A woman owns six vacation cottages that she rents out by the week. She supplies television sets at extra cost to those renters who request them. She estimates that the probability that a family who rents a cottage will want a television set is .50. She owns four sets. In a week in which all the cabins are rented, what is the probability that she will have enough sets to supply all the families that want one?

2-209   A card is drawn at random from a standard deck. Then it is replaced, the deck is shuffled, and another card is drawn. This procedure is followed five times in all. What is the probability of drawing a face card exactly three times out of the five?

2-210   Suppose that success in rolling two dice consists of getting a sum of either 7 or 11. What is the probability of exactly four successes in seven tries?

2-211   An animal trainer has attempted to train a cat to go to the larger of two boxes when given a choice. If the cat is as likely to go to either box each time it is given a choice, what is the probability that it would go to the larger box at least four times out of six?

2-212   Suppose that it is known that 48 members of the United States Senate are going to vote for a certain bill and 43 are sure to vote against it. The Vice-President is opposed to the bill, so a tie vote will defeat it. If the nine remaining senators are so uncertain of their opinion that each will decide how to vote by flipping a coin, what is the probability that the bill will pass the Senate?

2-213   A manufacturer of cardboard boxes tests five boxes chosen at random from the output of its factory to determine whether they will withstand a certain amount of pressure. If 15% of the boxes from the factory would fail the test, what is the probability that no more than one of the five boxes in the sample will fail the test?

2-9   Applications to Seismic Events

Let us now examine how we can use the formula we discovered in Section 2-8 to study the problem of detecting seismic events. Suppose that ten seismic events, earthquakes or nuclear explosions, of magnitude 4.75 or greater occur in some portion of the globe. If we assume that the probability of detection of each event is 9/10, what is the probability of detecting all ten events? Each seismic event is thought of as a separate experiment under identical conditions, with probability of success (detection) equal to 9/10. There are ten experiments and we wish to know the probability of achieving success in all ten. The answer, according to the general formula

$$p(k|n) = C_k^n p^k (1 - p)^{n-k}$$

is

$$p(10|10) = C_{10}^{10}\left(\frac{9}{10}\right)^{10}\left(\frac{1}{10}\right)^{0} = \left(\frac{9}{10}\right)^{10} = .35 \quad \text{(approximately)}$$

So, even though the probability of detecting each event is quite high, in a long sequence of events it is likely that some will be missed. Specifically, since the probability of detecting all ten events is about .35, then by the principles of probability, the probability of not detecting all, i.e., of overlooking at least one event, is about

$$1 - .35 = .65$$

If there are ten seismic events and the probability of detection of each event is 9/10, then the probability of detecting exactly nine events is, by the formula,

$$p(9|10) = C_9^{10}\left(\frac{9}{10}\right)^{9}\left(\frac{1}{10}\right)^{1} = .39 \quad \text{(approximately)}$$

while the probability of detecting exactly eight events is

$$p(8|10) = C_8^{10}\left(\frac{9}{10}\right)^{8}\left(\frac{1}{10}\right)^{2} = .19 \quad \text{(approximately)}$$

Under some circumstances, the detection of at least eight out of the ten events might be considered satisfactory. To find the probability of detecting at least eight events, think of the ten seismic events as a single experiment and consider the experiment a success if at least eight of the events are detected. Then, as we have seen,

$$p(\text{Detecting at least eight}) = p(8|10) + p(9|10) + p(10|10)$$
$$= .19 + .39 + .35 = .93 \quad \text{(approximately)}$$

Once a seismic event has been detected, it is important to know where the event took place, i.e., the location of the explosion or the epicenter of the earthquake. The instruments

Figure 2-6

of each seismological station will indicate the distance from the epicenter to that station. In order to locate the event, the distances to several stations are compared. For example, if there are three stations A, B, and C (Figure 2-6) which report being distances $d(A)$, $d(B)$, and $d(C)$, respectively, from the seismic event, then the intersection of the circles shown in Figure 2-7 should pinpoint the center of the event. In practice, the evidence of the distance to the source of the shock waves is not that clear-cut, and it takes a minimum of four stations reporting in order to get a reasonably accurate fix on the epicenter. Therefore, we wish to know, given a group of seismological stations, what is the probability that at least four of the stations will detect a seismic event?

The probability that at least four stations will detect an event should depend in some way on how many stations there are, since the more stations there are, the more likely it is that at least four will be successful. According to a British study published in 1970,

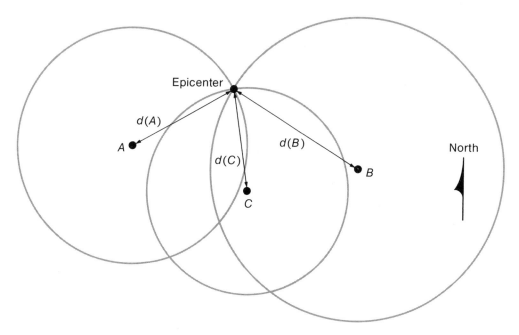

Figure 2-7

seven stations of the type required for monitoring a ban on underground nuclear testing were located near the Soviet Union at that time. The study proposed the construction of nineteen more stations for this purpose. Now, let us examine the probability of locating a seismic event of magnitude 4.75 or greater, first with the 1970 network of seven stations, and then with the proposed network of 26 stations.

Assume that there is a seven station network and a seismic event of magnitude at least 4.75 occurs in the Soviet Union, so that (according to the British report) there is a probability of 9/10 that it will be detected by the network. Further assume, for simplicity, that each station is as likely as any other to detect the event. In order to compute the probability that the network will be able to locate the event by detecting it at four stations or more, we need to know the probability that the event will be detected at each station. The computation depends on a trick. We know that the probability of detection at some, that means at least one, station is 9/10. According to a principle of probability, the probability that no station will detect the event is $1 - 9/10 = 1/10$. Our trick consists of finding a different formula for the probability of failure at all seven stations. Think of each station's attempt to detect a single event as a separate experiment with some probability p (which we do not yet know) of success. The probability that no station will detect the event is then the probability of failure in each of seven experiments. According to the general formula, that probability (when the probability of success in each experiment is p) is

$$p(0|7) = C_0^7 \, p^0 (1 - p)^7 = (1 - p)^7$$

But we already know that the probability of seven failures is 1/10, so we conclude that

$$(1 - p)^7 = \frac{1}{10}$$

To solve for p, the probability of detection at each station, we take the seventh root of both sides of the equation and get

$$1 - p = \sqrt[7]{\frac{1}{10}}$$

and therefore,

$$p = 1 - \sqrt[7]{\frac{1}{10}}$$

Solving for p on a calculator gives p equal to about .28.

We continue to think of the attempt to detect a single seismic event at seven stations as seven identical experiments. We have just computed that each experiment has probability of success close to .28. The probability of success at all seven stations is therefore

$$p(7|7) = C_7^7(.28)^7(.72)^0 = (.28)^7 = .0001 \quad \text{(approximately)}$$

By our formula, the probability of detection at four, five, or six stations is, respectively,

$$p(4|7) = C_4^7(.28)^4(.72)^3 = .080$$
$$p(5|7) = C_5^7(.28)^5(.72)^2 = .019 \quad \text{(approximate values)}$$
$$p(6|7) = C_6^7(.28)^6(.72)^1 = .0024$$

Locating the event was taken to mean detecting it at four stations at least, so that detection at exactly four stations counts as a success, as does detection at exactly five, or exactly six, or all seven stations. The appropriate principle tells us, then, that the probability of detection is computed by adding up the corresponding probabilities:

$$p(\text{Location}) = p(\text{Detection at four stations at least})$$
$$= p(4|7) + p(5|7) + p(6|7) + p(7|7)$$
$$= .080 + .019 + .0024 + .0001 = .10 \quad \text{(approximately)}$$

We have shown that if the probability of detecting a seismic event of magnitude 4.75 or more with a network of seven stations is 9/10, then the probability of its location is about 1/10.

We have computed the probability of detection of a seismic event of magnitude at least 4.75 to be .28 at each station of a seven station network. Assuming that each of the 19 new stations recommended by the British report would have the same probability of detecting such an event as do the existing stations, we will compute the probability of locating the event, that is, detecting it at four stations or more, using the proposed 26 station network. This assumption will tend to make the location probability lower than it actually would be, since the newer stations would have the latest seismographic equipment.

The probability of detection at four stations at least is the sum of the probability of detection at exactly four stations, at exactly five stations, and so on up to all 26 stations. It is both tedious and unnecessary to compute these 23 numbers and then add them up. Instead, we compute the probability of the opposite outcome, i.e., that fewer than four stations detect the event, and then subtract that number from one in order to obtain the probability we are really after. The probability that a single station will detect the event is .28, so the probability that all 26 will fail to detect the seismic event is

$$p(0|26) = C_0^{26}(.28)^0(.72)^{26} = (.72)^{26} = .0002 \qquad \text{(approximately)}$$

The probabilities of detection at exactly one, two, and three stations are, respectively,

$$p(1|26) = C_1^{26}(.28)^1(.72)^{25} = .0020$$

$$p(2|26) = C_2^{26}(.28)^2(.72)^{24} = .0096 \qquad \text{(approximate values)}$$

$$p(3|26) = C_3^{26}(.28)^3(.72)^{23} = .0299$$

and we compute that

$$p(\text{Failure to locate}) = p(\text{Detection at fewer than four stations})$$

$$= p(0|26) + p(1|26) + p(2|26) + p(3|26)$$

$$= .0002 + .0020 + .0096 + .0299 = .04 \qquad \text{(approximately)}$$

Therefore, the probability of locating an event of magnitude at least 4.75 using the network of 26 stations is

$$p(\text{Location}) = 1 - p(\text{Failure to locate})$$

$$= 1 - .04 = .96 \qquad \text{(approximately)}$$

Incidentally, we have also computed the probability that the 26 station network would detect an event of this magnitude. Failure to detect at any station, that is, $p(0|26)$, was computed to be about .0002, so the probability of detection is

$$1 - p(0|26) = 1 - .0002 = .9998 \qquad \text{(approximately)}$$

According to the British study referred to earlier, if all 26 seismological stations were put into operation, the system would have a 9/10 probability of detecting a seismic event of magnitude as small as 4.0, which corresponds to a nuclear explosion in hard rock equivalent to about 3 kilotons of conventional explosives.

2-31 **Example** Find the probability of locating a seismic event of magnitude at least 4.0 using a 26 station network if the probability that the network will detect is 9/10.

Solution The 26 stations constitute 26 experiments and, if the entire system has probability 9/10 of detecting an event of magnitude at least 4.0, then there is a probability of 1/10 that the entire system will miss the event—that all 26 experiments will fail. If the probability of detecting the event at a single station is called p, then the probability of failure at all stations is

$$p(0|26) = C_0^{26} p^0 (1 - p)^{26} = (1 - p)^{26}$$

But we know that the probability is 1/10, so we set

$$(1 - p)^{26} = \frac{1}{10}$$

take the twenty-sixth root of both sides of the equation,

$$1 - p = \sqrt[26]{\frac{1}{10}}$$

and find (with the aid of a calculator) that

$$p = 1 - \sqrt[26]{\frac{1}{10}} = .085 \qquad \text{(approximately)}$$

The efficient way to compute the location probability is to compute the probability that the event will not be located, that is, that fewer than four stations will detect the event. We already know the probability $p(0|26)$ of failure to detect the event at all 26 stations; it is 1/10. Using the probability figure of .085 that we just calculated for the detection of an event of magnitude at least 4.0 at each station, we compute the approximate probabilities of detection at exactly one, two, or three stations by the general formula:

$$p(1|26) = C_1^{26}(.085)^1(.915)^{25} = .24$$

$$p(2|26) = C_2^{26}(.085)^2(.915)^{24} = .28 \qquad \text{(approximate values)}$$

$$p(3|26) = C_3^{26}(.085)^3(.915)^{23} = .21$$

Then

$$p(\text{Failure to locate}) = p(\text{Detection at fewer than four stations})$$

$$= p(0|26) + p(1|26) + p(2|26) + p(3|26)$$

$$= .10 + .24 + .28 + .21 = .83 \qquad \text{(approximately)}$$

and thus

$$p(\text{Location}) = 1 - p(\text{Failure to locate})$$

$$= 1 - .83 = .17 \qquad \text{(approximately)}$$

We have found, although the 26 station network has a probability of 9/10 of detecting an event of magnitude at least 4.0, the probability that it will locate such an event is less than 2/10.

Summary

If there are n seismic events of a certain magnitude and a network of seismological stations has a probability p of detecting an event of that magnitude (i.e., at least one station in the network will detect), then the probability that the network will detect exactly k of the events is

$$p(k|n) = C_k^n p^k (1 - p)^{n-k}$$

Suppose there are r stations in a network and the network has probability P of detecting a seismic event of a certain magnitude. Assuming each station has the same probability of detecting an event and that detections by the various stations are independent, the probability p that an individual station will detect an event is computed by solving the following equation for p:

$$(1 - p)^r = 1 - P$$

If there are r stations in a network, each has probability p of detecting a seismic event of a certain magnitude, and at least four stations must detect in order to locate the event, then the probability of locating the event can be computed by

$$p(\text{Location}) = 1 - p(\text{Failure to locate})$$

where

$$p(\text{Failure to locate}) = p(0|r) + p(1|r) + p(2|r) + p(3|r)$$

The data calculated in this section is summarized in Table 2-8 (to two decimal places of accuracy).

	Table 2-8		
Magnitude (at least)	Number of stations	Detection probability	Location probability
4.75	7	.90 (assumed)	.10
4.75	26	.99	.96
4.00	26	.90 (assumed)	.17

Exercises

- *In Exercises 2-214 and 2-215, suppose there are five seismic events and there is a probability of detection of .95 for each.*

2-214   What is the probability of detecting all five events?

2-215   What is the probability of detecting at least four of the events?

2-216   If each seismological station in a five station network has a 30% probability of detecting a seismic event, what is the probability that at least two of the stations will detect the event?

- *In Exercises 2-217 and 2-218, suppose seismological equipment is improved in such a way that detection probabilities do not change, but detection of an event at three stations is sufficient to locate the event. The data in this section includes all the figures you will need to answer these questions.*

2-217   What would be the probability of location of events of magnitude at least 4.75 using the proposed 26 station network?

2-218   What would be the probability of location of events of magnitude at least 4.0 using the 26 station network? (See Example 2-31.)

2-219   If there are four seismic events, each with a probability of detection .90, what is the probability that at least one will escape detection?

2-220   If there is a probability of 4/5 of detecting a seismic event at one or two stations of a two station network and if both stations have the same probability of detecting the event, what is the probability of detection for each station?

2-221   Each station of a twelve station network has the same probability of detecting a seismic event. The system as a whole has an 85% probability of detecting an event of a certain magnitude. Write an equation to compute the probability of detection at each station, but do not attempt to solve the equation numerically.

2-222   The major South American seismological stations are located at Antofagasta, Bogota, Huancayo, La Paz, La Plata, Punta Arenas, and Santa Lucia. Suppose this network has a probability of .85 of detecting seismic events in South America of at least a certain magnitude. Then the probability of detection at each station (assuming all are the same) is .24. What is the probability of locating such an event, assuming that it requires at least four detections to locate an event? (If you do not have a calculator, just set up the computation and do not attempt to carry it out.)

2-10   Partitions

Two gamblers agree that each, after betting a certain sum, will draw a card from a standard deck and replace it in the deck. The higher card (with aces above kings) wins all the money.

The first gambler draws a seven and puts it back in the deck. Then, before the second gambler can draw a card, the first says that she is afraid her opponent will cheat, and to protect herself demands that the second gambler draw two cards from the deck and use the second card to determine the winner. The second gambler has no intention of cheating. She knows the probability is $28/52 = .5385$ (approximately) that she will beat the seven on her first draw (because there are 28 cards from eight to ace), but she wonders whether her opponent's demand might be a trick to reduce the probability that she will win. Thus, the second gambler wishes to know the probability that she will draw an eight or higher on her second card.

The new feature of this question is the fact that the probability of drawing an eight or higher after the first card is removed depends on what the first card is, but we do not know what it is. After the first card is removed, 51 cards remain. If the first card is an eight or higher, then only 27 cards at least as high as an eight remain in the deck, so the probability of getting at least an eight on the second card drops to $27/51 = .5294$ (approximately). On the other hand, if the first card is lower than an eight, then all 28 winning cards remain and the second gambler has a probability of $28/51 = .5490$ (approximately) that she will win. Neither calculation answers the gambler's question about the probability of winning, because she needs the answer *before* she knows what the first card is.

In order to analyze the gambler's problem, let A denote the winning outcomes. That is, A is the set of ordered pairs

(any card, eight or higher)

where the two cards in the pair are different because the first is kept out of the deck when the second is drawn. We will divide the successful outcomes, the pairs in A, into two types: those in which the first card of the pair is an eight or higher, and those in which it is not. To achieve this division we divide the sample space U of all the possible outcomes (ordered pairs of distinct cards), successful or not, in the same way. Thus, let B_1 mean that the first card in an ordered pair of distinct cards is an eight or higher and let B_2 mean that the first card is a seven or lower. We wish to divide A into outcomes of the type

(eight or higher, eight or higher)

and those of the type

(seven or lower, eight or higher)

The first type is both in A, because the second card is at least an eight, and in B_1, because the first card is also at least an eight. Thus, the first type of successful outcome is described by $(A$ and $B_1)$. By a similar argument, we can see that the other type of successful outcome is of the type $(A$ and $B_2)$.

No outcome can be both of type B_1 and of type B_2. Since every successful outcome is either in class (A and B_1) or class (A and B_2) and no successful outcome can belong to both, then the simplest form of the principle of probability in Section 2-5 [written there as $p(A \text{ or } B) = p(A) + p(B)$] applies, so

$$p(A) = p[(A \text{ and } B_1) \text{ or } (A \text{ and } B_2)] = p(A \text{ and } B_1) + p(A \text{ and } B_2)$$

Then, we can apply the appropriate form of the principle $p(A \text{ and } B) = p(A|B)p(B)$ worked out in Section 2-7 to produce the formula

$$p(A) = p(A|B_1)p(B_1) + p(A|B_2)p(B_2)$$

If we apply the formula to the gambler's problem, $p(B_1)$ just represents the probability of drawing an eight or higher on the first card, which is 28/52, while the probability of drawing a card lower than an eight is $p(B_2) = 24/52$. The symbol $p(A|B_1)$ means the probability of drawing an eight or higher on the second card when the first card is at least as high as an eight. We calculated before that $p(A|B_1) = 27/51$. We also calculated that $p(A|B_2)$, the probability of drawing an eight or higher on the second draw when the first produced a card below an eight, to be 28/51. By the formula we just developed,

$$p(\text{Winning with second card drawn}) = p(A) = p(A|B_1)p(B_1) + p(A|B_2)p(B_2)$$

$$= \left(\frac{27}{51}\right)\left(\frac{28}{52}\right) + \left(\frac{28}{51}\right)\left(\frac{24}{52}\right)$$

$$= \frac{(27 + 24)28}{(51)(52)} = \frac{(51)(28)}{(51)(52)} = \frac{28}{52}$$

which is the same as under the original rules of the game. So the gambler might just as well agree to using the second card to settle the bet, since it will make her opponent feel more secure.

A better way to understand the formula

$$p(A) = p(A|B_1)p(B_1) + p(A|B_2)p(B_2)$$

is to view the successful outcomes A as a subset of the sample space U of all outcomes (ordered pairs of distinct cards). Furthermore, B_1 and B_2 are subsets of outcomes: those in B_1 are pairs where the first member is at least as high as eight, all the rest are in B_2. Thus, U is divided into subsets B_1 and B_2 where no outcome is in both. We picture the situation as shown in Figure 2-8. Now, certainly $A = U \cap A$ and $U = B_1 \cup B_2$, so

$$A = U \cap A = (B_1 \cup B_2) \cap A$$

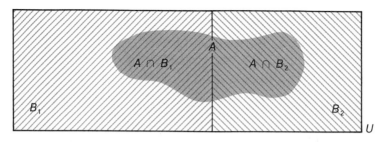

Figure 2-8

If an outcome is in $(B_1 \cup B_2) \cap A$, then it is either in B_1 or B_2 and at the same time, it is definitely in A. Thus, such an outcome is either in both B_1 and A or in both B_2 and A. In symbols,

$$(B_1 \cup B_2) \cap A = (B_1 \cap A) \cup (B_2 \cap A)$$

No outcome can be in both $B_1 \cap A$ and in $B_2 \cap A$, because it is impossible to be in B_1 and B_2 at the same time. Therefore, by the principles of probability,

$$
\begin{aligned}
p(A) &= p(U \cap A) \\
&= p[(B_1 \cup B_2) \cap A] \\
&= p[(B_1 \cap A) \cup (B_2 \cap A)] \\
&= p(B_1 \cap A) + p(B_2 \cap A) \\
&= p(A|B_1)p(B_1) + p(A|B_2)p(B_2)
\end{aligned}
$$

Table 2-9
Number of Heads

0	1	2	3	4	5
TTTTT	HTTTT	HHTTT	HHHTT	HHHHT	HHHHH
	THTTT	HTHTT	HHTHT	HHHTH	
	TTHTT	HTTHT	HHTTH	HHTHH	
	TTटHT	HTTTH	HTHHT	HTHHH	
	TTTTH	THHTT	HTHTH	THHHH	
		THTHT	HTTHH		
		THTTH	THHHT		
		TTHHT	THHTH		
		TTHTH	THTHH		
		TTTHH	TTHHH		

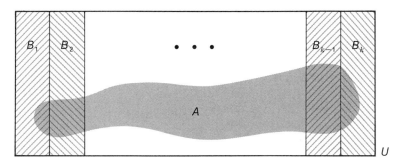

Figure 2-9

The division of U, the set of all outcomes, into the subsets B_1 and B_2, is a special case of a procedure which frequently occurs in studies of probability. A **partition** of the sample space U is a collection of subsets of U where each outcome in U belongs to exactly one of the subsets. A partition we have already encountered is the grouping of the outcomes of flipping five coins according to the number of heads that appear (see Table 2-9). Thus, if we let B_0 be the outcome with no heads, B_1 the outcomes with one head, B_2 those with two heads, and so on to the single outcome in B_5 in which all coins are heads, then B_0, B_1, B_2, B_3, B_4, B_5 is a partition of the set of outcomes of flipping five coins.

Now if say, B_1, B_2, . . . , B_{k-1}, B_k is a partition of a sample space U of outcomes, and A is a set of outcomes which are viewed as the successful outcomes of the experiment, then we visualize the situation as shown in Figure 2-9. An argument exactly like the one above, but involving more symbols, shows that

$$p(A) = p(A|B_1)p(B_1) + p(A|B_2)p(B_2) + \cdots + p(A|B_{k-1})p(B_{k-1}) + p(A|B_k)p(B_k)$$

2-32 **Example** A manufacturer has three assembly lines to produce the same product. Line 1, which uses old equipment, is so slow it produces only 5% of the total output and it is so inaccurate that 2% of the units produced are defective (say that line 1 has a defect rate of .02). Assembly line 2 produces 50% of the total output with a defect rate of only .005, while line 3 produces 45% of the output with a defect rate of .003. (*a*) What is the manufacturer's overall defect rate for this product, i.e., what is the probability that a unit chosen at random will be defective? (*b*) What would the defect rate become if the manufacturer were to close assembly line 1?

Solution (*a*) The outcomes are units of the product and we partition the units into subsets B_1, B_2, and B_3 according to which assembly line they come from. If a unit is chosen at random, then $p(B_1)$, the probability it comes from line 1, is .05. Also, $p(B_2) = .50$ and $p(B_3) = .45$ according to the statement of the problem. We wish to know $p(A)$, the probability that a unit is defective. The defect rate .02 for line 1 is the probability that a unit is defective under the condition that it comes from line 1, and so we can write $p(A|B_1) = .02$. In the same way, $p(A|B_2) = .005$ and $p(A|B_3) = .003$. By the formula,

$$p(A) = p(A|B_1)p(B_1) + p(A|B_2)p(B_2) + p(A|B_3)p(B_3)$$
$$= (.02)(.05) + (.005)(.50) + (.003)(.45) = .00485$$

(b) Closing line 1 means setting $p(B_1) = 0$ while lines 2 and 3 take a proportionately larger percentage of the total. The total production with line 1 closed is 95% of what it was before so now $p(B_2) = .50/.95 = .53$ (approximately) and $p(B_3) = .45/.95 = .47$ (approximately). Under these conditions, the defect rate is

$$p(A) = (.02)(0) + (.005)(.53) + (.003)(.47) = .00406 \quad \text{(approximately)}$$

The information we have just obtained would help the manufacturer decide whether the improved dependability of the product would compensate for the loss of 5% of the production as a result of closing line 1.

2-33   **Example**   A study of families with at least one blue-eyed child in some human population finds that 35% of them have two blue-eyed parents, 50% have exactly one blue-eyed parent, and the remaining 15% have neither parent with blue eyes. According to Mendel's law of genetics, if a couple is genetically capable of having blue-eyed children, then the probability that each child will be blue-eyed is 1.00 if both parents are blue-eyed, .50 if one parent is blue-eyed, and .25 if neither parent is blue-eyed. What is the probability that a child in this population from a family with at least one blue-eyed child will have blue eyes?

Solution   A child's probability of having blue eyes depends on the color of his parents eyes, so we partition the set of parents of families with at least one blue-eyed child into subsets B_1 where both parents have blue eyes, B_2 where one parent does, and B_3 where neither parent does. Let $p(A)$ be the probability that a child chosen at random from among those families with at least one blue-eyed child will have blue eyes. According to the formula above,

$$p(A) = p(A|B_1)p(B_1) + p(A|B_2)p(B_2) + p(A|B_3)p(B_3)$$

where the conditions of the problem imply that $p(B_1) = .35, p(B_2) = .50$, and $p(B_3) = .15$. The conditional probability $p(A|B_1)$ is the probability that a child from a family with at least one blue-eyed child (which might be that child himself) will have blue eyes, under the condition that both parents have blue eyes. Since it is known that the family has at least one blue-eyed child, we know—even without knowing anything about genetics— that the couple must be capable of having blue-eyed children. Therefore, Mendel's law tells us that $p(A|B_1) = 1.00, p(A|B_2) = .50$, and $p(A|B_3) = .25$. Substituting into the formula, we calculate

$$p(A) = (1.00)(.35) + (.50)(.50) + (.25)(.15) = .6375$$

Summary

A **partition** of the sample space U is a collection of subsets such that each outcome in U is in exactly one of the subsets.

If $B_1, B_2, \ldots, B_{k-1}, B_k$ is a partition of a sample space U and A is a subset of U, then

$$p(A) = p(A|B_1)p(B_1) + p(A|B_2)p(B_2) + \cdots + p(A|B_{k-1})p(B_{k-1}) + p(A|B_k)p(B_k)$$

Exercises

2-223   In a certain population, 15% of the men have difficulty hearing, while only 12% of the women have this problem. Assuming that there are as many men as women in the population, what is the probability that a person chosen at random from the population will have difficulty hearing?

2-224   A test for a certain disease will be positive for 80% of all carriers of the disease, but there is also a probability of .15 that the test will be positive even though the subject does not carry it. If 5% of the population are carriers of this disease, what is the probability that a person chosen at random will give a positive test?

2-225   An insurance company knows that 20% of the drivers covered by its policies are under 25 years of age and 35% are over 60. The company estimates that the accident rate for its drivers under 25 is .012; that is, there is a probability of .012 that a driver in this age group will have an accident during a given year. The accident rate for its over 60 drivers is .009, and for all other drivers is .005. Estimate the probability that a driver covered by this company's policies will have an accident in a given year.

2-226   The human blood types are A, B, AB, and O. Suppose that, in some population, 35% of the population is of type A, 42% of type B, 18% of type AB, and 5% of type O. Imagine there is a blood disease that is contracted depending on blood type. The probability of getting the disease is .001 for type A and type B individuals, .0005 for type AB, and .005 for type O. What is the probability that an individual from this population will contract the disease?

2-227   A factory works three shifts, with 1000 employees during the day shift (8 a.m. to 4 p.m.), 500 on the swing shift (4 p.m. to midnight), and 300 on the graveyard shift (midnight to 8 a.m.). The absentee rate, that is, the probability that an employee will fail to show up for work on a given day, is .02 for the day shift, .05 for the swing shift, and .07 for the graveyard shift. What is the absentee rate for the factory as a whole?

2-228   A Democrat and a Republican are running for mayor. The voters of the city are 50% Democrats, 30% Republicans, and 20% independents. If a poll indicates that 60% of the Democrats will vote for the Democratic candidate and the rest for the Republican, that 90% of the

Republicans will vote for their party's candidate and the rest for the Democrat, and that the independents are evenly divided between the candidates, who does the poll indicate will win the election?

- *In Exercises 2-229 through 2-233, suppose there is a hereditary birth defect that has a probability of .80 of affecting an offspring if both parents have the defective gene, .10 if one parent has it, and cannot affect the offspring if neither parent has the gene. Suppose that 1% of the adult population carries the defective gene and that mates are chosen independent of possession of the defective gene.*

2-229 What is the probability that both parents have the defective gene?

2-230 What is the probability that the father has the defective gene and the mother does not?

2-231 What is the probability that exactly one parent carries the defective gene?

2-232 What is the probability that neither parent has the defective gene?

2-233 What is the probability that an infant will be affected by the birth defect?

2-234 A game consists of rolling a die and drawing a card. If the player rolls an odd number on the die, then he wins the game only if he draws an ace. If he rolls an even number, then he wins if he draws a card with a value higher than the number of his roll, where aces are valued as one and face cards as ten. Thus, for example, a player who rolls a four will win if he draws a five, six, seven, eight, nine, ten, jack, queen, or king. What is the probability of winning in this game?

2-235 With respect to a particular gene, biologists classify individuals into three distinct genotypes: AA, Aa, and aa. Suppose that 30% of some population is of type AA, 60% of type Aa, and 10% of type aa. Further suppose that genotype with respect to this gene influences the probability that an infant will reach maturity as follows: 70% of type AA individuals will reach maturity, 60% of type Aa, and 30% of type aa. What is the probability that an individual chosen at random from this population at birth will reach maturity?

2-236 A chain of department stores has four branches. The shoplifting rates at the branches (amount per $1 of sales which will be lost because of shoplifting) and the annual sales volume is given in the table. How much money should the chain expect to lose because of shoplifting next year?

Branch	Shoplifting rate	Volume
A	.008	$300,000
B	.011	$200,000
C	.017	$250,000
D	.005	$250,000

2-11   Expected Value

A gambler who plays at a casino that offers both American roulette and a game called "keno" wishes to decide which game to play. If he understands probability, he knows that he can expect to lose money at both games. But if he views his losses as the cost of the entertainment provided by playing the game, he still might wonder which game, roulette or keno, is the cheaper form of entertainment.

Suppose the gambler always bets $1 in roulette that the winning number will be one of the first third of the numbers (numbers 1–12), and he also bets $1 that it will be in the last third (numbers 25–36). If the number that comes up is not in either of these groups, namely 0, 00, or 13–23, then he loses $2. If the number chosen is one of the first third of the numbers, then he loses the $1 bet on the last third, but makes $2 for the $1 bet on the first third. Thus, if one of the numbers 1–12 comes up, the gambler has a net gain of $1. Similarly, if a number from 25 through 36 is selected, he also comes out $1 ahead. Consequently, out of the 38 possible outcomes in roulette, 24 of them will cause the gambler to win $1 and the remaining 14 outcomes will cause him to suffer a loss of $2.

In the game of keno, the gambler selects a number from 1 through 80. Twenty numbers from 1 through 80 are chosen at random. If the gambler bets $1 and his number is one of the twenty selected, he wins $2. If his number is not selected, he loses $1.

In order to compare the games, the gambler can think of each play of a game as a single unit of entertainment and ask how much a single play of roulette costs compared to a single game of keno. Of course, the outcome of a single game of roulette is either the gain of $1 or the loss of $2, while the outcome from keno is either a win of $2 or a loss of $1. But what the gambler wants to know is the cost per game over the long run.

The measure of the cost of a game that we will use is the **expected value** of the game. To calculate the expected value of any gambling game, we divide up the possible outcomes according to the amount of the payoff, grouping together all outcomes with the same payoff (which may be negative if the outcomes result in a loss), and determine the probabilities that the result of the game will be an outcome in each group. The expected value of the game is calculated by multiplying the value of any outcome in a group by the probability that the result of the game will come from that group, and adding up the amounts calculated for all such groups.

The gambler's strategy in roulette was to bet $1 on the first third of the numbers and $1 on the last third. We have seen that if one of these numbers is selected, then the gambler wins $1. Since there are 24 winning numbers, the probability of winning $1 is 24/38. The fourteen remaining numbers each produce a loss of $2 if they are selected. A $2 loss is expressed as a payoff of −$2. The probability is 14/38 that this payoff will occur. By the rule we just stated for calculating expected value,

Expected value $= (\$1)(\text{Probability of a } \$1 \text{ payoff}) + (-\$2)(\text{Probability of a } -\$2 \text{ payoff})$

$$= (\$1)\left(\frac{24}{38}\right) + (-\$2)\left(\frac{14}{38}\right)$$

$$= -\frac{\$4}{38} = -10\tfrac{1}{2}\text{¢} \qquad (\text{approximately})$$

In the keno game, the probability of a $2 payoff is the probability 20/80 of winning. The probability of a loss is 60/80 and the amount lost is $1; this is a payoff of $-\$1$. Therefore,

$$\text{Expected value} = (\$2)\left(\frac{20}{80}\right) + (-\$1)\left(\frac{60}{80}\right) = -\frac{\$20}{80} = -25\text{¢}$$

Thus, we see that keno at $1 a game is more expensive than the proposed roulette strategy, even though roulette requires the risk of $2 on each spin of the wheel. Consequently, the gambler who wants as many plays as possible for his money would be better off betting on roulette than on keno.

2-34 **Example** The following gambling strategy in the game of roulette was discussed in Section 2-4: The gambler bets $1 on the first third of the numbers (1–12) and $1 on the last column (3, 6, 9, and so on to 36). The casino pays $2 for each $1 bet on the last column, which is the same payoff as for a win on the first third of the numbers. What is the expected value of roulette with this gambling strategy?

Solution If the number selected in the game is in the first third but not the last column (1, 2, 4, 5, 7, 8, 10, 11), then the gambler is $1 ahead, because he loses the $1 bet on the last column but wins $2 from the bet on the first third. The result is the same, a net gain of $1, if the number that comes up in the game is in the last column but not the first third (15, 18, 21, 24, 27, 30, 33, 36). Thus, there are sixteen numbers that pay the gambler $1, so the probability of this payoff is 16/38. If the number selected is 3, 6, 9, or 12, then the gambler wins $2 for *each* $1 bet, so the gain is $4. The remaining eighteen numbers cause a loss of $2 (payoff of $-\$2$). We calculate that

$$\text{Expected value} = (\$1)\left(\frac{16}{38}\right) + (\$4)\left(\frac{4}{38}\right) + (-\$2)\left(\frac{18}{38}\right)$$

$$= -\frac{\$4}{38} = -10\tfrac{1}{2}\text{¢} \qquad (\text{approximately})$$

Example 2-34 shows that the expected value of the roulette strategy of Section 2-4 is exactly the same as the expected value of the strategy of betting $1 on each of first third and last third. We now have defended our claim in Section 2-5 that neither is preferable to the other; the expected value is the same in either case.

The concept of expected value arises in many circumstances which have nothing to do with gambling, so we need to define it in a more abstract way. Let us, therefore, imagine a sample space U and a partition, say, $B_1, B_2, \ldots, B_k$ of U into subsets. Suppose we are given a numerical value v_1 which is associated with each element in B_1, a value v_2 for each outcome in B_2, and so on to a value v_k for the outcomes of B_k. In the gambling applications, the B_i are the outcomes with a certain payoff for the gambler and v_i is the amount of that payoff. Define the expected value of the probability problem represented by the partition and the given numbers v_i to be

$$\text{Expected value} = v_1 p(B_1) + v_2 p(B_2) + \cdots + v_k p(B_k)$$

where $p(B_i)$ is, as before, the probability that the result will come from the set B_i in the partition.

2-35 **Example**   Suppose the probability that a mature female of a certain animal species will not produce offspring in a given year is .35, that she will have one infant is .20, that she will have two is .30, three is .10, and four is .05. What is the expected value of the number of offspring in a given year for a mature female of this species?

Solution   The partition of outcomes is B_0 for no offspring, B_1 for one, and so on for B_2, B_3, and B_4. The value of an outcome is the number of offspring, so

$$\text{Expected value} = v_0 p(B_0) + v_1 p(B_1) + v_2 p(B_2) + v_3 p(B_3) + v_4 p(B_4)$$
$$= (0)(.35) + (1)(.20) + (2)(.30) + (3)(.10) + (4)(.05)$$
$$= 1.3 \text{ offspring}$$

Summary

Given a partition, say, $B_1, B_2, \ldots, B_k$ of a sample space U and a numerical value v_i for each subset B_i in the partition, the **expected value** is defined by

$$\text{Expected value} = v_1 p(B_1) + v_2 p(B_2) + \cdots + v_k p(B_k)$$

where $p(B_i)$ denotes the probability that the outcome will be in the set B_i.

Exercises

2-237   The European roulette wheel consists of the numbers 0, 1, 2, . . . , 36. A gambler wins
         $2 for a winning bet on the first third (1–12) and the same payoff for winning on the last
         column (3, 6, . . . , 36). If a gambler bets $1 on each of first third and last column, what
         is the expected value of the gambler's loss for the game? (Remember that European roulette
         uses only 0; it does not use 00.)

2-238   A gambler chooses a three digit number (from 000 to 999) for a lottery and bets 50¢. The
         lottery selects a three digit number at random and pays $450 to the holder of that number
         but does not return the winner's 50¢. What is the expected value of the gambler's gain
         (or loss) in the lottery?

2-239   Suppose the probability that a litter of a certain species of animal will consist of just one
         offspring is .08, the probability of two is .27, of three is .39, of four is .22, and of five is .04.
         What is the expected value of litter size for this species of animals?

2-240   An insurance company charges a woman a premium of $150 for an accident insurance
         policy for the coming year that will pay for at most one accident. If the policy holder
         suffers a major accident during the year, we estimate the company will pay her $5000,
         and, for a minor accident, $1000. The company does not return the $150 premium. If
         the probability is .08 that she will have a minor accident, .005 that she will have a major
         one, and thus .915 that she will have no accident at all, what is the expected value of the
         insurance policy to the company?

2-241   A gambler chooses a three digit number (000–999) for a lottery and bets 50¢. The lottery
         selects a three digit number at random. If the gambler has chosen the winning number,
         she receives $250 but does not get her 50¢ back. If the first two digits of the gambler's
         number are the same as the winning number (for example, if the winning number is 134
         and she holds 137), then she receives $25 but does not get her 50¢ back. What is the
         expected value of the lottery to the gambler?

2-242   The handsome prince has been searching all over his kingdom for the maiden who left
         her glass slipper at the ball. He has tried the slipper on all but three of the maidens in the
         kingdom without success. If one of the three remaining maidens is Cinderella, the owner
         of the slipper, and he tries the slipper on the three maidens in random order, what is the
         expected value of the number of maidens on whom the handsome prince must try the
         slipper in order to discover Cinderella?

2-243   In "6 spot keno," a gambler pays $1 and selects six numbers from 1 through 80. Twenty
         numbers from 1 through 80 are chosen at random as the winning numbers. If the gambler
         has selected two or fewer of the winning numbers, he loses the $1 bet. If the gambler has
         chosen three winning numbers, he gets his $1 back. If the gambler gets four winners, he
         receives $2 in addition to the $1 bet. If the gambler has five winners, he wins $89 in addition

to the $1; and if he has all six winning numbers, the payoff is $1799 plus the $1 bet. The approximate probability of choosing exactly three winners is .1298, four winners is .0285, five winners is .0031, and six winners is .0001. What is the approximate probability of losing $1? What is the expected value of this game to the gambler?

2-244   Cards are to be drawn from a standard deck, without replacement, until a card other than a black ace is drawn. What is the expected value of the number of cards drawn?

2-245   What is the expected value of the number of face cards in a three card hand drawn at random from a standard deck?

2-12   Bayes' Theorem

In Section 2-7, we found that the definition of conditional probability led to two different formulas for the probability $p(A \text{ and } B)$ that the result of an experiment will be both of type A and of type B. The formulas were

$$p(A \text{ and } B) = p(A|B)p(B)$$

$$p(A \text{ and } B) = p(B|A)p(A)$$

Since both of the products $p(A|B)p(B)$ and $p(B|A)p(A)$ are equal to $p(A \text{ and } B)$, we conclude that

$$p(B|A)p(A) = p(A|B)p(B)$$

Dividing both sides of the equation by $p(A)$ gives a fundamental relationship,

$$p(B|A) = \frac{p(A|B)p(B)}{p(A)}$$

between the two conditional probabilities $p(B|A)$ and $p(A|B)$.

In Example 2-22, we considered a school with a faculty of 20 men, 8 of whom had master's degrees, and 30 women, of whom 18 had master's degrees. We computed the probability $p(A|B)$ that a representative chosen at random from the faculty had a master's degree, under the condition that the representative was a woman. Here, A is the set of faculty members with master's degrees and B is the set of women faculty members. Consequently, $p(B|A)$ is the probability that the representative chosen is a woman, under the condition that the representative is known to possess a master's degree. To compute $p(B|A)$, we recall from Example 2-22 that $p(A) = 26/50$ and $p(A|B) = 18/30$. Since there are 50 teachers in all, of whom 30 are women, $p(B) = 30/50$. According to the formula above,

$$p(B|A) = \frac{p(A|B)p(B)}{p(A)} = \frac{\binom{18}{30}\binom{30}{50}}{\frac{26}{50}} = \frac{18}{26} = .69 \quad \text{(approximately)}$$

In this simple case, we could just have applied the counting definition of conditional probability,

$$p(B|A) = \frac{n(A \text{ and } B)}{n(A)} = \frac{18}{26}$$

but we need to see the general procedure.

2-36   **Example** Example 2-24 asks: If two dice are rolled, what is the probability $p(A|B)$ that the sum of the numbers is 6, given that the first die is a three. (*a*) What does $p(B|A)$ mean? (*b*) Compute $p(B|A)$.

Solution (*a*) From the way in which the question from Example 2-24 is asked, we know that A is the class of outcomes for which the sum is 6, while B is the class of outcomes in which the first die is a three. Therefore, the symbol $p(B|A)$ denotes the probability that the first die was a three, knowing the sum turned out to be 6.

(*b*) Example 2-24 calculated that $p(A|B) = 1/6$ and $p(A) = 5/36$. The six pairs in B were listed, and since there are 36 possibilities, $p(B) = 6/36 = 1/6$. [Alternatively, $p(B)$ is the probability that the first die is a three and since it does not matter what the second die does, $p(B)$ is really just the probability of rolling a three with one die—that certainly is 1/6.] By the formula,

$$p(B|A) = \frac{p(A|B)p(B)}{p(A)} = \frac{(1/6)(1/6)}{5/36} = \frac{1}{5} = .20$$

Now, we can also view Example 2-22 as a partition problem. Still letting A be the set of faculty members with master's degrees, let B_1 be the men teachers on the faculty and B_2 the women. Then B_1 and B_2 form a partition of the set of all faculty members and the problem above was the calculation of $p(B_2|A)$, the probability that the outcome was from set B_2 of the partition, under the condition that the outcome was successful—a teacher with a master's degree was chosen. A very common use of the formula

$$p(B|A) = \frac{p(A|B)p(B)}{p(A)}$$

is in partition questions of this sort. That is, given a partition $B_1, B_2, \ldots, B_k$ of a sample space and knowing the outcome of the experiment was successful in the sense of belonging to some set A, what is the probability $p(B_i|A)$ that the outcome belongs to some particular set B_i in the partition? The formula that answers the question is

$$p(B_i|A) = \frac{p(A|B_i)p(B_i)}{p(A)}$$

the previous formula with B_i in place of B.

In order to illustrate the use of the formula to determine the probability that a successful outcome belongs to a particular set in the partition, we recall Example 2-32. In that example, a manufacturer has three assembly lines 1, 2, and 3, producing the same product. Partitioning the factory's output into sets B_1, B_2, and B_3 according to which line produced the unit, the example stated that $p(B_1) = .05$, $p(B_2) = .50$, and $p(B_3) = .45$. For A the set of defective units, the example also gave the defect rates $p(A|B_1) = .02$, $p(A|B_2) = .005$, and $p(A|B_3) = .003$. In studying the effects of the inefficient line 1, the manufacturer might wish to know the proportion of the defective units from the factory for which line 1 is to blame. That is, the manufacturer wishes to know the probability $p(B_1|A)$ that a defective unit came from line 1. The solution to Example 2-32 showed that $p(A) = .00485$. Therefore, by the formula,

$$p(B_1|A) = \frac{p(A|B_1)p(B_1)}{p(A)} = \frac{(.02)(.05)}{.00485} = .21 \qquad \text{(approximately)}$$

Thus, line 1, which produces only 5% of the total output is responsible for more than 20% of the defective units.

2-37 **Example** The owner of a racehorse must decide whether to enter her horse in a race. The press reports that two horses, call them X and Y, which have sometimes defeated her horse, may also enter the race. The owner estimates on the basis of past experience that her horse's probability of winning is .60 if only horse X enters, .45 if only horse Y runs, .25 if both X and Y enter, but .85 if neither X nor Y is in the race. The reports state that there is a probability of .50 that X will enter and .25 that Y will enter. (*a*) What is the probability that both X and Y will enter, assuming that these events are independent? (*b*) What is the probability that neither X nor Y will enter? (*c*) What is the probability that the owner's horse will win the race? (*d*) If the owner's horse wins, what is the probability that the win will occur when neither X nor Y are entered?

Solution (*a*) Since we are to assume the events are independent,

$$p(\text{Both } X \text{ and } Y \text{ enter}) = p(X \text{ enters})p(Y \text{ enters}) = (.50)(.25) = .125$$

(*b*) Still assuming independence,

$$p(\text{Neither } X \text{ nor } Y \text{ enter}) = p(X \text{ does not enter})p(Y \text{ does not enter})$$

Since

$$p(X \text{ does not enter}) = 1 - p(X \text{ enters}) = 1 - .50 = .50$$

$$p(Y \text{ does not enter}) = 1 - p(Y \text{ enters}) = 1 - .25 = .75$$

then

$$p(\text{Neither } X \text{ nor } Y \text{ enters}) = (.50)(.75) = .375$$

(*c*) Let $p(A)$ denote the probability that the owner's horse will win. It may be that (1) only X will enter the race, (2) only Y will enter, (3) both X and Y enter, or (4) neither X nor Y enter. All these possibilities are distinct, yet one of them must happen, so let B_1, B_2, B_3, and B_4 be the partition of all possible races that result. Reasoning as in parts (*a*) and (*b*),

$$p(B_1) = p(X \text{ enters and } Y \text{ does not enter}) = p(X \text{ enters})p(Y \text{ does not enter})$$

$$= (.50)(.75) = .375$$

$$p(B_2) = p(X \text{ does not enter and } Y \text{ enters}) = p(X \text{ does not enter})p(Y \text{ enters})$$

$$= (.50)(.25) = .125$$

In part (*a*) we found that $p(B_3) = .125$, while part (*b*) showed that $p(B_4) = .375$. The problem gives us the conditional probability that the horse will win under each of the four possible conditions of the race, namely, $p(A|B_1) = .60$, $p(A|B_2) = .45$, $p(A|B_3) = .25$, and $p(A|B_4) = .85$. By the formula of Section 2-10,

$$p(A) = p(A|B_1)p(B_1) + p(A|B_2)p(B_2) + p(A|B_3)p(B_3) + p(A|B_4)p(B_4)$$

$$= (.60)(.375) + (.45)(.125) + (.25)(.125) + (.85)(.375) = .63125$$

(*d*) The question is to compute $p(B_4|A)$, the probability that neither X nor Y enters, assuming that the horse wins. By the formula above,

$$p(B_4|A) = \frac{p(A|B_4)p(B_4)}{p(A)} = \frac{(.85)(.375)}{.63125} = .505 \qquad \text{(approximately)}$$

Thus far, we have broken the problem of computing $p(B_i|A)$ into two parts. First we calculate

$$p(A) = p(A|B_1)p(B_1) + p(A|B_2)p(B_2) + \cdots + p(A|B_k)p(B_k)$$

and then we substitute the answer into

$$p(B_i|A) = \frac{p(A|B_i)p(B_i)}{p(A)}$$

If, instead, we substitute the formula for $p(A)$ into the formula for $p(B_i|A)$, we get a one-step formula for calculating $p(B_i|A)$:

$$p(B_i|A) = \frac{p(A|B_i)p(B_i)}{p(A|B_1)p(B_1) + p(A|B_2)p(B_2) + \cdots + p(A|B_k)p(B_k)}$$

This rather complicated formula is called **Bayes' theorem.**

2-38   **Example**   Scientists band individuals in a small population of animals of an endangered species for the purpose of studying their habits. The scientists band 10% of the population with plastic tags and 30% with metal tags. There is a 50% probability that a plastic tag will fall off an animal, but the metal tags have only a 5% probability of falling off. If an animal from that population is found without a tag, what is the probability that it once had a plastic tag?

Solution   Let A be the set of animals without tags and partition the animal population into B_1, those that were originally given plastic tags; B_2, those that were given metal tags; and B_3, those animals in the population that were never banded. Note that $p(B_1) = .10$, $p(B_2) = .30$, and $p(B_3) = .60$. According to the problem, the probability that an animal once wore a plastic tag and no longer has it, $p(A|B_1)$, is .50. Also, $p(A|B_2) = .05$, and, of course, $p(A|B_3) = 1.00$ since an animal that never had a tag can hardly acquire one by itself. The probability that an animal without a tag once wore a plastic one is $p(B_1|A)$, so Bayes' theorem tells us

$$p(B_1|A) = \frac{p(A|B_1)p(B_1)}{p(A|B_1)p(B_1) + p(A|B_2)p(B_2) + p(A|B_3)p(B_3)}$$

$$= \frac{(.50)(.10)}{(.50)(.10) + (.05)(.30) + (1.00)(.60)} = .075 \quad \text{(approximately)}$$

Summary

The conditional probabilities $p(B|A)$ and $p(A|B)$ are related by

$$p(B|A) = \frac{p(A|B)p(B)}{p(A)}$$

Let $B_1, B_2, \ldots, B_k$ be a partition of a sample space U and let A be a set of outcomes in U. Then **Bayes' theorem** states

$$p(B_i|A) = \frac{p(A|B_i)p(B_i)}{p(A|B_1)p(B_1) + p(A|B_2)p(B_2) + \cdots + p(A|B_k)p(B_k)}$$

Exercises

2-246   In a certain population, 10% of the men are color-blind and 2% of the women are color-blind. Assuming that there are as many men as women in the population, if a person chosen at random is color-blind, what is the probability that it is a man?

●   *For Exercises 2-247 and 2-248, the table below states accident rates (probabilities that a driver will have an accident in a given year) for drivers of various ages and the percentage of drivers in each age group covered by the policies of a certain insurance company.*

Age	Accident rate	Percentage of total
Under 25	.015	20%
25–35	.006	35%
36–50	.004	30%
Over 50	.011	15%

2-247   What percentage of accidents by drivers covered by the company's policies involve drivers under 25. That is, if a driver covered by the company is involved in an accident, what is the probability that the driver is under 25?

2-248   What percentage of accidents covered by the company involve drivers between the ages of 36 and 50?

2-249   A test for a certain disease will be positive for 90% of all carriers of the disease, but there is also a probability of .20 that a person who does not have the disease will have a positive reaction to the test. If 5% of the population carries the disease, what is the probability that a person who has a positive reaction to the test really has the disease?

2-250   In a certain city, 80% of all households have television sets. Of the households with television sets, 20% own stereos, while only 5% of the households without television sets have stereos. If a household has a stereo, what is the probability that it also has a television set?

2-251   The table below lists human blood types, the percentage of some population with that type of blood, and the probability that each type will contract a certain blood disease. What is the probability that a person who gets the disease has type O blood?

Type	Percentage of population	Probability of disease
A	35	.001
B	40	.001
AB	20	.0002
O	5	.007

2-252   A Democrat and a Republican are running for mayor. The voters of the city are 40% Democrats, 40% Republicans, and 20% independents. Suppose a poll indicates that 70% of the Democrats will vote for the candidate of their party and 30% against her. Furthermore, 80% of the Republicans will vote for their party's candidate and the rest will vote for the Democrat. The independents support the Democratic candidate 60% to 40%. How much of the Democratic candidate's support is coming from her own party; that is, if a person who will vote for the Democratic candidate is chosen at random, what is the probability that the person will be a Democrat?

●   *Biologists classify individuals with respect to a particular gene into genotypes AA, Aa, and aa. For Exercises 2-253 and 2-254, suppose that with respect to a gene which affects hair color, 40% of the population is of type AA, 50% is of type Aa, and the rest is of type aa. Suppose further that no individual of genotype AA can have blond hair, that 30% of the type Aa individuals have blond hair, and that all the type aa individuals are blond.*

2-253   What is the probability that a member of this population has blond hair?

2-254   If an individual in the population has blond hair, what is the probability that this person's genotype is aa?

2-255   A department store chain has three branches. The shoplifting rates (proportion of $1 of sales lost due to shoplifting) and the annual sales volumes for the branches are presented in the table. How much of the chain's shoplifting losses are attributable to store *C*? That is, what is the probability that stolen merchandise worth $1 was taken from branch *C*?

Branch	Shoplifting rate	Volume
A	.01	$200,000
B	.006	$100,000
C	.03	$200,000

2-256   A manufacturer of sports cars buys its speedometers from companies that specialize in their manufacture. One small company produces speedometers with a high degree of precision; only one unit in 1000 is defective (the defect rate is .001). However, this company can supply only 80% of the car maker's requirements. The rest of the speedometers needed must be bought from a large company whose mass-production methods give rise to a defect rate of .05. What proportion of the defective speedometers the car manufacturer installs comes from the less dependable manufacturer, i.e., what is the probability that a defective speedometer came from the large speedometer manufacturer?

2-257   The table below lists the number of employees in each shift at a factory and the absentee rate for each shift (the probability that any particular employee in that shift will fail to show up for work on any given day). What proportion of the absenteeism at the factory is attributable to the graveyard shift? In other words, what is the probability that an absent employee, chosen at random, was supposed to be working during the graveyard shift?

Shift	Number of employees	Absentee rate
Day (8 a.m. to 4 p.m.)	500	.01
Swing (4 p.m. to midnight)	500	.03
Graveyard (midnight to 8 a.m.)	300	.07

2-258   Suppose a poker player has the betting strategy that he will always raise his bet if he has a good hand and that he will raise his bet on 30% of his poor hands in hopes of bluffing the other players into believing he has a good hand. If one out of four hands the player is dealt is good and if the player raises his bet on a hand, what is the probability that he is bluffing?

2-259   Wheat farms are classified as poor, fair, or good on the basis of the quality of the soil. In a particular state, 25% of the farms have poor soil, 45% fair, and 30% good. A new variety of wheat is used in all the farms in the state that will reportedly increase the yield per acre of 80% of all farms with poor soil, 40% of farms rated fair, and 5% of farms with good soil. How much of the improvement in wheat yields from the new variety is the result of increasing the yield of farms with poor soil? That is, if a farm from the state whose yield per acre has been improved is chosen at random, what is the probability that it has poor soil?

2-260   Suppose that 40% of the men in a certain city went to college and that 2% of them refuse to admit they did. Of the men in the same city who never went to college, 3% claim that they did. If a man from the city tells you he went to college, what is the probability that he is telling the truth? That is, what is the probability that he did go to college?

2-261   Suppose that in some country, 25% of the people have parents who are divorced. Suppose also that there is a probability of .30 that a person will be divorced if his or her parents were divorced and a probability of .20 if they were not. If a person is divorced, what is the probability that his or her parents were also divorced?

2-262   Two dice are rolled and the sum is 4. What is the probability that the first die was a one?

2-13   An Application of Bayes' Theorem

One way out of the impasse caused by the differences between the Soviet and American positions on an underground test ban treaty for nuclear weapons would open up if the United States were to feel so confident in its ability to monitor the treaty by means of remote seismological stations that it no longer insisted on on-site inspections. The problem of measuring confidence in a monitoring system is a very complicated one; it will be discussed further in our concluding essay. For now, we restrict ourselves to a single, relatively simple, part of that problem.

Suppose a monitoring system detects a number of seismic events in some region of the earth over a period of time. How confident can one be that all the events that occurred there (or, more realistically, the great majority of them) have been detected? In the language of this chapter, one wishes to know the probability that all, or most, of the events that took place were detected. If k events have been detected, what is the probability that n actually occurred? The symbol $p(n|k \text{ detected})$ will be used to represent this probability. For example, if during some period of time, five seismic events were observed in some region of the world and if we wish to determine the probability that an additional, undetected, seismic event happened, then we wish to calculate $p(6|5 \text{ detected})$, the conditional probability that there were six seismic events, given that five such events were recorded by seismological equipment.

A different probability question concerning seismic events plays an important role in the determination of $p(n|k \text{ detected})$: Suppose that in fact n seismic events occurred in some region of the world during a certain period of time, what is the probability that k of the events will be detected? This question was discussed in Section 2-9. There, we viewed each event as an experiment conducted under identical conditions with the same probability of success, detection of the event, each time the experiment takes place. The probability of k detections out of n events was represented by the symbol $p(k|n)$, because it is a probability of the type analyzed in Section 2-8.

Although the probabilities $p(k|n)$ and $p(n|k$ detected$)$ are different, they are intimately related. To state the relationship, we need one other type of information: the probability that a certain number of seismic events will take place. Letting $p(n)$ represent the probability that n seismic events will occur, we require numbers for $p(0)$, $p(1)$, $p(2)$, and so on until n is so large that there is, essentially, no likelihood that so many events would take place.

On one level, the determination of the numbers $p(n)$ can be viewed as a straightforward exercise in estimating probabilities on the basis of past experience. For an example, we will use the number of shallow earthquakes of magnitude at least 6.0 in the Mongolian region (34° to 42° N and 75° to 120° E) from 1930 through 1939:

Year	1930	1931	1932	1933	1934	1935	1936	1937	1938	1939
Number of events	1	4	3	2	1	2	2	1	1	3

From this table, we count the number of years in which each number of events occurred:

Number of events	0	1	2	3	4
Number of years	0	4	3	2	1

Thus, since one event per year occurred in 4 of the 10 years (1930, 1934, 1937, 1938) for which we have information, we estimate $p(1)$, the probability that there will be one such event in the Mongolian region in any given year, at $p(1) = 4/10 = .40$. In the same way, we estimate $p(2) = .30$, $p(3) = .20$, $p(4) = .10$, and $p(n) = 0$ for all other n. The reader may object that this example is not very relevant, since the test-monitoring problem is concerned with seismic events of magnitudes much smaller than 6.0. However, data on smaller events in many regions is not available, and since, according to seismologists, there is some relationship between the number of large earthquakes and the number of small ones, this large earthquake data still is quite useful. As in our discussion of the computation of detection probabilities, data of this kind is collected on a continuing basis, not just for a decade. With enough experience, figures close to the correct probabilities can be obtained unless conditions change.

However, in practice, there are a number of difficulties involved in computing the probabilities $p(n)$. If we had been able to determine the number of seismic events with complete accuracy in the past, we would not now be seeking a way to find the probability that some were overlooked. In fact, even the record of large events in the example above may not be complete. Also, if the seismic events recorded in the past in some part of the world included both earthquakes and nuclear explosions, the frequency of events in the future will depend in part on whether it is expected that nuclear testing in that area will cease (perhaps because of a ban), remain at its present level, or increase. A reasonable approach in such a case would be to use various sets of values for the $p(n)$, depending on hypotheses about future political decisions, and then perform the probability calculations below many times (on a computer, of course), once for each choice of hypothesis. In this way, the decision-makers can be shown how the probability of overlooking events would depend on various possible levels of testing activity.

We will assume for simplicity that a single set of probabilities $p(n)$, one for each $n = 0, 1, 2$, and so on, has somehow been arrived at. The probability $p(k|n)$ of k detections out of n events can be computed by the formula

$$p(k|n) = C_k^n p^k (1 - p)^{n-k}$$

assuming that the detection probability p for a single event is known, it remains constant over time, and events are independent.

If k events are detected, then the probability $p(n|k$ detected$)$ that there were really n events can, we claim, be computed by Bayes' theorem. We partition the possible levels of seismic activity into sets B_0, B_1, B_2, . . . , B_r, where B_0 represents 0 events, B_1 represents one event, and so on to B_r, where r is chosen so large that there is no possibility of more than r seismic events ever occurring in the given time period. Let A mean that exactly k events were detected; then $p(n|k$ detected$)$ is, in our new terminology, the conditional probability $p(B_n|A)$. Bayes' theorem tells us that

$$p(B_n|A) = \frac{p(A|B_n)p(B_n)}{p(A|B_0)p(B_0) + p(A|B_1)p(B_1) + \cdots + p(A|B_r)p(B_r)}$$

Changing to the notation we used earlier in this section, $p(B_0) = p(0)$, $p(B_1) = p(1)$, and so on to $p(B_r) = p(r)$. Furthermore, $p(A|B_0) = p(k|0)$, $p(A|B_1) = p(k|1)$, and on to $p(A|B_n) = p(k|n)$, continuing until $p(A|B_r) = p(k|r)$. Substituting into Bayes' theorem produces

$$p(n|k \text{ detected}) = \frac{p(k|n)p(n)}{p(k|0)p(0) + p(k|1)p(1) + \cdots + p(k|r)p(r)}$$

A further simplification takes place when we observe what, for example, the symbol $p(k|0)$ means when k is greater than zero (note that the formula for $p(k|n)$ only makes sense when n is at least as large as k). The symbol $p(k|0)$ denotes the probability of detecting k events when in fact none occurred. But the seismographs cannot detect events that never happen, so there is no chance that k events will be detected if none take place; this means that $p(k|0) = 0$. More generally, $p(k|j) = 0$ whenever k is larger than j. Consequently, the terms $p(k|0)p(0)$, $p(k|1)p(1)$, and the rest are all zero, at least until we reach the expression $p(k|k)p(k)$ in the sum in the denominator of the formula for $p(n|k$ detected$)$. Thus, the formula we require can be written

$$p(n|k \text{ detected}) = \frac{p(k|n)p(n)}{p(k|k)p(k) + p(k|k + 1)p(k + 1) + \cdots + p(k|r)p(r)}$$

As an example, we take the probability of detecting a single event to be, as before, $9/10 = .90$ and use the figures for the $p(n)$ based on the Mongolian region data:

Number of events	1	2	3	4
Number of years	4	3	2	1

Supposing that two seismic events were detected during a year, what is the probability that there were, in fact, only two events and no more? We wish to compute $p(2|2 \text{ detected})$. Since from the table above $p(n) = 0$ for n greater than four, the formula gives us

$$p(2|2 \text{ detected}) = \frac{p(2|2)p(2)}{p(2|2)p(2) + p(2|3)p(3) + p(2|4)p(4)}$$

We calculate from $p(k|n) = C_k^n p^k (1 - p)^{n-k}$ that

$$p(2|2) = C_2^2(.90)^2(.10)^0 = .81$$

$$p(2|3) = C_2^3(.90)^2(.10)^1 = .24 \qquad \text{(approximate values)}$$

$$p(2|4) = C_2^4(.90)^2(.10)^2 = .05$$

From the table above we have the probabilities $p(2) = .30$, $p(3) = .20$, and $p(4) = .10$. We therefore obtain

$$p(2|2 \text{ detected}) = \frac{p(2|2)p(2)}{p(2|2)p(2) + p(2|3)p(3) + p(2|4)p(4)}$$

$$= \frac{(.81)(.30)}{(.81)(.30) + (.24)(.20) + (.05)(.10)}$$

$$= \frac{.243}{.296} = .82 \qquad \text{(approximately)}$$

We can also ask, given the same data and still assuming that two events were detected, what the probability is that three events actually took place, but one was not detected. The answer, according to the formula and the calculations above, is

$$p(3|2 \text{ detected}) = \frac{p(2|3)p(3)}{p(2|2)p(2) + p(2|3)p(3) + p(2|4)p(4)}$$

$$= \frac{(.24)(.20)}{.296} = \frac{.016}{.296} = .16 \quad \text{(approximately)}$$

(Notice that the denominator of the formula for $p(n|2 \text{ detected})$ is the same no matter what n is.) Thus, if two events were detected, there is about an 82% probability that no event was overlooked and about a 16% probability that there was on additional event.

2-39 **Example** Earthquake activity of magnitude about 4.0 and greater in the Scandinavian region (north of 55° to the Arctic Ocean and from 0° to 45° E) from 1936 to 1955 is given by the following table:

Year	1936	1937	1938	1939	1940	1941	1942	1943	1944	1945
Number of events	4	2	1	2	0	2	2	1	0	1

Year	1946	1947	1948	1949	1950	1951	1952	1953	1954	1955
Number of events	1	0	3	3	0	0	2	1	0	2

Suppose there is a 95% probability of detecting a seismic event of magnitude 4.0 and greater in this region and that, in some year, no seismic activity is recorded. What is the probability that one undetected earthquake occurred?

Solution Since there were no earthquakes in 6 of the 20 years from 1936 to 1955, we estimate $p(0)$ to be $6/20 = .30$. Similarly, we find that $p(1) = 5/20 = .25$, $p(2) = .30$, $p(3) = .10$, and $p(4) = .05$. Since no earthquake was observed and we wish to know the probability that one actually took place, we are trying to compute $p(1|0 \text{ detected})$. From the formula, this is

$$p(1|0 \text{ detected}) = \frac{p(0|1)p(1)}{p(0|0)p(0) + p(0|1)p(1) + p(0|2)p(2) + p(0|3)p(3) + p(0|4)p(4)}$$

If no earthquake takes place, we are certain to detect none, so $p(0|0)$ is always one. Applying the formula for $p(k|n)$, we find that, since we were given $p = .95$,

$$p(0|1) = C_0^1(.95)^0(.05)^1 = .05$$

$$p(0|2) = C_0^2(.95)^0(.05)^2 = .0025$$

$$p(0|3) = C_0^3(.95)^0(.05)^3 = .000125$$

The probability $p(0|4)$ is so small that we will take it to be zero. Substituting into the formula,

$$p(1|0 \text{ detected}) = \frac{(.05)(.25)\cdot}{(1)(.30) + (.05)(.25) + (.0025)(.30) + (.000125)(.10) + 0}$$

$$= \frac{.0125}{.3132625} = .04 \quad \text{(approximately)}$$

This means there is about a 4% probability that a seismic event was overlooked.

If a political leader wants to know how much confidence to place in a seismic detection system, the leader is asking, in part, how many events the system is likely to overlook. It may be that the leader should be willing to accept a system which is quite imperfect in this regard on the grounds that the events missed may well be earthquakes and because even some probability of clandestine nuclear weapons testing might be tolerable — if that probability were small enough. If acceptable estimates of the probabilities $p(n)$ of different levels of seismic activity could be obtained and a probability p of detecting each event agreed upon, then, as we have seen, Bayes' theorem could aid the leader by calculating the probabilities of various numbers of undetected events.

Summary

Let $p(k|n)$ denote the probability that if n events occur, then k events will be detected. If there is a probability p that each event will be detected, then

$$p(k|n) = C_k^n p^k (1 - p)^{n-k}$$

Let $p(n)$ be the probability that n events will take place and let $p(n|k \text{ detected})$ be the probability that if k events were detected, then n events actually occurred. Then Bayes' theorem implies

$$p(n|k \text{ detected}) = \frac{p(k|n)p(n)}{p(k|k)p(k) + p(k|k+1)p(k+1) + \cdots + p(k|r)p(r)}$$

where r is chosen so large that more than r events cannot occur.

Exercises

- *For Exercises 2-263 and 2-264, use the information from Example 2-39. Suppose that one event was detected.*

2-263   Compute the probability that no other event occurred.

2-264   Compute the probability that there was one more event.

2-265   Suppose in Example 2-39 that the probability of detecting a seismic event of magnitude over 4.0 were only .80. If no such event is detected, what is the probability that one such event actually took place in the Scandinavian region?

- *For Exercises 2-266 and 2-267, recall that the estimated probabilities of numbers of events of magnitude at least 6.0 in the Mongolian region were $p(1) = .40$, $p(2) = .30$, $p(3) = .20$, and $p(4) = .10$. Suppose the probability of detection of such an event is .90, and one event is detected.*

2-266   What is the probability that there was no other event?

2-267   What is the probability that two events took place?

2-268   For the years 1936–1955 in the Scandinavian region, two seismic events of magnitude above 5.25 were observed in 1936 and exactly one such event in each of the years 1942, 1943, and 1949. No events of such large magnitude were detected in the other years. If the probability of detecting an event of magnitude above 5.25 in the Scandinavian region is .95, compute the probability that one such event actually occurred in a year in which none is detected.

2-269   The number of seismic events of magnitude 5.0 and above in California between 1926 and 1945 is given in the table below, which is based on a standard, but incomplete, list. If there is a probability of .85 that an earthquake of this magnitude will appear on the list and three events are entered for a given year, calculate the probability that four actually took place.

Year	1926	1927	1928	1929	1930	1931	1932	1933	1934	1935
Number of events	4	4	1	1	3	3	3	6	5	3
Year	1936	1937	1938	1939	1940	1941	1942	1943	1944	1945
Number of events	3	2	3	3	4	10	2	0	4	2

2-270   The number of large shallow earthquakes (magnitude at least 7.75) for the years 1918–
1942 throughout the world is given in the table below. While there is little possibility that
such a powerful seismic event would escape detection, its magnitude might be under-
estimated so that it fails to be included in a list of large earthquakes. Assume a 95% probability that an earthquake of magnitude 7.75 and larger will be recognized as such. In a
year in which two events are entered in the list, what is the probability that there were
just two events of this magnitude?

Year	1918	1919	1920	1921	1922	1923	1924	1925	1926
Number of events	4	2	3	0	1	2	2	0	0

Year	1927	1928	1929	1930	1931	1932	1933	1934	1935
Number of events	2	2	2	0	4	3	1	2	2

Year	1936	1937	1938	1939	1940	1941	1942
Number of events	0	0	2	4	1	3	4

Concluding Essay

We have purposely oversimplified the problem of monitoring an underground nuclear test ban to keep both our discussion and the mathematics employed in it at a reasonably elementary level. We have looked at two parts of the process of monitoring such a ban, the detection and the location of seismic events, but we omitted the most important and most difficult part of the process, namely, the identification of the event. In other words, is an event an earthquake or a nuclear explosion? It is possible that an explosion will appear to be an earthquake, and thus a violation of the treaty will be overlooked. But that is not the only problem. The Swedish representative to the Committee on Disarma-

ment has warned of "the risk that 'freaks of nature,' in the form of earthquakes similar in appearance to explosions would induce unwarranted political accusations."

The identification problem is both politically sensitive and technically difficult. A Swedish report made in 1970 declared that with the seven station seismic detection network then in existence, the correct identification of nuclear explosions of at least 90 kilotons in hard rock would be 90% probable. With the proposed 26 station network the predicted level for identification, with 90% probability, would drop to 12 kilotons. The United States position has been that it requires substantial assurance of detection, location, and identification for seismic events of magnitudes as small as 4.0, equal to an explosion of approximately 8 kilotons. Since this magnitude is lower than is likely to be achieved by the proposed 26 station network, the consequence is what is inevitably known as the "magnitude gap."

Another difficulty we have passed over concerns the magnitude computations we used for explosions; all of them assumed the explosion takes place in hard rock. If a country intended to make detection and identification difficult, it could explode the device in soft rock and thereby produce a seismic event of much lower magnitude than the same explosion would produce in hard rock. For example, it has been estimated that a 10 kiloton explosion which produces a magnitude of about 4.75 in hard rock could be "decoupled" (to use the technical term) to produce an event of magnitude less than 4.0. Thus, even if the magnitude gap can be bridged, it is still not clear that a treaty could be monitored in a manner that all parties would find satisfactory.

The mathematics being employed to aid the search for effective means of monitoring a treaty is considerably more sophisticated than is the material in this chapter, but it is based on essentially the same principles. What you have seen, then, is a valid, if considerably simplified, approach to the problems that arise in this area. It seems likely that a complete solution to the underground test-ban problem—a solution that leads to an effective, enforceable treaty—will require substantial improvements in seismological theory and instrumentation, in mathematics, and, most crucially, in the world political climate.

The problems in the monitoring of an underground nuclear test-ban treaty are certainly not the only applications of probability theory. If you scan the exercises, you will see that one of the main areas of application for probability is in gambling. In fact, the scientific theory of probability began with a question from gambling. A seventeenth century Frenchman, the Chevalier de Mere, asked the noted philosopher and mathematician Blaise Pascal a question that had bothered gamblers for centuries: How should the pot be divided in a dice game that has to be discontinued? Pascal consulted another famous mathematician, Pierre de Fermat, and it was from their correspondence that the theory of probability emerged.

Another major application of probability theory is the mathematics used by insurance companies to determine the probability of various kinds of accidents and thus the rates to be charged for insurance. Other applications include the frequency patterns of telephone calls and the resulting demands on telephone traffic and switching equipment, and the sampling techniques used for estimating the quality of many products or the opinions of many individuals (see Chapter 7). Probability has been applied to disciplines as diverse as physics, biology, and history. Indeed, the probability concept occurs in a significant portion of applied mathematics.

Supplementary Exercises

• *One skill required in order to solve probability problems is the ability to determine what part of the theory applies to a particular problem. The exercise sets of the previous sections consisted primarily of problems to which the material of that section applied. These supplementary exercises are of the same level of difficulty as the previous ones, but they have been randomly mixed so the reader has no external clues as to which portion of probability theory is appropriate to the question. Good luck!*

2-271   The last time a certain mathematics class was given, 19 students received a grade of A, 25 of B, 16 of C, and 14 lower than C. On the other hand, when a statistics course was last given, 30 students received an A, 34 a B, 23 a C, and 25 below C. A student can choose to take either the mathematics class or the statistics class and decides to take the class in which there appears to be a higher probability of receiving an A. On the basis of the information given, which class should this student take?

2-272   Three dice are rolled. What is the probability of getting no more than two sixes?

2-273   Suppose the probability that a child will learn to play a musical instrument is .85 if both its parents play instruments, .70 if one parent does, and .45 if neither parent plays. Suppose that, of the children at a certain school, 15% come from families in which both parents play musical instruments, 25% from families in which one parent plays, and the rest from families in which neither parent plays. What is the probability that a child from this school will learn to play a musical instrument?

2-274   Five officers and eight enlisted men volunteer for a secret military mission. The names of the four soldiers who will actually carry out the mission are drawn at random. What is the probability that at least three out of the four chosen will be enlisted men?

2-275   A pinochle card deck is made up of the nines, tens, jacks, queens, kings, and aces from two standard decks. Thus, for example, there are two kings of spades in a pinochle deck, but no sixes at all. If a five card hand is dealt from a pinochle deck, what is the probability that it will contain exactly three aces?

2-276   An icosahedron is a regular twenty sided solid. Suppose one face of the icosahedron has one dot on it, another two dots, and so on to a face with twenty dots on it. If the icosahedron is rolled like a die, then each face is equally likely to end up topmost. What is the probability that the topmost face shows a prime number of dots, that is, a number greater than one which is evenly divisible only by itself and one.

2-277   The students in a finite mathematics class consist of 33 men and 26 women. Of the men, 19 are social science majors, 5 have chosen other majors, and the rest are undecided as to a major. Of the women, 11 are social science majors, 8 have chosen other majors, and the rest are undecided as to a major. A member of the class is chosen at random. What is the probability that the person chosen is either a man or a person undecided as to a major?

2-278   The price of the stock of an oil company will rise if the company's oil explorations uncover a large amount of oil or if the government lowers the taxes on oil company profits. The president of the company estimates the probability that the company will discover large amounts of oil to be .10 and the probability that the government will lower the taxes on oil company profits to be .65. If the president considers the two events to be independent, what is the estimated probability that the price of stock in the company will rise?

2-279   Overall, a high jumper has won her event in 60% of the track meets she has entered. But in the meets in which the world record holder has competed, she has won only 15% of the time. The probability that the world record holder will compete in a meet along with this high jumper is .45. Use this information to estimate the probability that this high jumper will win her event if the world record holder does not compete.

2-280   Suppose that in a species of animals only individuals of genotype aa with respect to a certain gene can be albino and the probability is .40 for an individual of genotype aa to be albino. If the probability is .15 that a member of this species will be of genotype aa, what is the probability that an individual of this species chosen at random will be albino?

2-281   The major seismological stations in and near the western United States are located at Berkeley, Boulder, Boulder City, Eureka, Hungry Horse, Mt. Hamilton, Pasadena, Shasta, Tucson, and Victoria. If this network has a 98% probability of detecting a seismic event of magnitude at least 3.0 in this area, then the probability of detection at each station (assumed all the same) is about .32. What is the probability of locating the event if that requires detection at four stations or more? (Set up the computation, but do not attempt to carry it out.)

2-282   An organization with 100 members, of whom six are officers, plans to elect delegates to attend a convention. There are to be two delegates; one must be an officer, and the other cannot be an officer. In addition, an alternate delegate, either an officer or not, will be elected and will attend if one of the regular delegates is unable to do so. How many different outcomes can this election have?

2-283   In a study of 200 married couples, it is found that in 70 couples at least one has a college degree, in 40 couples two members are college graduates, and that 50 of the men in the study are college graduates. How many of the women in the study are college graduates?

2-284   A player bets $4, rolls a single die, and receives $1.10 times the number on the die. What is the expected value of this game to the player?

2-285   Suppose that of all hurricanes originating in the Caribbean Sea 35% reach the United States. What is the probability that at least two of the next seven Caribbean hurricanes will reach the United States?

2-286   Suppose that each time a baseball player comes to bat in his next game the probability is .25 that he will get a hit. Suppose also that the probability that he will get only three chances at bat in the game is .15, that he will get four chances is .60, and that he will get five chances is .25. What is the probability that the baseball player will get exactly three hits in his next game?

2-287   A pinochle card deck is made up of the nines, tens, jacks, queens, kings, and aces from two standard decks. Thus, for example, there are two kings of spades, but no sixes at all. If a card is drawn at random from a pinochle deck, what is the probability that it is a face card (jack, queen, or king)?

2-288   A new system of watering fruit trees is tried out on ten trees. If there is a probability of .40 that a tree's yield of fruit will increase significantly as a result of the new system, then the approximate probability that exactly k of the ten trees will produce higher yields is given by the following table:

k	0	1	2	3	4	5	6	7	8	9	10
$p(k)$	.01	.04	.12	.22	.25	.20	.11	.04	.01	.00	.00

What is the approximate probability that at least five of the trees will produce higher yields of fruit?

2-289   A man who owns two cars, one new and one old, will get to work on time if at least one of the cars will start in the morning. He estimates that the probability is .95 that the new car will start and .80 that the old car will. If he estimates the probability that he will get to work on time as .99, what does he consider to be the probability that both cars will start? Is he assuming that the outcomes (New car starts) and (Old car starts) are independent?

2-290   A motel has 120 rooms; 80 have color television sets, while the rest have only black and white sets but also correspondingly lower room rates. If on a typical night, 60 of the rooms with color television sets and 30 of the rooms with black and white sets are rented, does the evidence suggest that the public prefers its motel rooms to be equipped with color television sets, even if it has to pay higher room rents for this feature? Explain your answer.

2-291   Suppose the probability is .70 that the value of the stock of a company listed on the New York Stock Exchange will increase next year. An investor decides on an investment strategy which consists of rolling three dice and buying stocks of as many companies as the total number of dots on the dice; the companies are to be chosen at random from those listed on the Exchange. Thus, for example, if she rolled a two on each of the dice, she would select six companies at random and purchase their stock. What is the probability that, of the stocks purchased by this investor, those of seventeen companies will increase in value next year? [*Hint:* If the sum on the dice is less than 17, there is no chance the investor will purchase seventeen stocks with increasing value.]

2-292 A game consists of flipping a coin and rolling a die. Success in the game occurs if the coin comes up heads and the die shows any number except three; any other outcome is a failure. If this game is played four times, what is the probability of at least two successes?

2-293 In a chess tournament, a player receives 1 point for a win, 1/2 point for a draw, and 0 points for a loss. Based on her past performance, a player estimates there is no chance that she will earn fewer than 3 points out of 5 possible in an upcoming tournament. The probability that she will receive each possible score between 3 and 5 points is given by the following table:

Points	3	3½	4	4½	5
Probability	.15	.20	.30	.25	.10

What is the expected value of the number of points she will earn in the tournament?

2-294 One type of bet at horse races is called the "daily double." A bettor chooses one horse in each of the first two races of the day and if both win, the bettor wins some money. If there are seven horses in the first race and nine in the second, how many different daily double bets are there?

2-295 An architect will receive a substantial bonus from her firm either if her design for a large building is chosen by the owners as the one to be constructed or if she wins an award for a building she designed previously. She estimates the probability that her design for the new building will be chosen as .30, that the building she designed previously will win an award as .05, and that both will happen as .015. What does she estimate to be the probability that she will receive a bonus?

2-296 The proud father of a newborn baby girl believes that the probability is .45 that she will earn a private pilot's license when she grows up, .20 that she will become a physician, and .15 that she will write a novel. If the father considers these events all to be independent, what does he believe to be the probability that, someday, he could sit in a plane reading a novel written by the pilot—his daughter, the doctor?

2-297 Suppose that 20% of the drivers covered by an automobile insurance company are under 25 years of age and that 1% of the company's policyholders are drivers under the age of 25 who have had an accident this year. What is the accident rate for drivers under 25 years of age covered by this company? That is, if we choose at random, what is the probability that a driver under 25 years of age covered by this company is one who had an accident this year?

2-298   Suppose a customer in a restaurant wishes to choose a main dish, two vegetables, and a dessert from a menu which lists six main dishes, five different vegetables, and four desserts. How many different meals can this customer choose?

2-299   A stock market analyst has a theory that predicts stock price behavior during any period in which stock prices are declining. The theory is that if a stock's price rose on a certain day, the probability is .60 that it will rise again the next day; if the stock's price dropped on that day, the probability is .25 that it will rise the next day; and if the price remained constant on a particular day, the probability is .30 that it will rise the next day. Suppose the period of May 28 and 29 was one of declining stock prices and that, of 487 stocks traded on the Pacific Stock Exchange on May 28, 196 had a price rise, 209 had prices drop, and the remaining 82 were stocks for which the price stayed constant. Suppose that the analyst's theory is correct. At random we select a stock which was traded on May 28 and whose price rose on May 29. What is the probability that its price rose on May 28?

2-300   The coach of a professional basketball team estimates the probability that his team will win more games than it loses this season as .70 and the probability that his annual contract will be renewed at the end of the season as .75. If he thinks the probability that at least one of these desirable events will take place is .80, is the coach assuming that the events are independent?

2-301   A game consists of drawing a card from a standard deck and then drawing a second card without replacing the first. The player wins if at least one card is an ace. What proportion of wins occur because of an ace on the first card, that is, what is the probability that the reason for a win was an ace on the first draw?

2-302   An octahedron is a regular solid with eight faces. Suppose a die is constructed in the shape of an octahedron with one dot on a face, two dots on another face, and so on to a face with eight dots. If the octahedron is rolled, then each face is equally likely to end up topmost. A game consists of flipping a coin and rolling the octahedral die. A player who flips heads wins the game if the topmost face on the die has more than three dots. A player who flips tails must roll eight dots on the die to win. What is the probability of winning this game?

2-303   An instructor puts a very difficult problem on an examination and estimates the probability is .35 that an A student will solve it, .15 that a B student will solve it, .05 that a C student will solve it, and no chance that a student earning less than a C will solve the problem. Suppose 15% of the students in the class are A students, 30% B students, 40% C students, and the rest are students earning grades below C. What is the probability that a student who solved the problem is, nevertheless, a C student?

2-304   Oregon weather forecasters are very good at predicting rainy days, because they get so much practice doing it. In fact, weather forecasters in Oregon correctly predict the weather 75% of the time when the day is to be rainy, but only 60% of the time for clear days. Suppose that 45% of all days in Oregon are rainy. For a day chosen at random, what is the probability that an Oregon weather forecast will be correct?

2-305   A California automobile license plate consists of a three digit number followed by three letters. Combinations of three letters that the state considers obscene or improper in some way are not used. Suppose there are 200 forbidden letter combinations. How many different license plates can the state of California use?

2-306   A mathematics class consisting of 35 men and 65 women decides to select a committee of four members from the class at random to discuss student complaints with the instructor. What is the probability that the committee will consist of one man and three women?

2-307   Suppose a game consists of betting $1 and drawing a card at random from a standard deck. The player gets 50¢ back (and so has a net loss of 50¢) if a spade is drawn that is not a face card. The player is given $5 (for a net gain of $4) if a face card is drawn that is not a spade. The payment for the jack, queen, or king of spades combines both other payments, so the player receives $5.50 (for a net gain of $4.50). What is the expected value of this game to the player?

2-308   An expert wine taster is asked to judge the wines of a certain variety at a county fair. There are eleven entries and the taster is to select a first, second, and third prize winner, with no ties allowed. How many different outcomes can there be?

●   In Exercises 2-309 and 2-310, the students in a calculus class are divided by major into mathematics, engineering, and other majors, and according to whether they are freshmen or not:

	Mathematics	Engineering	Other
Freshman	32	19	6
Not freshman	5	8	24

A member of the class is chosen at random.

2-309   If the person selected is a mathematics major, what is the probability that this person is a freshman?

2-310   Show that the outcomes (mathematics major) and (freshman) are not independent.

2-311   Taking a baseball player's batting average as an estimate of whether he will get a hit each time he comes up to bat, suppose a player with a .310 average (probability of a hit is .31) has four turns at bat in his next game. What is the probability that he will fail to get a hit?

References
Kurt Kreith's article, "Mathematics and Arms Control," in *SIAM Review*, Vol. 8 (1966), gave us the idea of illustrating probability by means of the problem of the limitation of underground nuclear testing.

All the actual information and the quotations from the talks come from *Documents on Disarmament,* United States Arms Control and Disarmament Agency. These documents are issued annually by the U.S. Government Printing Office, Washington, D.C.

Interesting nontechnical chapters on probability can be found in the Life Science Library book *Mathematics* by David Bergamini and the Editors of Life (New York: Time Inc., 1963) and in *Mathematics and the Imagination* by Edward Kasner and James R. Newman (New York: Simon and Schuster, 1940). Somewhat more technical articles are those, both titled "Probability," by Warren Weaver from the October 1950 *Scientific American* and by Mark Kac from the September 1964 issue of the same magazine. These articles are reproduced in *Mathematics in the Modern World* (San Francisco: W. H. Freeman, 1968).

For those who wish to learn more about the mathematics of probability, a more complete treatment of the subject can be found in John E. Freund's *Introduction to Probability* (Los Angeles: Dickenson, 1973).

A good starting point for the study of earthquakes would be *Elementary Seismology* by Charles F. Richter (San Francisco: W. H. Freeman, 1958).

Appendix:
The Standard
Deck of Cards

The standard deck of playing cards consists of 52 different cards. Each card is identified by its **value** and its **suit.** There are four cards of each value and thirteen possible values:

$$A(\text{ace}),\ 2,\ 3,\ 4,\ 5,\ 6,\ 7,\ 8,\ 9,\ 10,\ J(\text{jack}),\ Q(\text{queen}),\ K(\text{king})$$

The values are ordered from left to right. Thus, an eight is higher than a three, a jack is higher than an eight, and a king is higher than a jack. However, in many card games, aces are the highest rather than the lowest value, that is, aces are higher than kings.

The **face cards** are the jacks, queens, and kings, so there are twelve face cards in all.

Each of the four cards of a fixed value is of a different suit. The suits are:

$$\text{spades } (\spadesuit),\ \text{hearts } (\heartsuit),\ \text{diamonds } (\diamondsuit),\ \text{clubs } (\clubsuit)$$

Spades and clubs are printed with black ink and hearts and diamonds are printed with red ink, so half the cards in the standard deck are black and the other half are red. Thus, a red seven means either the seven of hearts ($7\heartsuit$) or the seven of diamonds ($7\diamondsuit$).

Chapter 3
Economic Planning and the Arithmetic of Matrices

Introductory Essay

In 1973, Wassily Leontief won the Nobel Prize in Economic Science for his development of input–output economics. Leontief, who was born in Leningrad in 1906 and received his doctorate from the University of Berlin in 1928, came to the United States in 1931 as a research associate of the National Bureau of Economic Research in New York. He later joined the Harvard faculty, where he became a professor of economics in 1946.

Leontief's ideas on the workings of the economy were first formulated in a paper written while he was still a graduate student in Berlin. His basic work, *The Structure of the American Economy, 1919–1929,* was published in 1941 by Harvard University Press, Cambridge, Mass. The theory was not put to use until World War II when, with W. Duane Evans and Marvin Hoffenberg at the U.S. Bureau of Labor Statistics, Leontief developed useful input–output tables for the American economy.

In an article that appeared in *Scientific American* in 1951 (see References), Leontief defined input–output economics as

a method of analysis that takes advantage of the relatively stable pattern of the flow of goods and services among the elements of our economy to bring a much more detailed statistical picture of the system into the range of manipulation by economic theory. As such, the method has had to await the modern high-speed computing machine as well as the present propensity of government and private agencies to accumulate mountains of data.

The central concept of Leontief's theory is that there is a fundamental relationship between the volume of the output from an industry and the size of the inputs to it. These fundamental relationships are expressed in input–output analysis as the ratio of each input to the total output of which it becomes part. For example, for each $5000 of production, the automobile industry uses so many hundred dollars worth of iron and steel, industrial and heating equipment, electrical equipment, and so on. The dollar amounts rise or fall with changing prices, but the actual *amount* of steel going into

a car varies only when the technology changes. The ratio of input of material to output of cars remains relatively stable over a period of several years. Barring any drastic technological change, the needs of the industry can be predicted quite accurately, assuming we know how many cars will be produced in a given year.

The relationships among all the industries can be presented in a systematic way by displaying them in tabular form in what is called an "input–output table." The input–output ratio that describes the relationship between two industries (known as "sectors" in this theory) occupies a specific location in the table. As long as the ratios remain the same, that is, as long as each sector (such as the automobile industry) continues to use the same percentage of a material (such as glass), a change in the number of units (automobiles) produced in a year does not change the basic input–output table. However, when the technology of an industry changes, the ratios of the inputs in the table change and predictions based on the table must be recomputed. As we shall see in this chapter, although the mathematics involved is not difficult, it requires complicated arithmetic. Also, the input–output models have grown larger as the economy has become more complex. Fortunately, the electronic computer has become much more efficient, so that while it took the old Mark II computer 56 hours to manipulate the 42 sector table for the 1939 economy, the IBM 7090 was able to perform comparable operations for the 81 sector table of the 1958 economy in only 3 minutes.

One of the most important uses of input–output economics is to predict the future needs of the economy. A study of the postwar economy was conducted in 1945 based on a list of economic demands assuming full employment in 1950. The demand was inserted into an input–output model based on a table of ratios for the year 1939, the last year of normal, peacetime production. The model predicted, for example, a need for 98 million ingot tons of steel. Steel production—at full capacity—in 1950 was in fact 96.8 million tons, and there was a shortage.

Another use for input–output economics is producing models of the economies of developing nations. These models give planners a clearer picture of the strengths and weaknesses of the developing economy. Also, input–output theory is useful in managing the environment. This application involves associating the amount of pollutants generated (foul air, polluted water, solid waste, noise, etc.) to each unit of output of each economic sector.

Individual corporations have also constructed input–output tables to aid in planning. From this kind of analysis, corporations discover indirect costs of which they had been previously unaware. For example, the Celanese Plastics Company found that in order to turn out products that sell for $100 they use $43.20 in chemicals bought directly from a manufacturer. But, in producing their $100 worth of products they actually consume $61.30 in chemicals. The extra $18.10 represents chemicals bought by the packaging concerns, printers, and other companies whose products are used in the manufacture of plastics. The importance of this indirect cost for the plastics company is that if the price of chemicals changes, the change must be figured on the indirect cost of chemicals as well as the direct cost when estimating the necessary price changes for the plastics.

The value of input–output economics in planning, whether it is used by a single company, a nation, or an ecologist studying global problems, depends on the accuracy of the data put into the model and the skill and intelligence of those who interpret the information the models yield. Thus, the proper use of input–output theory requires skills from business and economics as well as mathematics. In this chapter, we will discuss the mathematics on which Leontief's model depends.

3-1   Matrices

The Leontief economic model assumes there is a simple relationship between the amount produced by each sector of an economy and the amount of input required from other sectors. We illustrate the relationship by means of Leontief's analysis of the 1958 American economy. (See his article in the April 1965 issue of *Scientific American*, pp. 26–27.) Leontief divided the economy into 81 sectors and grouped them into six families of related sectors. In order to keep the discussion reasonably simple, we will treat each family of sectors as a single sector and so, in effect, work with a six sector model. The sectors are listed in Table 3-1.

	Table 3-1
Sector	Examples
Final nonmetal (FN)	Furniture, processed food
Final metal (FM)	Household appliances, motor vehicles
Basic metal (BM)	Machine-shop products, mining
Basic nonmetal (BN)	Agriculture, printing
Energy (E)	Petroleum, coal
Services (S)	Amusements, real estate

The workings of the American economy in 1958 are described in the **input–output table** (Table 3-2) based on Leontief's figures. We will demonstrate the meaning of Table 3-2 by considering the left-hand column. The numbers in this column mean that 1 unit of final nonmetal production requires the consumption of 0.170 unit of (other) final nonmetal production, 0.003 unit of final metal output, 0.025 unit of basic metal products, and so on down the column. Since the unit of measurement that Leontief used for this table is millions of dollars, we conclude that the production of $1 million worth of final nonmetal production consumes $0.017 million, or $170,000, worth of other final nonmetal products, $3000 of final metal production, $25,000 of basic metal products, and so on. Similarly, the entry in the column headed FM and opposite S of 0.074 means that $74,000 worth of output from the service industries goes into the production of $1 million worth of final metal products, and the number 0.358 in the column headed E and opposite E means that $358,000 worth of energy must be consumed to produce $1 million worth of energy.

	Table 3-2					
	FN	FM	BM	BN	E	S
FN	0.170	0.004	0	0.029	0	0.008
FM	0.003	0.295	0.018	0.002	0.004	0.016
BM	0.025	0.173	0.460	0.007	0.011	0.007
BN	0.348	0.037	0.021	0.403	0.011	0.048
E	0.007	0.001	0.039	0.025	0.358	0.025
S	0.120	0.074	0.104	0.123	0.173	0.234

By the underlying assumption of Leontief's model, the production of n units (n = any number) of final nonmetal production consumes $0.170n$ unit of final nonmetal output, $0.003n$ unit of final metal output, $0.025n$ unit of basic metal production, and so on. Thus, production of $50 million worth of products from the final nonmetal section of the 1958 American economy required $(0.170)(50) = 8.5$ units ($8.5 million) worth of final nonmetal output, $(0.003)(50) = 0.15$ unit of final metal output, $(0.025)(50) = 1.25$ units of basic metal production, and so on.

3-1   **Example**   According to the simplified input–output table for the 1958 American economy, how many dollars worth of final metal products, basic nonmetal products, and services are required to produce $120 million worth of basic metal products?

Solution   Each unit ($1 million worth) of basic metal products requires 0.018 unit of final metal products because the number in the BM column of the table opposite FM is 0.018. Thus, $120 million, or 120 units, requires $(0.018)(120) = 2.16$ units, or $2.16 million worth of final metal products. Similarly, 120 units of basic metal production uses $(0.021)(120) = 2.52$ units of basic nonmetal production and $(0.104)(120) = 12.48$ units of services, or $2.52 million and $12.48 million worth of basic nonmetal output and services, respectively.

The Leontief model also involves a **bill of demands**, that is, a list of requirements for units of output beyond that required for its inner workings as described in the input–output table. These demands represent exports, surpluses, government and individual consumption, and the like. The bill of demands for the simplified version of the 1958 American economy we have been using was (in millions)

FN	$99,640
FM	$75,548
BM	$14,444
BN	$33,501
E	$23,527
S	$263,985

A problem which the Leontief model attempts to solve is: How many units of output from each sector are needed in order to run the economy and fill the bill of demands? The reason the problem is somewhat complicated is that each sector must produce not only the units required by the bill of demands, but also units of input to the other sectors so that they, in turn, can produce what the first sector needs to operate. Notice that the more each sector produces, the more input it needs, but then the more input from it the other sectors need in order to satisfy its demands. This interdependence among the various sectors of the economy is at the heart of the input–output economic theory.

To solve the problem of determining the production required from each sector, we must develop the appropriate mathematical tools. Let us return to the 1958 input–output table (Table 3-2). The mathematical properties of the table do not depend on the meaning of the numbers in it (e.g., there is no mathematical significance to the fact that the 0.120 in the lower left corner determines the number of units of services required to produce 1 unit of final nonmetal output). If we delete the labels in the input–output table, we have

$$\begin{bmatrix} 0.170 & 0.004 & 0 & 0.029 & 0 & 0.008 \\ 0.003 & 0.295 & 0.018 & 0.002 & 0.004 & 0.016 \\ 0.025 & 0.173 & 0.460 & 0.007 & 0.011 & 0.007 \\ 0.348 & 0.037 & 0.021 & 0.403 & 0.011 & 0.048 \\ 0.007 & 0.001 & 0.039 & 0.025 & 0.358 & 0.025 \\ 0.120 & 0.074 & 0.104 & 0.123 & 0.173 & 0.234 \end{bmatrix}$$

In other words, the input–output table is, mathematically, just a rectangular array of numbers. A rectangular array of numbers is called a **matrix.**

In the general mathematical theory, there is no restriction on the size of the matrix or on the numbers in it. Thus, all of the following are matrices:

$$\begin{bmatrix} 2 & 3 & -1 \\ 2 & 1 & 0 \\ 0 & 0 & 1 \end{bmatrix} \quad \begin{bmatrix} -1 & -1 & 0 \\ 0 & 0 & 0 \end{bmatrix} \quad \begin{bmatrix} 2 \\ 3 \\ 2 \\ -5 \\ 1 \\ -\dfrac{1}{2} \end{bmatrix} \quad \begin{bmatrix} 1 & 1 & 1 & 1 \end{bmatrix} \quad \begin{bmatrix} 0.47 & 0.12 \\ -3.1 & -6 \end{bmatrix}$$

If we remove the labels from the bill of demands on p. 127, we obtain another matrix:

$$\begin{bmatrix} 99,640 \\ 75,548 \\ 14,444 \\ 33,501 \\ 23,527 \\ 263,985 \end{bmatrix}$$

A matrix is described by means of **rows** and **columns.** The input–output matrix has 6 rows, always numbered from top to bottom; they are

First row: [0.170   0.004   0   0.029   0   0.008]
Second row: [0.003   0.295   0.018   0.002   0.004   0.016]
Third row: [0.025   0.173   0.460   0.007   0.011   0.007]
Fourth row: [0.348   0.037   0.021   0.403   0.011   0.048]
Fifth row: [0.007   0.001   0.039   0.025   0.358   0.025]
Sixth row: [0.120   0.074   0.104   0.123   0.173   0.234]

It also has 6 columns, which are always numbered from left to right, thus:

First column:	Second column:	Third column:	Fourth column:	Fifth column:	Sixth column:
0.170	0.004	0	0.029	0	0.008
0.003	0.295	0.018	0.002	0.004	0.016
0.025	0.173	0.460	0.007	0.011	0.007
0.348	0.037	0.021	0.403	0.011	0.048
0.007	0.001	0.039	0.025	0.358	0.025
0.120	0.074	0.104	0.123	0.173	0.234

The bill of demands matrix has 6 rows also:

First row: [99,640]
Second row: [75,548]
Third row: [14,444]
Fourth row: [33,501]
Fifth row: [23,527]
Sixth row: [263,985]

but only 1 column

$$\begin{bmatrix} 99,640 \\ 75,548 \\ 14,444 \\ 33,501 \\ 23,527 \\ 263,985 \end{bmatrix}$$

Similarly, the matrix

$$\begin{bmatrix} -1 & -1 & 0 \\ 0 & 0 & 0 \end{bmatrix}$$

has 2 rows and 3 columns, and the matrix

$$[1 \quad 1 \quad 1 \quad 1]$$

has 1 row and 4 columns.

The number of rows and columns of a matrix is called its **size.** The number of rows is always given first and the numbers are separated with the word "by." Thus, the input–output matrix is a 6 by 6 matrix (abbreviated 6 × 6), and the bill of demands matrix is 6 by 1 (or 6 × 1). Similarly,

$$\begin{bmatrix} -1 & -1 & 0 \\ 0 & 0 & 0 \end{bmatrix}$$

is a 2 × 3 matrix.

For the economic theory, we will be concerned with input–output matrices which necessarily have the same number of rows (inputs) as columns (outputs). A matrix with an equal number n of rows and columns is called an $n \times n$ or **square** matrix. The bill of demands matrix has only a single column, i.e., is an $n \times 1$ matrix. These are the only sizes of matrices that appear in our problem, but other sizes are important in the many other problems to which matrix theory is applied so we will not restrict ourselves as we discuss the mathematics.

A number in a matrix has a **location** described by stating the row and column in which it lies. In the matrix

$$\begin{bmatrix} 0.170 & 0.004 & 0 & 0.029 & 0 & 0.008 \\ 0.003 & 0.295 & 0.018 & 0.002 & 0.004 & 0.016 \\ 0.025 & 0.173 & 0.460 & 0.007 & 0.011 & 0.007 \\ 0.348 & 0.037 & 0.021 & 0.403 & 0.011 & 0.048 \\ 0.007 & 0.001 & 0.039 & 0.025 & 0.358 & 0.025 \\ 0.120 & 0.074 & 0.104 & 0.123 & 0.173 & 0.234 \end{bmatrix}$$

the location of the number 0.039 is specified by saying that it is in the fifth row and the third column. Just as with the convention for size, row location is always given before column location. In the same matrix, the number in the second row and the sixth column is 0.016.

3-2 **Example**   Given the matrix

$$\begin{bmatrix} 1 & -1 & 2 & 1 & 3 \\ 4 & 1 & 2 & 2 & \dfrac{1}{2} \\ 1 & 2 & 0 & 1 & 0 \\ 3 & 4 & 1 & 2 & 2 \end{bmatrix}$$

what is (a) the second row, (b) the third column, (c) the size of the matrix, (d) the number in the fourth row and the second column?

Solution

(a) $\begin{bmatrix} 4 & 1 & 2 & 2 & \dfrac{1}{2} \end{bmatrix}$

(b) $\begin{bmatrix} 2 \\ 2 \\ 0 \\ 1 \end{bmatrix}$

(c) 4×5

(d)   4

A matrix is characterized not only by its size and the numbers in it, but by their location. Thus, two matrices are **equal** only if they are of the same size and have equal numbers in every corresponding location.

3-3 **Example**   Are the following matrices equal?

(a) $\begin{bmatrix} 2 & 1 & 3 \\ 4 & -1 & 0 \end{bmatrix}$ and $\begin{bmatrix} 1+1 & 1 & 4-1 \\ 4 & -1 & 2-2 \end{bmatrix}$

(b) $\begin{bmatrix} -1 & 0 \\ 1 & 1 \end{bmatrix}$ and $\begin{bmatrix} 0 & -1 \\ 1 & 1 \end{bmatrix}$

Solution   (a) Yes, because both matrices are 2×3 and $1 + 1 = 2$, $4 - 1 = 3$, and $2 - 2 = 0$.

(b) No, because $-1 \neq 0$ and therefore the matrices differ, for example, in the numbers located in the first row and the first column.

3-4 **Example** Solve for x, y, and z:

$$\begin{bmatrix} x & -1 \\ 0 & y+1 \\ 2z & 2 \end{bmatrix} = \begin{bmatrix} 1 & 1-2 \\ 0 & 2 \\ 4 & 1+1 \end{bmatrix}$$

Solution The matrices will be equal only if they agree in every position. Thus, x must equal 1 so that the numbers located in the first row and first column of each matrix will be the same. For the same reason, it must be that $y + 1 = 2$, or $y = 1$, and that $2z = 4$, or $z = 2$.

The solution to the problem of determining the production required in each sector of the economy in order both to run the economy and satisfy the bill of demands depends, as we shall see, on the ability to perform arithmetic operations with matrices. Addition of two matrices to produce a third matrix is possible only if the two matrices are the same size, in which case their sum will be that size also. Otherwise, the sum of matrices is just not defined, i.e., there is no such thing. To add matrices of the same size, the simple rule is: Add the numbers in the corresponding locations in the two matrices and put their sum in that corresponding location of the sum matrix.

3-5 **Example** Perform the indicated addition, if possible.

(a) $\begin{bmatrix} 1 & 2 & -1 \\ 3 & 2 & 0 \\ 1 & 1 & 2 \end{bmatrix} + \begin{bmatrix} 3 & 1 & 3 \\ -1 & -1 & -2 \\ 1 & 1 & -1 \end{bmatrix}$

(b) $\begin{matrix} [1 & -1 & 2] \\ +[0 & 1 & 1] \end{matrix}$

(c) $\begin{bmatrix} -1 & 0 \\ 1 & 2 \\ 2 & 0 \end{bmatrix} + \begin{bmatrix} 1 \\ 1 \\ 3 \end{bmatrix}$

Solution

(a) $\begin{bmatrix} 1 & 2 & -1 \\ 3 & 2 & 0 \\ 1 & 1 & 2 \end{bmatrix} + \begin{bmatrix} 3 & 1 & 3 \\ -1 & -1 & -2 \\ 1 & 1 & -1 \end{bmatrix} = \begin{bmatrix} 1+3 & 2+1 & -1+3 \\ 3+(-1) & 2+(-1) & 0+(-2) \\ 1+1 & 1+1 & 2+(-1) \end{bmatrix}$

$$= \begin{bmatrix} 4 & 3 & 2 \\ 2 & 1 & -2 \\ 2 & 2 & 1 \end{bmatrix}$$

(b) $\begin{array}{rrr} [1 & -1 & 2] \\ +[0 & 1 & 1] \\ \hline [1 & 0 & 3] \end{array}$

(c)   The sum is not defined, because the left-hand matrix is 3 × 2 while the right-hand matrix is a different size, 3 × 1.

There are matrices that act like 0 in ordinary arithmetic. That is, just as $1 + 0 = 1$ and $0 + 14 = 14$, so when these matrices are added to a matrix, the matrix does not change. These matrices have a 0 in every location; they are called, naturally, **zero matrices.**

3-6   **Example**   Compute the sums.

(a) $\begin{bmatrix} 2 & -1 \\ 0 & 0 \\ 2 & 1 \end{bmatrix} + \begin{bmatrix} 0 & 0 \\ 0 & 0 \\ 0 & 0 \end{bmatrix}$

(b) $\begin{array}{rrrr} [0 & 0 & 0 & 0] \\ +[2 & 2 & 1 & -1] \end{array}$

Solution

(a) $\begin{bmatrix} 2 & -1 \\ 0 & 0 \\ 2 & 1 \end{bmatrix} + \begin{bmatrix} 0 & 0 \\ 0 & 0 \\ 0 & 0 \end{bmatrix} = \begin{bmatrix} 2+0 & (-1)+0 \\ 0+0 & 0+0 \\ 2+0 & 1+0 \end{bmatrix} = \begin{bmatrix} 2 & -1 \\ 0 & 0 \\ 2 & 1 \end{bmatrix}$

(b) $\begin{array}{rrrr} [0 & 0 & 0 & 0] \\ +[2 & 2 & 1 & -1] \\ \hline [2 & 2 & 1 & -1] \end{array}$

Thus, in matrix theory, there are many zeros, one for each size of matrix.

Addition of matrices behaves just like addition of ordinary numbers, in particular, it does not matter in what order matrices are added.

3-7   **Example**   Perform the additions.

(a) $\begin{bmatrix} -1 & 0 \\ 2 & 1 \end{bmatrix} + \begin{bmatrix} 0 & 2 \\ 2 & 2 \end{bmatrix}$

(b) $\begin{bmatrix} 0 & 2 \\ 2 & 2 \end{bmatrix} + \begin{bmatrix} -1 & 0 \\ 2 & 1 \end{bmatrix}$

Solution

(a) $\begin{bmatrix} -1 & 0 \\ 2 & 1 \end{bmatrix} + \begin{bmatrix} 0 & 2 \\ 2 & 2 \end{bmatrix} = \begin{bmatrix} -1+0 & 0+2 \\ 2+2 & 1+2 \end{bmatrix} = \begin{bmatrix} -1 & 2 \\ 4 & 3 \end{bmatrix}$

(b) $\begin{bmatrix} 0 & 2 \\ 2 & 2 \end{bmatrix} + \begin{bmatrix} -1 & 0 \\ 2 & 1 \end{bmatrix} = \begin{bmatrix} 0+(-1) & 2+0 \\ 2+2 & 2+1 \end{bmatrix} = \begin{bmatrix} -1 & 2 \\ 4 & 3 \end{bmatrix}$

It might be unnecessary to emphasize the similarity between addition of matrices and addition of ordinary numbers were it not for the fact that, in contrast, matrix multiplication does not behave like ordinary multiplication, as we shall see in Section 3-2.

Subtraction of matrices, however, does not pose any problems. The rule is the one we might expect: Matrices are subtracted, provided they are of the same size, by subtracting the numbers in the corresponding location and placing their difference in that corresponding location of the difference matrix.

3-8 **Example** Subtract:

$$\begin{bmatrix} 2 & 3 & 2 \\ -1 & 4 & 2 \end{bmatrix} - \begin{bmatrix} 3 & 0 & 1 \\ -2 & -2 & 2 \end{bmatrix}$$

Solution

$$\begin{bmatrix} 2 & 3 & 2 \\ -1 & 4 & 2 \end{bmatrix} - \begin{bmatrix} 3 & 0 & 1 \\ -2 & -2 & 2 \end{bmatrix} = \begin{bmatrix} 2-3 & 3-0 & 2-1 \\ (-1)-(-2) & 4-(-2) & 2-2 \end{bmatrix} = \begin{bmatrix} -1 & 3 & 1 \\ 1 & 6 & 0 \end{bmatrix}$$

The operation of subtraction is not defined between matrices of different sizes.

With these rules of arithmetic for matrices, it is possible to solve simple equations.

3-9 **Example** Solve the following equation for the matrix X:

$$\begin{bmatrix} 1 & 2 \\ 3 & -1 \end{bmatrix} + X = \begin{bmatrix} 0 & 2 \\ 2 & 0 \end{bmatrix}$$

Solution The problem is to find a matrix X, which necessarily must be of size 2×2, so that when it is added to the matrix on its left, we get the matrix to the right of the equal sign. Just as the ordinary equation $3 + x = 2$ is solved by subtracting 3 from both sides of the equation to produce $x = 2 - 3$, so the matrix equation is solved by subtracting the left-hand matrix from both sides:

$$X = \begin{bmatrix} 0 & 2 \\ 2 & 0 \end{bmatrix} - \begin{bmatrix} 1 & 2 \\ 3 & -1 \end{bmatrix} = \begin{bmatrix} -1 & 0 \\ -1 & 1 \end{bmatrix}$$

We may check the solution by verifying that in fact

$$\begin{bmatrix} 1 & 2 \\ 3 & -1 \end{bmatrix} + \begin{bmatrix} -1 & 0 \\ -1 & 1 \end{bmatrix} = \begin{bmatrix} 0 & 2 \\ 2 & 0 \end{bmatrix}$$

Summary

A **matrix** is any rectangular array of numbers. **Rows** are numbered from top to bottom; **columns** from left to right. A **location** in a matrix is specified by giving the row number and the column number in that order. The **size** of a matrix is $m \times n$ if it has m rows and n columns. Two matrices are **equal** provided they are the same size and have equal numbers in each location.

Two matrices can be added (subtracted) only if they are the same size, in which case the numbers in the same location of the two matrices are added (subtracted) to produce the number in that same location of their sum (difference), a matrix of that same size.

A **zero matrix** is a matrix with 0 in every location. A zero matrix, when added to another matrix, does not change the matrix.

Exercises

3-1 A much simplified version of an input–output table for the 1958 Israeli economy divides that economy into three sectors: agriculture, manufacturing, and energy, with the following result:*

	Agriculture	Manufacturing	Energy
Agriculture	0.293	0	0
Manufacturing	0.014	0.207	0.017
Energy	0.044	0.010	0.216

(a) How many units of agricultural production are required to produce 1 unit of agricultural output? (b) How many units of agricultural production are required to produce 200,000 units of agricultural output? (c) How many units of agricultural product go into the produc-

*Wassily Leontief, *Input–Output Economics* (New York: Oxford University Press, 1966), pp. 54–57.

tion of 50,000 units of energy? (*d*) How many units of energy go into the production of 50,000 units of agricultural products?

3-2   The interdependence of certain basic industries of the 1947 American economy is presented in the following input–output table for chemicals (C), petroleum (P), rubber products (R), and metals (M) (units are billions of dollars):

	C	P	R	M
C	0.184	0.015	0.213	0.010
P	0.023	0.353	0.004	0.048
R	0	0	0.014	0
M	0.014	0.001	0.004	0.369

(*a*) How many units of metal products are required to produce 1 unit of chemicals? (*b*) How many units of rubber products are required to produce 5 units of petroleum products? (*c*) How many units of chemical output are needed to produce 5 units of rubber products? (*d*) How many dollars worth of petroleum go into producing $1 billion worth of metal products? (*e*) How many dollars worth of petroleum go into producing $10 billion worth of metal products? (*f*) How many dollars worth of chemicals are consumed in the production of $7 billion worth of chemical products?

3-3   The interdependence among the motor vehicle industry and other basic industries in the 1958 American economy is described by the following input–output table for motor vehicles (V), steel (S), glass (G), and rubber and plastics (R).

	V	S	G	R
V	0.298	0.002	0	0
S	0.088	0.212	0	0.002
G	0.010	0	0.050	0.006
R	0.029	0.003	0.004	0.030

The final demand for these products in millions of dollars is

V	5444
S	3276
G	119
R	943

Express the input–output table as a matrix and then write (a) the first row, (b) the second column, (c) the fourth row, (d) the fifth column, (e) the number in the first row and the third column, (f) the number in the third row and the second column. Express the bill of demands as a matrix. (g) What is the size of the matrix? (h) What is the number in the third row and the first column of the matrix? (i) What is the number in the first row and the second column of the matrix?

3-4   The following table gives the distances in miles between the cities listed:

	Boston	New York	Chicago	Denver	San Francisco
Boston	0	208	980	2025	3186
New York	208	0	850	1833	3049
Chicago	980	850	0	1038	2299
Denver	2025	1833	1038	0	1270
San Francisco	3186	3049	2299	1270	0

(a) Write the corresponding matrix. (b) What is the first column? (c) What is the fourth row? (d) What number is in the third row and the fourth column?

3-5   A polling organization gives the results of asking the preferences of a sample of 1089 voters in a coming election for governor in the following table:

	Democratic candidate	Republican candidate	Other candidates	Undecided
Democrats	415	91	6	77
Republicans	65	281	4	63
Independents	31	19	8	29

(a) Write the corresponding matrix. (b) What numbers form the third column? (c) What number is in the second row and the fourth column? (d) What number is in the fourth row and the second column?

3-6   The cost (in cents) of ingredients for a recipe of each of two types of desserts is listed in the table:

	Brownies	Cookies
Sugar	25	8
Butter	13	20
Eggs	0	7
Flour	3	6
Vanilla	3	3
Chocolate	28	0

(*a*) If a bakery makes 50 recipes of brownies, how much money will it spend for chocolate? (*b*) If a bakery makes 30 recipes of brownies and 40 of cookies, how much in all will it spend for butter? (*c*) If a bakery makes 50 recipes of brownies and 20 of cookies, how much will it spend for sugar? (*d*) Write the matrix corresponding to the table. (*e*) What size is the matrix? (*f*) What numbers form the sixth row of the matrix? (*g*) What number is in the third row and the first column?

3-7   Give examples of matrices of the following sizes: (*a*) 1×2, (*b*) 3×4, (*c*) 2×1, (*d*) 5×2.

●   *In Exercises 3-8 through 3-12, are the matrices equal? If they are not equal, in what way do they fail to be equal?*

3-8 $\begin{bmatrix} 2 & (2)^2 \\ -1 & 3+2 \end{bmatrix}$ and $\begin{bmatrix} 3-1 & 4 \\ -1 & 5 \end{bmatrix}$

3-9 $\begin{bmatrix} \dfrac{3}{2} & -2 & 0.1 \end{bmatrix}$ and $\begin{bmatrix} 1+\dfrac{1}{2} & 1-3 & \dfrac{1}{10} \end{bmatrix}$

3-10 $\begin{bmatrix} 2 \\ 0 \end{bmatrix}$ and $\begin{bmatrix} 2 & 0 \end{bmatrix}$

3-11 $\begin{bmatrix} 2 & 1 \\ 0 & 0 \\ 0 & 1 \end{bmatrix} + \begin{bmatrix} -1 & -1 \\ 0 & -1 \\ 2 & 2 \end{bmatrix}$ and $\begin{bmatrix} 1 & 0 \\ 0 & -1 \\ 2 & 3 \end{bmatrix}$

3-12 $\begin{bmatrix} 2 & 0 \\ 1 & 0 \end{bmatrix}$ and $\begin{bmatrix} 2 & 0 \\ 2-1 & -3 \end{bmatrix}$

●   *In Exercises 3-13 through 3-17, solve for x and y, if possible.*

3-13 $\begin{bmatrix} -1 & x \\ 2 & y \end{bmatrix} = \begin{bmatrix} -1 & 4 \\ 1+1 & -1 \end{bmatrix}$

3-14 $\begin{bmatrix} 1 & x \\ x & 2y \end{bmatrix} = \begin{bmatrix} 1 & 4 \\ 3+2 & 6 \end{bmatrix}$

3-15 $\begin{bmatrix} x \\ y \end{bmatrix} = \begin{bmatrix} 2x+1 \\ -y+2 \end{bmatrix}$

3-16 $[x+1 \quad y-2] = [2 \quad -3] + [x \quad 1-y]$

3-17 $\begin{bmatrix} x \\ y \\ 0 \end{bmatrix} = \begin{bmatrix} 2x \\ 4 \\ 2 \end{bmatrix} + \begin{bmatrix} 1 \\ 1 \\ -2 \end{bmatrix}$

- In Exercises 3-18 through 3-22, perform the arithmetic, if possible.

3-18 $\begin{bmatrix} -1 & 0 \\ 0 & 0 \end{bmatrix} - \begin{bmatrix} 2 & 2 \\ -1 & 2 \end{bmatrix}$

3-19 $\begin{bmatrix} 2 & \dfrac{1}{2} & -1 \end{bmatrix}$
$+ [-1 \quad -1 \quad -1]$

3-20 $\begin{bmatrix} 0 \\ 1 \\ 1 \end{bmatrix} + \left(\begin{bmatrix} 2 \\ -1 \\ -1 \end{bmatrix} - \begin{bmatrix} 3 \\ 3 \\ 1 \end{bmatrix} \right)$

3-21 $\begin{bmatrix} 2 & -1 \\ 1 & 1 \end{bmatrix} + \begin{bmatrix} 0 \\ 0 \end{bmatrix}$

3-22 $\begin{bmatrix} -1 & 2 \\ 1 & 1 \\ 1 & 0 \end{bmatrix} - \left(\begin{bmatrix} 2 & 1 \\ 3 & 1 \\ 0 & 0 \end{bmatrix} + \begin{bmatrix} 0 & 0 \\ 0 & 0 \\ 0 & 0 \end{bmatrix} \right)$

- In Exercises 3-23 through 3-28, find the unknown matrix X.

3-23 $X + \begin{bmatrix} 2 \\ 1 \end{bmatrix} = \begin{bmatrix} -1 \\ -1 \end{bmatrix}$

3-24 $\begin{bmatrix} -1 & 0 \\ 0 & 1 \\ 1 & 1 \end{bmatrix} - X = \begin{bmatrix} 2 & 1 \\ 0 & 0 \\ 0 & 0 \end{bmatrix}$

3-25 $([2 \quad 1 \quad 4] + X) + [-3 \quad 1 \quad 1] = [0 \quad 0 \quad 2]$

3-26 $\begin{bmatrix} 2 & -1 & 1 \\ 3 & 1 & 2 \\ 1 & 2 & 3 \end{bmatrix} + X = \begin{bmatrix} 0 & 0 & 0 \\ 0 & 0 & 0 \\ 0 & 0 & 0 \end{bmatrix}$

3-27 $\left(\begin{bmatrix} 2 \\ 1 \\ 2 \end{bmatrix} + X \right) + \begin{bmatrix} 0 \\ 0 \\ 0 \end{bmatrix} = \begin{bmatrix} -2 \\ -2 \\ -2 \end{bmatrix}$

3-28 $\begin{bmatrix} 2 & 1 \\ 3 & 2 \end{bmatrix} - X = \begin{bmatrix} 0 & 0 \\ 0 & 0 \end{bmatrix}$

3-2   Matrix Multiplication

We stated in Section 3-1 that two matrices could be added together only if they were the same size. In symbols, letting A and B represent matrices, their sum $A + B$ exists provided that A and B are the same size and, in that event, $A + B$ is of that size as well. Similarly, the difference $A - B$ only exists if A and B are of the same size. Thus, if A and B are matrices of different sizes, neither the symbol $A + B$ nor $A - B$ has any meaning.

In view of the restrictions on taking sums and differences of matrices, it should not surprise you to learn that the **product** of matrices A and B, written AB, is defined only for matrices A and B of the appropriate sizes. In this case, however, there is a different size restriction. In order to form the product AB, the number of columns of A, the matrix on the left, must equal the number of rows of B, the matrix on the right. Thus, A must be of size $m \times k$ and B of size $k \times n$, for some numbers m, k, and n.

3-10   **Example**   Is the product AB defined in each of the following cases?

(a) $A = [1 \quad 2 \quad -1] \qquad B = \begin{bmatrix} 0 \\ -2 \\ 3 \end{bmatrix}$

(b) $A = \begin{bmatrix} 2 & 1 & -1 & 2 \\ 0 & 0 & 1 & 2 \end{bmatrix} \qquad B = \begin{bmatrix} 2 & 1 & 0 \\ 1 & 2 & -1 \\ 0 & 2 & -1 \\ 0 & 1 & 0 \end{bmatrix}$

(c) $A = [2 \quad 1 \quad 2] \qquad B = [3 \quad 2 \quad 1]$

Solution   (*a*) Yes, because *A* has 3 columns and *B* has 3 rows.

(*b*) Yes, because *A* is of size 2 × 4 and *B* is 4 × 3, so *A* has 4 columns and *B* has 4 rows.

(*c*) No, because *A* and *B* are both of size 1 × 3, so *A* has 3 columns but *b* has only 1 row.

We noted in Section 3-1 that addition of matrices behaves just like addition of ordinary numbers. If $A + B$ makes sense, so does $B + A$ and, in fact, $A + B = B + A$. However, returning to Example 3-10, part (*b*) just above, which was

$$A = \begin{bmatrix} 2 & 1 & -1 & 2 \\ 0 & 0 & 1 & 2 \end{bmatrix} \qquad B = \begin{bmatrix} 2 & 1 & 0 \\ 1 & 2 & -1 \\ 0 & 2 & -1 \\ 0 & 1 & 0 \end{bmatrix}$$

we saw that AB exists. But we claim that BA is not defined in this example, because the left-hand matrix (now *B*) has 3 columns, while the number of rows in the right-hand matrix (*A*) is only 2. Thus, the existence of AB does not automatically imply the existence of BA, and we conclude that it is important to keep track of the order in which matrices are multiplied.

To explain the rules for multiplying matrices of the correct sizes, we begin with the simplest case. This case is illustrated by the definition of AB when

$$A = \begin{bmatrix} 4 & -2 & 1 \end{bmatrix} \qquad B = \begin{bmatrix} 0 \\ -1 \\ -3 \end{bmatrix}$$

Again, we emphasize that AB is defined because the number of columns of *A* (3) is equal to the number of rows of *B*. The rule of multiplication is:

1. Start at the left of *A* and at the top of *B*.

2. Moving to the right in *A* and down in *B*, multiply corresponding (first, second, etc.) numbers together.

3. Add up the result.

Applying the rule to the example gives

$$AB = (4)(0) + (-2)(-1) + (1)(-3) = 0 + 2 + (-3) = -1$$

Thus, the product of the 1 × 3 matrix *A* and the 3 × 1 matrix *B* is a number, −1. We may, however, write −1 as [−1] and think of it as a matrix, with only 1 row and 1 column. In this way, the product of these matrices may be thought of either as a number or as another matrix. We shall have occasion to make use of both points of view.

3-11 **Example** Find the product AB when

$$A = [2 \quad -1 \quad 3 \quad -2 \quad 4] \qquad B = \begin{bmatrix} 5 \\ 4 \\ 1 \\ -3 \\ -2 \end{bmatrix}$$

Solution The rule of multiplication requires us to start at the left of A (with 2) and at the top of B (with 5), multiply corresponding numbers, and add up the products. Thus,

$$AB = (2)(5) + (-1)(4) + (3)(1) + (-2)(-3) + (4)(-2)$$
$$= 10 + (-4) + 3 + 6 + (-8) = 7$$

By now, you may find it convenient to have a formula in addition to the verbal rule of multiplication we have been using:

A general $1 \times n$ matrix A can be written thus:

$$A = [a_1 \quad a_2 \quad a_3 \cdots a_n]$$

and a general $n \times 1$ matrix B thus:

$$B = \begin{bmatrix} b_1 \\ b_2 \\ b_3 \\ \cdot \\ \cdot \\ \cdot \\ b_n \end{bmatrix}$$

Then the rule of multiplication for such matrices is

$$AB = a_1 b_1 + a_2 b_2 + a_3 b_3 + \cdots + a_n b_n$$

which may be thought of as a 1×1 matrix.

Although the rule for multiplying a $1 \times n$ matrix times an $n \times 1$ matrix may look formidable, fortunately the rules for multiplying in general (when it is defined) can be reduced to performing this special case over and over. In order to formulate these general rules, we are first required to know the size of the matrix AB. Matrix AB has as many rows as A and as many columns as B. That is, if A is of size $m \times k$ and B is of size $k \times n$, then AB is of size $m \times n$. Notice that when we multiplied a $1 \times n$ matrix by an $n \times 1$ matrix above, the product was a 1×1 matrix, in agreement with the rule just stated.

3-12   **Example**   (*a*) What is the size of AB when

$$A = [-1 \quad -2 \quad 0] \qquad B = \begin{bmatrix} 1 & 2 \\ 0 & 0 \\ 3 & 1 \end{bmatrix}$$

(*b*) If A is an 8×7 matrix and B is 7×6, what size is AB?

Solution   (*a*) Since A is a 1×3 matrix and B is 3×2, then AB is 1×2.

(*b*) 8×6.

The rule for the size of a product indicates that, even if AB and BA are both defined, they will not in general be equal (in contrast to the situation for addition), because they will not necessarily even be of the same size.

3-13   **Example**   What sizes are AB and BA when

$$A = [1 \quad 2 \quad -1] \qquad B = \begin{bmatrix} 0 \\ -3 \\ 2 \end{bmatrix}$$

Solution   Since A is 1×3 and B is 3×1, then AB is a 1×1 matrix. When we consider BA, then B is the left-hand matrix and A is on the right. The product BA is defined, because B has as many columns (1) as A has rows. Now BA has as many rows as B, 3, and as many columns as A, also 3. We conclude that while AB is a 1×1 matrix, BA is a 3×3 matrix so certainly $AB \neq BA$.

If you feel comfortable with the development of matrix multiplication thus far, and wish to skip directly to the general definition, proceed to p. 147. Since some people will benefit from a more leisurely treatment of the subject, at this point we have included an intermediate level of generality between the $1 \times n$ times $n \times 1$ case above and the most general situation for matrix multiplication.

As a first step toward understanding how the general rules of matrix multiplication reduce to the rule for multiplying a matrix with 1 row times a matrix with 1 column, consider the example

$$A = [-3 \quad -2 \quad 0] \qquad B = \begin{bmatrix} 4 & 2 \\ -1 & 1 \\ 3 & 5 \end{bmatrix}$$

Since A has 1 row and B has 2 columns, we know that AB must be a 1×2 matrix, that is, of the form

$$AB = [x_1 \quad x_2]$$

The problem is to define the numbers x_1 and x_2. The procedure is to think of each column of

$$B = \begin{bmatrix} 4 & 2 \\ -1 & 1 \\ 3 & 5 \end{bmatrix}$$

as a 3×1 matrix in its own right and multiply that matrix by A according to the rule we used before. That is, let

$$B_1 = \begin{bmatrix} 4 \\ -1 \\ 3 \end{bmatrix} \qquad \text{and} \qquad B_2 = \begin{bmatrix} 2 \\ 1 \\ 5 \end{bmatrix}$$

and compute

$$AB_1 = [-3 \quad -2 \quad 0] \begin{bmatrix} 4 \\ -1 \\ 3 \end{bmatrix} = (-3)(4) + (-2)(-1) + (0)(3) = -10$$

and

$$AB_2 = [-3 \quad -2 \quad 0] \begin{bmatrix} 2 \\ 1 \\ 5 \end{bmatrix} = (-3)(2) + (-2)(1) + (0)(5) = -8$$

In this case, the definition is

$$x_1 = AB_1 = -10 \qquad x_2 = AB_2 = -8$$

So,

$$AB = [-3 \quad -2 \quad 0] \begin{bmatrix} 4 & 2 \\ -1 & 1 \\ 3 & 5 \end{bmatrix} = [-10 \quad -8]$$

The method for multiplying a $1 \times n$ matrix A by an $n \times r$ matrix B is a straightforward extension of what we have just done. Think of each column of B (there are r of these) as an $n \times 1$ matrix and multiply it by A. The number that results goes into the place in the $1 \times r$ matrix AB that corresponds to the column of B, e.g., the product of A and the third column from the left in B is the third number from the left in AB.

3-14   **Example**   Compute AB when

$$A = [2 \quad -1] \qquad B = \begin{bmatrix} 3 & 1 & 2 & 1 & 3 \\ 0 & 1 & -2 & 2 & 1 \end{bmatrix}$$

Solution   Write

$$B = \left[\begin{pmatrix} 3 \\ 0 \end{pmatrix} \begin{pmatrix} 1 \\ 1 \end{pmatrix} \begin{pmatrix} 2 \\ -2 \end{pmatrix} \begin{pmatrix} 1 \\ 2 \end{pmatrix} \begin{pmatrix} 3 \\ 1 \end{pmatrix} \right]$$

and compute

$$[2 \quad -1] \begin{bmatrix} 3 \\ 0 \end{bmatrix} = 6 \qquad [2 \quad -1] \begin{bmatrix} 1 \\ 1 \end{bmatrix} = 1 \qquad [2 \quad -1] \begin{bmatrix} 2 \\ -2 \end{bmatrix} = 6$$

$$[2 \quad -1] \begin{bmatrix} 1 \\ 2 \end{bmatrix} = 0 \qquad [2 \quad -1] \begin{bmatrix} 3 \\ 1 \end{bmatrix} = 5$$

Then the rule rells us that

$$AB = [2 \quad -1] \begin{bmatrix} 3 & 1 & 2 & 1 & 3 \\ 0 & 1 & -2 & 2 & 1 \end{bmatrix} = [6 \quad 1 \quad 6 \quad 0 \quad 5]$$

We come now to the most general situation for matrix multiplication. Let us illustrate the procedure by means of an example:

$$AB = \begin{bmatrix} 0 & 1 & 4 \\ -1 & 3 & -2 \end{bmatrix} \begin{bmatrix} 1 & -2 \\ -3 & 5 \\ -1 & 0 \end{bmatrix}$$

Since A is a 2 × 3 matrix and B is 3 × 2, we know that AB must be a 2 × 2 matrix, that is,

$$AB = \begin{bmatrix} x & y \\ z & w \end{bmatrix}$$

We think of B as a collection of columns thus:

$$B = \left[\overset{B_1}{\begin{pmatrix} 1 \\ -3 \\ -1 \end{pmatrix}} \quad \overset{B_2}{\begin{pmatrix} -2 \\ 5 \\ 0 \end{pmatrix}} \right]$$

In a similar manner, we consider A to be a collection of rows:

$$A = \begin{bmatrix} (0 & 1 & 4) \\ (-1 & 3 & -2) \end{bmatrix}$$

and we give the rows names,

$$A_1 = \begin{bmatrix} 0 & 1 & 4 \end{bmatrix}$$
$$A_2 = \begin{bmatrix} -1 & 3 & -2 \end{bmatrix}$$

Now, as above, each row of A, a 1 × 3 matrix, can be multiplied by each column of B, a 3 × 1, to produce a number. There are four possible such calculations in this example, they are

$$A_1 B_1 = \begin{bmatrix} 0 & 1 & 4 \end{bmatrix} \begin{bmatrix} 1 \\ -3 \\ -1 \end{bmatrix} = 0 + (-3) + (-4) = -7$$

$$A_1 B_2 = \begin{bmatrix} 0 & 1 & 4 \end{bmatrix} \begin{bmatrix} -2 \\ 5 \\ 0 \end{bmatrix} = 0 + 5 + 0 = 5$$

$$A_2 B_1 = \begin{bmatrix} -1 & 3 & -2 \end{bmatrix} \begin{bmatrix} 1 \\ -3 \\ -1 \end{bmatrix} = (-1) + (-9) + 2 = -8$$

$$A_2 B_2 = \begin{bmatrix} -1 & 3 & -2 \end{bmatrix} \begin{bmatrix} -2 \\ 5 \\ 0 \end{bmatrix} = 2 + 15 + 0 = 17$$

As you would expect, these are the four numbers we need for the 2×2 matrix AB. The only question is: Where does each number go in AB?

> The general rule for the multiplication of matrices is: To find the number in the matrix AB which is in the ith row and the jth column, multiply the entire ith row of A by the entire jth column of B.

Recall that we wrote

$$AB = \begin{bmatrix} x & y \\ z & w \end{bmatrix}$$

Since x is in the first row and the first column of AB, then x is the product of the first row (A_1) of A and the first column (B_1) of B. Thus,

$$x = A_1 B_1 = [0 \quad 1 \quad 4]\begin{bmatrix} 1 \\ -3 \\ -1 \end{bmatrix} = -7$$

Similarly, since y is in the first row and the second column of AB, then $y = A_1 B_2 = 5$. Also, $z = A_2 B_1 = -8$ and $w = A_2 B_2 = 17$. We conclude that

$$AB = \begin{bmatrix} 0 & 1 & 4 \\ -1 & 3 & -2 \end{bmatrix}\begin{bmatrix} 1 & -2 \\ -3 & 5 \\ -1 & 0 \end{bmatrix} = \begin{bmatrix} -7 & 5 \\ -8 & 17 \end{bmatrix}$$

3-15 **Example** Compute AB when

(a) $A = \begin{bmatrix} -1 & 1 \\ 2 & 0 \\ 3 & 2 \end{bmatrix}$ $B = \begin{bmatrix} 1 & 0 \\ 1 & -1 \end{bmatrix}$

(b) $A = \begin{bmatrix} 0 \\ 2 \\ 1 \end{bmatrix}$ $B = [1 \quad 2 \quad -1]$

Solution (a) Since A is 3×2 and B is 2×2, then AB is 3×2. That is,

$$AB = \begin{bmatrix} -1 & 1 \\ 2 & 0 \\ 3 & 2 \end{bmatrix}\begin{bmatrix} 1 & 0 \\ 1 & -1 \end{bmatrix} = \begin{bmatrix} (-1 \ 1) \\ (\ 2 \ 0) \\ (\ 3 \ 2) \end{bmatrix}\left[\begin{pmatrix} 1 \\ 1 \end{pmatrix}\begin{pmatrix} 0 \\ -1 \end{pmatrix}\right] = \begin{bmatrix} \cdot & \cdot \\ \cdot & \cdot \\ \cdot & \cdot \end{bmatrix}$$

Multiplying rows by columns, we have

$$[-1 \quad 1]\begin{bmatrix} 1 \\ 1 \end{bmatrix} = -1 + 1 = 0 \qquad [-1 \quad 1]\begin{bmatrix} 0 \\ -1 \end{bmatrix} = 0 + (-1) = -1$$

$$[2 \quad 0]\begin{bmatrix} 1 \\ 1 \end{bmatrix} = 2 + 0 = 2 \qquad [2 \quad 0]\begin{bmatrix} 0 \\ -1 \end{bmatrix} = 0 + 0 = 0$$

$$[3 \quad 2]\begin{bmatrix} 1 \\ 1 \end{bmatrix} = 3 + 2 = 5 \qquad [3 \quad 2]\begin{bmatrix} 0 \\ -1 \end{bmatrix} = 0 + (-2) = -2$$

We conclude that

$$\begin{bmatrix} -1 & 1 \\ 2 & 0 \\ 3 & 2 \end{bmatrix} \begin{bmatrix} 1 & 0 \\ 1 & -1 \end{bmatrix} = \begin{bmatrix} 0 & -1 \\ 2 & 0 \\ 5 & -2 \end{bmatrix}$$

(b) Here, A is 3×1 and B is 1×3, so AB must be 3×3. Each number in A is a distinct row and each number in B is a distinct column. The rule for multiplying implies that

$$AB = \begin{bmatrix} 0 \\ 2 \\ 1 \end{bmatrix} [1 \quad 2 \quad -1] = \begin{bmatrix} 0 & 0 & 0 \\ 2 & 4 & -2 \\ 1 & 2 & -1 \end{bmatrix}$$

Summary

Matrices A and B can be multiplied to form the product AB only if the number of columns of A is equal to the number of rows in B. If A is of size $m \times k$ and B is a $k \times n$ matrix, then AB is $m \times n$. To obtain the number in the ith row and the jth column of the matrix AB, multiply the entire ith row of A by the entire jth column of B. If the ith row of A is

$$[a_1 \quad a_2 \quad a_3 \cdots a_k]$$

and the jth column of B is

$$\begin{bmatrix} b_1 \\ b_2 \\ b_3 \\ \cdot \\ \cdot \\ b_k \end{bmatrix}$$

then·their product is defined to be the number

$$a_1b_1 + a_2b_2 + a_3b_3 + \cdots + a_kb_k$$

Exercises

● *In each of Exercises 3-29 through 3-36, decide whether AB or BA exist, and if they do, what size they must be. Do not actually carry out the multiplication.*

3-29 $A = \begin{bmatrix} -1 & 2 & \frac{1}{2} \end{bmatrix}$ $B = \begin{bmatrix} 2 & 0 & 0 & 1 \\ 3 & 0 & 1 & -1 \\ 2 & \frac{1}{2} & 0 & 0 \end{bmatrix}$

3-30 $A = \begin{bmatrix} 2 \\ 1 \end{bmatrix}$ $B = \begin{bmatrix} -1 & 0 & 0 \\ 1 & 1 & 2 \\ 0 & 0 & 0 \end{bmatrix}$

3-31 $A = \begin{bmatrix} 2 & 1 \\ 1 & -1 \end{bmatrix}$ $B = \begin{bmatrix} 0 & 3 \\ -2 & 1 \end{bmatrix}$

3-32 $A = \begin{bmatrix} 1 & -1 & 1 & 2 \end{bmatrix}$ $B = \begin{bmatrix} 3 \end{bmatrix}$

3-33 $A = \begin{bmatrix} 1 & 0 & 1 & -1 \\ 2 & -1 & 1 & 1 \\ 2 & 0 & 1 & 2 \end{bmatrix}$ $B = \begin{bmatrix} 1 & 2 & 1 \end{bmatrix}$

3-34 $A = \begin{bmatrix} 1 & 2 & 2 & -1 \\ 0 & 0 & 1 & 0 \end{bmatrix}$ $B = \begin{bmatrix} -1 & -1 \\ 0 & 0 \\ 1 & -2 \\ 2 & -1 \end{bmatrix}$

3-35 $A = \begin{bmatrix} 0 & 1 & 2 \end{bmatrix}$ $B = \begin{bmatrix} 1 & 1 & 1 \\ -1 & 2 & 0 \end{bmatrix}$

3-36 $A = \begin{bmatrix} 1 \\ 2 \end{bmatrix}$ $B = \begin{bmatrix} 2 & -1 & 1 & 2 \end{bmatrix}$

● *In Exercises 3-37 through 3-48, compute AB.*

3-37 $A = \begin{bmatrix} 2 & 1 \end{bmatrix}$ $B = \begin{bmatrix} 0 \\ 1 \end{bmatrix}$

3-38 $A = \begin{bmatrix} 0 & 1 & -1 & -1 \end{bmatrix}$ $B = \begin{bmatrix} -2 \\ 1 \\ 0 \\ 0 \end{bmatrix}$

3-39 $A = \begin{bmatrix} -1 & 2 & 1 \end{bmatrix}$ $B = \begin{bmatrix} x \\ y \\ z \end{bmatrix}$

3-40 $A = \begin{bmatrix} -1 & -1 \end{bmatrix}$ $B = \begin{bmatrix} 0 & 1 & -1 & 2 \\ 0 & 0 & -1 & 0 \end{bmatrix}$

3-41 $A = \begin{bmatrix} 0 & \dfrac{1}{2} & 2 & -1 & -3 & -2 \end{bmatrix}$ $B = \begin{bmatrix} 1 & \dfrac{1}{2} \\ -1 & 1 \\ 0 & -1 \\ 0 & 0 \\ 2 & 1 \\ -1 & 2 \end{bmatrix}$

3-42 $A = \begin{bmatrix} a & b & c \end{bmatrix}$ $B = \begin{bmatrix} 2 & 1 & -1 \\ 1 & 0 & 0 \\ -1 & 1 & 0 \end{bmatrix}$

3-43 $A = \begin{bmatrix} 1 & 0 \\ 0 & 1 \end{bmatrix}$ $B = \begin{bmatrix} -1 & 3 & 1 \\ \dfrac{1}{2} & -1 & 2 \end{bmatrix}$

3-44 $A = \begin{bmatrix} 1 \\ 2 \\ -1 \\ -2 \end{bmatrix}$ $B = \begin{bmatrix} 1 & 0 & 2 \end{bmatrix}$

3-45 $A = \begin{bmatrix} \dfrac{1}{2} & \dfrac{1}{2} & 2 & 1 \\ 1 & 2 & \dfrac{1}{2} & 2 \end{bmatrix}$ $B = \begin{bmatrix} -1 \\ 2 \\ -1 \\ 0 \end{bmatrix}$

3-46 $A = \begin{bmatrix} 1 \\ 0 \\ -1 \\ -1 \\ 2 \\ 3 \end{bmatrix}$ $B = [2 \quad 1 \quad 0 \quad 0 \quad 0 \quad 1 \quad 3 \quad 4]$

3-47 $A = \begin{bmatrix} -1 & 2 & 0 \\ 3 & 0 & 0 \\ -1 & 0 & -1 \end{bmatrix}$ $B = \begin{bmatrix} x \\ y \\ z \end{bmatrix}$

3-48 $A = \begin{bmatrix} 1 & 2 & -1 \\ 0 & 1 & 1 \end{bmatrix}$ $B = \begin{bmatrix} 1 & 0 & 0 \\ 0 & 1 & 0 \\ 0 & 0 & 1 \end{bmatrix}$

● In Exercises 3-49 through 3-51, compute both AB and BA.

3-49 $A = \begin{bmatrix} -1 & 2 \\ 0 & 1 \end{bmatrix}$ $B = \begin{bmatrix} 0 & 2 \\ 0 & 0 \end{bmatrix}$

3-50 $A = \begin{bmatrix} -4 & 0 \\ 0 & 3 \end{bmatrix}$ $B = \begin{bmatrix} -2 & -1 \\ 1 & 3 \end{bmatrix}$

3-51 $A = \begin{bmatrix} 4 & 2 \\ -2 & 0 \end{bmatrix}$ $B = \begin{bmatrix} -3 & 0 \\ 0 & -3 \end{bmatrix}$

● In Exercises 3-52 through 3-55, compute AB + C.

3-52 $A = [1 \quad 2 \quad 3 \quad 1]$ $B = \begin{bmatrix} \frac{1}{2} & 2 & 1 \\ 0 & 1 & 1 \\ 0 & 0 & 0 \\ 2 & -2 & 0 \end{bmatrix}$ $C = [-1 \quad 0 \quad 2]$

3-53 $A = \begin{bmatrix} 0 & 0 & 3 \\ 0 & 0 & 2 \end{bmatrix}$ $B = \begin{bmatrix} 2 & 3 \\ -1 & 1 \\ 0 & 0 \end{bmatrix}$ $C = \begin{bmatrix} 2 & 1 \\ 0 & 1 \end{bmatrix}$

3-54 $A = \begin{bmatrix} -1 & 1 \\ 2 & 0 \end{bmatrix}$ $B = \begin{bmatrix} -1 \\ -1 \end{bmatrix}$ $C = \begin{bmatrix} 1 \\ 1 \end{bmatrix}$

3-55 $A = \begin{bmatrix} 0 & 1 & 1 \\ 1 & -1 & 1 \end{bmatrix}$ $B = \begin{bmatrix} -1 & 0 & -2 \\ 1 & 2 & 3 \\ 3 & 2 & 1 \end{bmatrix}$ $C = \begin{bmatrix} 0 & 0 & 0 \\ 0 & 0 & 0 \end{bmatrix}$

● In Exercises **3-56** and **3-57**, compute $A - (BC)$.

3-56 $A = [-2]$ $B = [1 \quad 0 \quad 2]$ $C = \begin{bmatrix} 2 \\ 1 \\ -2 \end{bmatrix}$

3-57 $A = \begin{bmatrix} 2 \\ -3 \end{bmatrix}$ $B = \begin{bmatrix} 0 & 2 & -1 & 0 \\ 1 & 1 & 0 & 1 \end{bmatrix}$ $C = \begin{bmatrix} -2 \\ 0 \\ 0 \\ 1 \end{bmatrix}$

● In Exercises **3-58** and **3-59**, compute $AB + CD$.

3-58 $A = C = \begin{bmatrix} -1 & 2 \\ 1 & 0 \end{bmatrix}$ $B = \begin{bmatrix} -1 \\ 2 \end{bmatrix}$ $D = \begin{bmatrix} 2 \\ 0 \end{bmatrix}$

3-59 $A = \begin{bmatrix} 0 & 0 \\ 2 & 1 \\ 0 & 0 \\ \frac{1}{2} & 2 \\ 0 & 0 \\ 1 & 0 \end{bmatrix}$ $B = \begin{bmatrix} 0 & -1 & 1 & 0 & 2 \\ 0 & 0 & 0 & 0 & 0 \end{bmatrix}$

$C = \begin{bmatrix} 1 \\ 2 \\ 3 \\ 4 \\ 5 \\ 6 \end{bmatrix}$ $D = [-2 \quad -1 \quad 0 \quad 1 \quad 2]$

3-3 Properties of Multiplication

The rather complicated definition of matrix multiplication in Section 3-2 may impress you as being somewhat unnatural. There are, however, mathematical problems which are quite readily and usefully expressed in the language of matrix multiplication.

Let us suppose that a person is planning to buy the following staple food items: milk, bread, coffee, flour, and sugar. Further suppose that the costs of the items are:

Milk	$0.75 per half-gallon
Bread	$0.50 a loaf
Coffee	$2.25 a pound
Flour	$1.00 for 5 pounds
Sugar	$1.50 for 5 pounds

If the person purchases four half-gallons of milk, the cost of the milk is ($0.75)(4) = $3.00. Similarly, suppose other purchases are two loaves of bread for ($0.50)(2), 3 pounds of coffee for ($2.25)(3), two 5-pound sacks of flour for ($1.00)(2), and one 5-pound bag of sugar for ($1.50)(1). Let c stand for the total cost of all the purchases. Then

$$c = (\$0.75)(4) + (\$0.50)(2) + (\$2.25)(3) + (\$1.00)(2) + (\$1.50)(1).$$

In order to express this equation in matrix form, we define a 1×5 matrix P of prices:

$$P = [\$0.75 \quad \$0.50 \quad \$2.25 \quad \$1.00 \quad \$1.50]$$

and a 5×1 matrix Q of quantities:

$$Q = \begin{bmatrix} 4 \\ 2 \\ 3 \\ 2 \\ 1 \end{bmatrix}$$

Notice that the food items are in the order: milk, bread, coffee, flour, sugar from left to right in P and from top to bottom in Q. By the definition of matrix multiplication,

$$PQ = [\$0.75 \quad \$0.50 \quad \$2.25 \quad \$1.00 \quad \$1.50] \begin{bmatrix} 4 \\ 2 \\ 3 \\ 2 \\ 1 \end{bmatrix}$$

$$= (\$0.75)(4) + (\$0.50)(2) + (\$2.25)(3) + (\$1.00)(2) + (\$1.50)(1)$$

$$= c$$

Thus, the formula for the total cost of the groceries becomes the simple matrix equation $c = PQ$.

The same procedure can be applied to any list of purchases. Letting P be the matrix of prices and Q the matrix of quantities purchased, then $c = PQ$ still determines the total cost. Thus, if P is known, several possible shopping lists (matrices for Q) can be multiplied by P to determine, for example, what combinations of purchases are possible within a fixed budget (i.e., keeping c less than a given amount).

3-16   **Example**   A morning's work at an automobile service station consists of four cars and a truck that need oil changes, two cars requiring tune-ups, and one truck with a defective starter. If it takes 1/3 hour to change the oil in a car, 1/2 hour to do the same for a truck, 1/2 hour to tune up a car and 1 hour to repair the starter, how many man-hours are required to complete the morning's work?

Solution   The "price" in this example is calculated in man-hours. Taking the order

$$\text{oil car} \qquad \text{oil truck} \qquad \text{tune-up} \qquad \text{starter}$$

then the matrix of "prices" is

$$P = \begin{bmatrix} \dfrac{1}{3} & \dfrac{1}{2} & \dfrac{1}{2} & 1 \end{bmatrix}$$

The matrix of quantities is

$$Q = \begin{bmatrix} 4 \\ 1 \\ 2 \\ 1 \end{bmatrix}$$

so the total "cost" is

$$c = PQ = \begin{bmatrix} \dfrac{1}{3} & \dfrac{1}{2} & \dfrac{1}{2} & 1 \end{bmatrix} \begin{bmatrix} 4 \\ 1 \\ 2 \\ 1 \end{bmatrix}$$

$$= \frac{4}{3} + \frac{1}{2} + 1 + 1 = 3\tfrac{5}{6} \text{ man-hours}$$

A more complicated situation is exemplified by a baker who bakes one type of cake and one kind of cookie. Each cake requires $1\tfrac{1}{2}$ cups of sugar and 2 eggs, while each batch of cookies uses 1/2 cup of sugar and 1 egg. If the baker plans to bake 15 cakes and 10 batches of cookies, how much sugar and how many eggs will be required? Since the baker uses $1\frac{1}{2}$ cups of sugar for each cake, the baker needs $(1\frac{1}{2})(15) = 22\frac{1}{2}$ cups for all the cakes. The sugar requirement for the 10 batches of cookies is $(1/2)(10) = 5$ cups, so the total sugar need is

$$\text{Sugar} = (1\frac{1}{2})(15) + \left(\frac{1}{2}\right)(10) = 27\frac{1}{2} \text{ cups}$$

Similarly, the egg demand is

$$\text{Eggs} = (2)(15) + (1)(10) = 40$$

The baker's requirements for sugar and eggs per unit of product can be expressed in a table as follows:

		Product	
		Cake	Cookies
Ingredient	Sugar	$1\frac{1}{2}$	$\frac{1}{2}$
	Eggs	2	1

We hope this table reminds you of the input–output table of Section 3-1. As in that section, we can write a corresponding matrix, here of size 2 × 2:

$$\begin{bmatrix} 1\frac{1}{2} & \frac{1}{2} \\ 2 & 1 \end{bmatrix}$$

We call this matrix A. We also form a 2 × 1 matrix Q of the quantities of cakes and cookies being produced:

$$Q = \begin{bmatrix} 15 \\ 10 \end{bmatrix} \begin{matrix} \text{(Cakes)} \\ \text{(Batches of cookies)} \end{matrix}$$

Finally, we let N be the matrix of quantities of each ingredient needed:

$$N = \begin{bmatrix} \text{Sugar} \\ \text{Eggs} \end{bmatrix}$$

We claim that the two equations above,

$$\text{Sugar} = (1\tfrac{1}{2})(15) + \left(\frac{1}{2}\right)(10)$$

$$\text{Eggs} = (2)(15) + (1)(10)$$

become the single matrix equation

$$N = \begin{bmatrix} \text{Sugar} \\ \text{Eggs} \end{bmatrix} = \begin{bmatrix} 1\tfrac{1}{2} & \frac{1}{2} \\ 2 & 1 \end{bmatrix} \begin{bmatrix} 15 \\ 10 \end{bmatrix} = AQ$$

The equation $N = AQ$ *means* (by the definition of equality of matrices) that the top number in N (Sugar) is equal to the top number in AQ, and the bottom number in N (Eggs) equals the bottom number of AQ. The top number of the 2×1 matrix AQ is

$$\begin{bmatrix} 1\tfrac{1}{2} & \frac{1}{2} \end{bmatrix} \begin{bmatrix} 15 \\ 10 \end{bmatrix} = (1\tfrac{1}{2})(15) + \left(\frac{1}{2}\right)(10) = 27\tfrac{1}{2}$$

and the bottom number is

$$\begin{bmatrix} 2 & 1 \end{bmatrix} \begin{bmatrix} 15 \\ 10 \end{bmatrix} = (2)(15) + (1)(10) = 40$$

which verifies our claim.

More generally, a manufacturer who uses a number of raw materials to produce a number of finished products can form a table of the amount of each material that goes into the manufacture of each unit of product.

	Finished products
Raw materials	

Again there is a corresponding matrix we call A. The matrix A has as many rows as there are raw materials (say m of them) and as many columns as there are finished products (let us say n of these). If the manufacturer plans to produce a certain quantity of each of the finished products, the manufacturer can express these quantities in an $n \times 1$ matrix Q (in the same order from top to bottom as the columns of A from left to right). The $n \times 1$ matrix $N = AQ$ is a shopping list of the total needs for each raw material, for that choice of Q, in the same order as the rows of A. Thus, in each time period, the manufacturer can compute the material requirements using that same A, once the production amounts Q are established.

3-17 **Example** Suppose a manufacturer of motor vehicles uses steel, glass, and rubber (measured in tons) for each unit of one model of automobile and one model of pickup truck according to the following table:

	Car	Truck
Steel	2	3
Glass	0.08	0.05
Rubber	0.06	0.07

If the manufacturer plans to produce 2000 cars and 500 trucks, how much of each raw material will be required?

Solution

$$AQ = \begin{bmatrix} 2 & 3 \\ 0.08 & 0.05 \\ 0.06 & 0.07 \end{bmatrix} \begin{bmatrix} 2000 \\ 500 \end{bmatrix} = \begin{bmatrix} 5500 \\ 185 \\ 155 \end{bmatrix} = N$$

That is, the manufacturer will use 5500 tons of steel, 185 tons of glass, and 155 tons of rubber.

Returning to the more abstract features of matrices, we recall that a zero matrix, a matrix all of whose entries are zeros, behaves like the number 0 in ordinary arithmetic—when it is added to a matrix of the same size, the matrix remains unchanged. There are also matrices that behave like the number 1. Just as $1 \cdot x = x \cdot 1 = x$ for any number x, so when these matrices are multiplied by any matrix, the matrix remains unchanged. These matrices, called **identity matrices,** are all square, that is, they have the same number of rows and columns. There is one such matrix of each possible size. The first few identity matrices, in order of size are

$$[1]$$

$$\begin{bmatrix} 1 & 0 \\ 0 & 1 \end{bmatrix}$$

$$\begin{bmatrix} 1 & 0 & 0 \\ 0 & 1 & 0 \\ 0 & 0 & 1 \end{bmatrix}$$

In general, the identity matrices are defined to be those square matrices that have a 1 in every position where the row number and the column number are the same (the diagonal from upper left to lower right) and 0s in all other positions. Identity matrices are usually identified by the letter I. They have the property that $AI = A$ and $IB = B$ for all matrices

A and B such that multiplication is defined. For example, in $AI = A$, the size of I is determined by the number of columns in A.

To see that the identity matrices behave as we claim, let us multiply

$$\begin{bmatrix} 2 & -1 & 1 \\ 3 & 0 & 1 \end{bmatrix} \begin{bmatrix} 1 & 0 & 0 \\ 0 & 1 & 0 \\ 0 & 0 & 1 \end{bmatrix}$$

Then

$$[2 \quad -1 \quad 1] \begin{bmatrix} 1 \\ 0 \\ 0 \end{bmatrix} = 2 + 0 + 0 = 2$$

$$[2 \quad -1 \quad 1] \begin{bmatrix} 0 \\ 1 \\ 0 \end{bmatrix} = 0 + (-1) + 0 = -1$$

$$[2 \quad -1 \quad 1] \begin{bmatrix} 0 \\ 0 \\ 1 \end{bmatrix} = 0 + 0 + 1 = 1$$

and, similarly,

$$[3 \quad 0 \quad 1] \begin{bmatrix} 1 \\ 0 \\ 0 \end{bmatrix} = 3$$

$$[3 \quad 0 \quad 1] \begin{bmatrix} 0 \\ 1 \\ 0 \end{bmatrix} = 0$$

$$[3 \quad 0 \quad 1] \begin{bmatrix} 0 \\ 0 \\ 1 \end{bmatrix} = 1$$

So,

$$\begin{bmatrix} 2 & -1 & 1 \\ 3 & 0 & 1 \end{bmatrix} \begin{bmatrix} 1 & 0 & 0 \\ 0 & 1 & 0 \\ 0 & 0 & 1 \end{bmatrix} = \begin{bmatrix} 2 & -1 & 1 \\ 3 & 0 & 1 \end{bmatrix}$$

3-18 **Example** Perform the multiplication.

$$\begin{bmatrix} 1 & 0 \\ 0 & 1 \end{bmatrix}\begin{bmatrix} 2 & -1 & 1 \\ 3 & 0 & 1 \end{bmatrix}$$

Solution

$$[1 \quad 0]\begin{bmatrix} 2 \\ 3 \end{bmatrix} = 2 + 0 = 2 \qquad [1 \quad 0]\begin{bmatrix} -1 \\ 0 \end{bmatrix} = -1 + 0 = -1$$

$$[1 \quad 0]\begin{bmatrix} 1 \\ 1 \end{bmatrix} = 1 + 0 = 1 \qquad [0 \quad 1]\begin{bmatrix} 2 \\ 3 \end{bmatrix} = 0 + 3 = 3$$

$$[0 \quad 1]\begin{bmatrix} -1 \\ 0 \end{bmatrix} = 0 + 0 = 0 \qquad [0 \quad 1]\begin{bmatrix} 1 \\ 1 \end{bmatrix} = 0 + 1 = 1$$

and therefore

$$\begin{bmatrix} 1 & 0 \\ 0 & 1 \end{bmatrix}\begin{bmatrix} 2 & -1 & 1 \\ 3 & 0 & 1 \end{bmatrix} = \begin{bmatrix} 2 & -1 & 1 \\ 3 & 0 & 1 \end{bmatrix}$$

Another important property of matrix arithmetic, which it shares with ordinary arithmetic, is called the **distributive law.** We know that given the arithmetic problem

$$(3 \cdot 2) + (3 \cdot 4)$$

we can either solve directly,

$$(3 \cdot 2) + (3 \cdot 4) = 6 + 12 = 18$$

or we can first factor out the 3, thus:

$$(3 \cdot 2) + (3 \cdot 4) = 3(2 + 4) = 3(6) = 18$$

The algebraic formula for factoring is the distributive law; it is

$$ab + ac = a(b + c)$$

Using capital letters to denote matrices, the same distributive law works for matrices, that is

$$AB + AC = A(B + C)$$

3-19   **Example**   Compute $AB + AC$ and $A(B + C)$ when

$$A = \begin{bmatrix} 1 & -1 \\ 2 & 3 \\ 0 & 1 \end{bmatrix} \qquad B = \begin{bmatrix} -1 & -2 & -3 \\ 0 & 2 & 4 \end{bmatrix} \qquad C = \begin{bmatrix} 2 & -2 & 0 \\ 1 & -1 & 2 \end{bmatrix}$$

Solution

$$AB = \begin{bmatrix} 1 & -1 \\ 2 & 3 \\ 0 & 1 \end{bmatrix} \begin{bmatrix} -1 & -2 & -3 \\ 0 & 2 & 4 \end{bmatrix} = \begin{bmatrix} -1 & -4 & -7 \\ -2 & 2 & 6 \\ 0 & 2 & 4 \end{bmatrix}$$

and

$$AC = \begin{bmatrix} 1 & -1 \\ 2 & 3 \\ 0 & 1 \end{bmatrix} \begin{bmatrix} 2 & -2 & 0 \\ 1 & -1 & 2 \end{bmatrix} = \begin{bmatrix} 1 & -1 & -2 \\ 7 & -7 & 6 \\ 1 & -1 & 2 \end{bmatrix}$$

So,

$$AB + AC = \begin{bmatrix} -1 & -4 & -7 \\ -2 & 2 & 6 \\ 0 & 2 & 4 \end{bmatrix} + \begin{bmatrix} 1 & -1 & -2 \\ 7 & -7 & 6 \\ 1 & -1 & 2 \end{bmatrix} = \begin{bmatrix} 0 & -5 & -9 \\ 5 & -5 & 12 \\ 1 & 1 & 6 \end{bmatrix}$$

On the other hand,

$$B + C = \begin{bmatrix} -1 & -2 & -3 \\ 0 & 2 & 4 \end{bmatrix} + \begin{bmatrix} 2 & -2 & 0 \\ 1 & -1 & 2 \end{bmatrix} = \begin{bmatrix} 1 & -4 & -3 \\ 1 & 1 & 6 \end{bmatrix}$$

and therefore

$$A(B + C) = \begin{bmatrix} 1 & -1 \\ 2 & 3 \\ 0 & 1 \end{bmatrix} \begin{bmatrix} 1 & -4 & -3 \\ 1 & 1 & 6 \end{bmatrix} = \begin{bmatrix} 0 & -5 & -9 \\ 5 & -5 & 12 \\ 1 & 1 & 6 \end{bmatrix}$$

which is the same as $AB + AC$ calculated above.

Thus, we can factor matrices just as we do numbers. As the example illustrates, factoring has computational advantages for matrices. That is, matrix multiplication is difficult, so factoring to eliminate a multiplication step is an attractive computational shortcut.

One must be careful, however, in using the distributive law in matrix theory. The arithmetic problem

$$(3 \cdot 2) + (4 \cdot 3)$$

allows for factoring because

$$(3 \cdot 2) + (4 \cdot 3) = (3 \cdot 2) + (3 \cdot 4) = 3(2 + 4)$$

but, in general, $AB + CA$ cannot be factored because it may be that $CA \neq AC$ (remember that AC may not even be defined). However, the forms of factoring

$$BA + CA = (B + C)A$$
$$AB - AC = A(B - C)$$

and

$$BA - CA = (B - C)A$$

are all permitted in matrix arithmetic just as they are in ordinary arithmetic.

3-20   **Example** Compute $BA - CA$ when

$$A = \begin{bmatrix} -1 & -2 \\ 0 & 2 \end{bmatrix} \qquad B = \begin{bmatrix} 1 & 2 \\ -1 & -2 \\ 0 & 1 \end{bmatrix} \qquad C = \begin{bmatrix} 2 & 3 \\ -1 & 2 \\ 1 & -1 \end{bmatrix}$$

Solution   The efficient procedure is to use the formula $BA - CA = (B - C)A$. Now,

$$B - C = \begin{bmatrix} 1 & 2 \\ -1 & -2 \\ 0 & 1 \end{bmatrix} - \begin{bmatrix} 2 & 3 \\ -1 & 2 \\ 1 & -1 \end{bmatrix} = \begin{bmatrix} -1 & -1 \\ 0 & -4 \\ -1 & 2 \end{bmatrix}$$

and so

$$BA - CA = (B - C)A = \begin{bmatrix} -1 & -1 \\ 0 & -4 \\ -1 & 2 \end{bmatrix} \begin{bmatrix} -1 & -2 \\ 0 & 2 \end{bmatrix} = \begin{bmatrix} 1 & 0 \\ 0 & -8 \\ 1 & 6 \end{bmatrix}$$

As a final remark on factoring, we can factor A out of

$$AB + A$$

to write the expression as a product of matrices. The method, as usual, is suggested by ordinary arithmetic. To write $(3 \cdot 2) + 3$ as a product of numbers, we would write

$$(3 \cdot 2) + 3 = (3 \cdot 2) + (3 \cdot 1) = 3(2 + 1) = 3 \cdot 3$$

Thus, we write

$$AB + A = AB + AI = A(B + I)$$

Notice that in this problem B must necessarily be a square matrix, since otherwise AB and A would be different sizes and so $AB + A$ would not be defined.

3-21 **Example** Write as a product of two matrices and then multiply.

$$\begin{bmatrix} -1 & 2 & 0 \\ 1 & 1 & 1 \\ 2 & 1 & 2 \end{bmatrix} \begin{bmatrix} 2 \\ 1 \\ 3 \end{bmatrix} + \begin{bmatrix} 2 \\ 1 \\ 3 \end{bmatrix}$$

Solution The problem is of the form $BA + A$, where B is 3×3. We can use the formula

$$BA + A = BA + IA = (B + I)A$$

where I is the 3×3 identity matrix.

$$B + I = \begin{bmatrix} -1 & 2 & 0 \\ 1 & 1 & 1 \\ 2 & 1 & 2 \end{bmatrix} + \begin{bmatrix} 1 & 0 & 0 \\ 0 & 1 & 0 \\ 0 & 0 & 1 \end{bmatrix} = \begin{bmatrix} 0 & 2 & 0 \\ 1 & 2 & 1 \\ 2 & 1 & 3 \end{bmatrix}$$

So,

$$\begin{bmatrix} -1 & 2 & 0 \\ 1 & 1 & 1 \\ 2 & 1 & 2 \end{bmatrix} \begin{bmatrix} 2 \\ 1 \\ 3 \end{bmatrix} + \begin{bmatrix} 2 \\ 1 \\ 3 \end{bmatrix} = BA + A = (B + I)A = \begin{bmatrix} 0 & 2 & 0 \\ 1 & 2 & 1 \\ 2 & 1 & 3 \end{bmatrix} \begin{bmatrix} 2 \\ 1 \\ 3 \end{bmatrix} = \begin{bmatrix} 2 \\ 7 \\ 14 \end{bmatrix}$$

Summary

An **identity matrix** I is a square ($n \times n$) matrix with 1s in locations where the row number and the column number are the same and 0s elsewhere. It has the property $AI = A$ and $IB = B$ for any matrices A, B for which the product is defined.

Matrices obey the **distributive law,**

$$AB + AC = A(B + C)$$

and its variants: $BA + CA = (B + C)A$, $AB - AC = A(B - C)$, and $BA - CA = (B - C)A$. The distributive law implies that $AB + A = A(B + I)$.

Exercises

3-60 An investor plans to buy 100 shares of telephone stock, 200 shares of oil stock, 400 shares of automobile stock, and 100 shares of airline stock. The telephone stock is selling for $46 a share, the oil stock for $34 a share, the automobile stock for $15 a share, and the airline stock for $10 a share. ($a$) Express the numbers of shares as a matrix with 1 row. (b) Express the prices of the stocks as a matrix with 1 column. (c) Use matrix multiplication to compute the total cost of the investor's purchases.

3-61 A manufacturer of custom-designed jewelry has orders for two rings, three pairs of earrings, five pins, and one necklace. The manufacturer estimates that it takes 1 man-hour of labor to make a ring, $1\frac{1}{2}$ man-hours to make a pair of earrings, 1/2 man-hour for each pin, and 2 man-hours to make a necklace. (a) Express the manufacturer's orders as a matrix with 1 row. (b) Express the man-hour requirements for the various types of jewelry as a matrix with 1 column. (c) Use matrix multiplication to calculate the total number of man-hours it will require to complete all the orders.

3-62 A company pays its executives a salary and gives them shares of its stock as an annual bonus. Last year, the president of the company received $80,000 and 50 shares of stock, each of the three vice-presidents were paid $45,000 and 20 shares of stock, and the treasurer was paid $40,000 and 10 shares of stock. (a) Express the payments to the executives in money and stocks by means of a 2×3 matrix. (b) Express the number of executives of each rank by means of a matrix with 1 column. (c) Use matrix multiplication to calculate the total amount of money and the total number of shares of stock the company paid these executives last year.

3-63 A tourist returns from a European trip with the following foreign currency: 1000 Austrian schillings, 20 British pounds, 100 French francs, 5000 Italian lire, and 50 German marks. In American money, a schilling is worth $0.055, the pound $1.80, the franc $0.20, the lira

$0.001, and the mark $0.40. ($a$) Express the quantity of each currency by means of a matrix with 1 row. (b) Express the value of each currency in American money by means of a matrix with 1 column. (c) Use matrix multiplication to compute how much the tourist's foreign currency is worth in American money.

3-64   One day, an ice cream shop sold 80 ice cream sodas and 150 milk shakes. The ingredients in a soda are 1 ounce of syrup, 1 ounce of milk, 3 ounces of soda water, and 4 ounces of ice cream. The recipe for a milk shake is to blend together 1 ounce of syrup, 4 ounces of milk, and 3 ounces of ice cream. (a) Express the syrup, milk, and ice cream in each end product by means of a matrix. (b) Express the number of sodas and milk shakes sold that day as a matrix with 1 column. (c) Use matrix multiplication to compute the amount of syrup, milk, and ice cream used by the shop that day.

3-65   A family consists of two adults, one teenager, and three young children. Each adult consumes 1/5 loaf of bread, no milk, 1/10 pound of coffee, and 1/8 pound of cheese in an average day. The teenager eats 2/5 loaf of bread, drinks 1 quart of milk but no coffee, and eats 1/8 pound of cheese. Each child eats 1/5 loaf of bread, drinks 1/2 quart of milk and no coffee, and eats 1/16 pound of cheese. (a) Express the daily consumption of bread, milk, coffee, and cheese by the various types of family members using a matrix. (b) Express the number of family members of the various types by means of a matrix with 1 column. (c) Use matrix multiplication to calculate the total amount of bread, milk, coffee, and cheese consumed by this family in an average day.

3-66   Write the $n \times n$ identity matrix when (a) n is 4, (b) n is 5, (c) n is 6.

● *In Exercises 3-67 and 3-68, carry out the indicated multiplication, showing your work in detail.*

3-67 $\begin{bmatrix} 1 & 0 & 0 \\ 0 & 1 & 0 \\ 0 & 0 & 1 \end{bmatrix} \begin{bmatrix} 1 & 2 \\ -1 & 0 \\ 2 & 3 \end{bmatrix}$

3-68 $\begin{bmatrix} 2 & -1 \end{bmatrix} \begin{bmatrix} 1 & 0 \\ 0 & 1 \end{bmatrix}$

● *For the matrices of Exercises 3-69 through 3-72, compute both $AB + AC$ and $A(B + C)$.*

3-69 $A = \begin{bmatrix} 2 & -1 & 0 \\ 1 & 2 & 3 \end{bmatrix}$ $B = \begin{bmatrix} -1 & -1 \\ 0 & 1 \\ 0 & 2 \end{bmatrix}$ $C = \begin{bmatrix} 0 & -4 \\ 3 & 2 \\ 1 & 0 \end{bmatrix}$

3-70 $A = \begin{bmatrix} 2 \\ 1 \\ 3 \end{bmatrix}$ $B = \begin{bmatrix} -1 & 0 \end{bmatrix}$ $C = \begin{bmatrix} 1 & 2 \end{bmatrix}$

3-71 $A = \begin{bmatrix} -1 & \dfrac{1}{2} \\ 2 & 0 \\ 1 & 1 \end{bmatrix}$ $B = \begin{bmatrix} 1 & 2 \\ -1 & 1 \end{bmatrix}$ $C = \begin{bmatrix} 0 & 0 \\ 0 & 0 \end{bmatrix}$

3-72 $A = \begin{bmatrix} 3 & -1 & 1 \\ 0 & \dfrac{1}{2} & 1 \\ 1 & 2 & 0 \end{bmatrix}$ $B = \begin{bmatrix} 1 \\ 2 \\ -1 \end{bmatrix}$ $C = \begin{bmatrix} -1 \\ 2 \\ 1 \end{bmatrix}$

●   In Exercises 3-73 through 3-78, compute as efficiently as possible.

3-73 $\begin{bmatrix} -1 & 0 \\ 2 & 3 \end{bmatrix} \begin{bmatrix} -1 & 2 & 1 \\ 1 & 0 & 1 \end{bmatrix} + \begin{bmatrix} 0 & 1 \\ -1 & -2 \end{bmatrix} \begin{bmatrix} -1 & 2 & 1 \\ 1 & 0 & 1 \end{bmatrix}$

3-74 $\begin{bmatrix} 2 & -1 & 1 & 2 \\ 3 & 1 & 4 & 0 \end{bmatrix} \begin{bmatrix} 1 \\ 2 \\ 0 \\ 1 \end{bmatrix} - \begin{bmatrix} 2 & -1 & 1 & 2 \\ 3 & 1 & 4 & 0 \end{bmatrix} \begin{bmatrix} 2 \\ 1 \\ 1 \\ 2 \end{bmatrix}$

3-75 $\begin{bmatrix} 1 \\ -1 \end{bmatrix} \begin{bmatrix} 2 & -1 & 0 \end{bmatrix} + \begin{bmatrix} 1 \\ -1 \end{bmatrix} \begin{bmatrix} -2 & 1 & 0 \end{bmatrix}$

3-76 $\begin{bmatrix} 2 & -1 & \dfrac{1}{2} \end{bmatrix} \begin{bmatrix} 2 \\ 1 \\ 0 \end{bmatrix} - \begin{bmatrix} 2 & -1 & \dfrac{1}{2} \end{bmatrix} \begin{bmatrix} 3 \\ 0 \\ 2 \end{bmatrix}$

3-77 $\begin{bmatrix} 0 & 1 & 1 & 3 \\ -1 & 1 & 2 & 1 \end{bmatrix} \begin{bmatrix} 2 & -1 \\ 1 & 2 \\ 0 & 1 \\ 0 & 0 \end{bmatrix} + \begin{bmatrix} 0 & 1 & 1 & 3 \\ -1 & 1 & 2 & 1 \end{bmatrix} \begin{bmatrix} 0 & 0 \\ 1 & 2 \\ -1 & 1 \\ 0 & 0 \end{bmatrix}$

3-78 $\begin{bmatrix} 2 & -1 \\ 0 & 2 \\ 1 & 3 \\ -1 & 2 \end{bmatrix} \begin{bmatrix} 1 & -1 \\ 2 & 0 \end{bmatrix} - \begin{bmatrix} 0 & 0 \\ 0 & 0 \\ 0 & 0 \\ 0 & 0 \end{bmatrix} \begin{bmatrix} 1 & -1 \\ 2 & 0 \end{bmatrix}$

●   In Exercises 3-79 through 3-82, rewrite as a product of matrices and then multiply.

3-79 $\begin{bmatrix} -1 & 2 & 3 \end{bmatrix} \begin{bmatrix} 2 & 1 & -1 \\ 0 & 1 & 2 \\ 0 & 0 & 1 \end{bmatrix} + \begin{bmatrix} -1 & 2 & 3 \end{bmatrix}$

3-80 $\begin{bmatrix} 3 & 2 \\ -1 & 0 \end{bmatrix}\begin{bmatrix} -1 & 0 \\ 0 & 1 \end{bmatrix} + \begin{bmatrix} 3 & 2 \\ -1 & 0 \end{bmatrix}$

3-81 $\begin{bmatrix} -2 & 1 \\ 3 & 4 \end{bmatrix}\begin{bmatrix} -2 \\ 1 \end{bmatrix} - \begin{bmatrix} -2 \\ 1 \end{bmatrix}$

3-82 $\begin{bmatrix} 2 & 1 & 3 \\ 1 & -1 & 0 \end{bmatrix} - \begin{bmatrix} 2 & 1 & 3 \\ 1 & -1 & 0 \end{bmatrix}\begin{bmatrix} -1 & 0 \\ 0 & -1 \end{bmatrix}$

3-4 The Inverse of a Matrix

We may view an ordinary division problem as a multiplication problem. Instead of dividing 3 into 9, writing $9 \div 3$ or $9/3$, we may multiply 9 by $1/3$; that is,

$$\frac{9}{3} = \left(\frac{1}{3}\right)9$$

The number $1/3$ is called the **reciprocal** or **inverse** of 3. Arithmetically, the relationship between the number 3 and its inverse is given by

$$3\left(\frac{1}{3}\right) = \left(\frac{1}{3}\right)3 = 1$$

For a general algebraic formula, we write

$$b \div a = (a^{-1})b$$

where a^{-1} (read "a inverse") is that number for which

$$a(a^{-1}) = (a^{-1})a = 1$$

Thus, when a is 3, then a^{-1} is $1/3$.

We can, to some extent, carry this point of view over to the subject of matrices. Specifically, a matrix A is said to have an inverse, written A^{-1}, if the matrix A^{-1} has the property that

$$A(A^{-1}) = (A^{-1})A = I$$

(Recall that the identity matrix I behaves like the number 1.)

There is, however, no concept of division in matrix arithmetic. In ordinary arithmetic, we can define $b \div a$ either as $(a^{-1})b$ or as $b(a^{-1})$, because $(a^{-1})b = b(a^{-1})$. Such a definition cannot be used with matrices, as the following example shows: Let

$$A = \begin{bmatrix} 2 & 1 \\ 1 & 1 \end{bmatrix}$$

Then

$$A^{-1} = \begin{bmatrix} 1 & -1 \\ -1 & 2 \end{bmatrix}$$

You may check this by multiplying

$$\begin{bmatrix} 2 & 1 \\ 1 & 1 \end{bmatrix}\begin{bmatrix} 1 & -1 \\ -1 & 2 \end{bmatrix} \qquad \text{and} \qquad \begin{bmatrix} 1 & -1 \\ -1 & 2 \end{bmatrix}\begin{bmatrix} 2 & 1 \\ 1 & 1 \end{bmatrix}$$

Now, let

$$B = \begin{bmatrix} 2 & -1 \\ -2 & 0 \end{bmatrix}$$

Then

$$(A^{-1})B = \begin{bmatrix} 1 & -1 \\ -1 & 2 \end{bmatrix}\begin{bmatrix} 2 & -1 \\ -2 & 0 \end{bmatrix} = \begin{bmatrix} 4 & -1 \\ -6 & 1 \end{bmatrix}$$

while

$$B(A^{-1}) = \begin{bmatrix} 2 & -1 \\ -2 & 0 \end{bmatrix}\begin{bmatrix} 1 & -1 \\ -1 & 2 \end{bmatrix} = \begin{bmatrix} 3 & -4 \\ -2 & 2 \end{bmatrix}$$

We conclude that, in general, $(A^{-1})B$ and $B(A^{-1})$ need not be the same matrix.

It can even happen that one of the products $(A^{-1})B$ and $B(A^{-1})$ is defined and the other is not. For example, if A^{-1} is a 2×2 matrix like the one above, and B is a 2×4 matrix, then $(A^{-1})B$ makes sense but $B(A^{-1})$ does not.

Although, as we have seen, an attempt to define division of matrices by analogy with ordinary arithmetic fails, we will find that the concept of the inverse of a matrix is still very useful. Consequently, we must next learn more about it.

Since A^{-1} must satisfy the equations

$$A(A^{-1}) = (A^{-1})A = I$$

then both of the products must be defined. If A is an $m \times n$ matrix, then, for $A(A^{-1})$ to exist, A^{-1} must be an $n \times s$ matrix for some s. On the other hand, for $(A^{-1})A$ to exist, s must be the same as m, which means that A^{-1} is an $n \times m$ matrix. We conclude that $A(A^{-1})$ is $n \times n$, while $(A^{-1})A$ is $m \times m$. Finally, the equation $A(A^{-1}) = (A^{-1})A$ tells us that $A(A^{-1})$ and $(A^{-1})A$ are the same size, so in fact $m = n$ and we see that both A and A^{-1} must be $n \times n$ matrices. We have demonstrated that A can have an inverse only if A is square.

However, just as the equations

$$a(a^{-1}) = (a^{-1})a = 1$$

have no solution in ordinary arithmetic when a is zero, so A^{-1} does not exist even for some square matrices. It should not come as a surprise that there is no inverse for A when A is a square zero matrix. In fact, the product of a zero matrix and any other matrix is again a zero matrix (compare $0 \cdot x = 0$ for all numbers x). Thus, when A is a zero matrix, $A(A^{-1}) = I$ is impossible for any matrix A^{-1}, because the product of A and any matrix is a zero matrix, not an identity matrix. Moreover, even some matrices that do not contain any zeros, such as

$$A = \begin{bmatrix} 1 & 2 \\ 3 & 6 \end{bmatrix}$$

can be shown not to possess an inverse. Such matrices are called **singular,** while a matrix that does have an inverse is called a **nonsingular** matrix.

One use of division in ordinary arithmetic is in the solution of an equation like

$$4x = 16$$

where division of both sides of the equation by 4 will produce x. Thus, we have $x = 16/4$, or, in our preferred terminology,

$$x = \left(\frac{1}{4}\right)16 = (4)^{-1}(16)$$

More generally, we can solve equations of the form

$$ax = b$$

in ordinary arithmetic, provided $a \neq 0$, to obtain the solution

$$x = (a^{-1})b$$

In just the same way, if A is a nonsingular matrix and we are given a matrix B with the same number of rows as A, we can solve an equation of the form

$$AX = B$$

for an unknown matrix X. Multiplying both sides of the equation on the left by A^{-1}, we obtain

$$(A^{-1})AX = (A^{-1})B$$

We know that $(A^{-1})A = I$, so

$$(A^{-1})AX = IX = X$$

by the property of I. Thus, we have the solution

$$X = (A^{-1})B$$

3-22 **Example** Solve the equation $AX = B$ when

$$A = \begin{bmatrix} 2 & -1 \\ 1 & 1 \end{bmatrix} \quad \text{and} \quad B = \begin{bmatrix} 0 & 2 & -1 \\ 1 & -2 & 1 \end{bmatrix}$$

given that

$$A^{-1} = \begin{bmatrix} \dfrac{1}{3} & \dfrac{1}{3} \\ -\dfrac{1}{3} & \dfrac{2}{3} \end{bmatrix}$$

Solution

$$X = (A^{-1})B = \begin{bmatrix} \dfrac{1}{3} & \dfrac{1}{3} \\ -\dfrac{1}{3} & \dfrac{2}{3} \end{bmatrix} \begin{bmatrix} 0 & 2 & -1 \\ 1 & -2 & 1 \end{bmatrix} = \begin{bmatrix} \dfrac{1}{3} & 0 & 0 \\ \dfrac{2}{3} & -2 & 1 \end{bmatrix}$$

We should check that

$$AX = \begin{bmatrix} 2 & -1 \\ 1 & 1 \end{bmatrix} \begin{bmatrix} \dfrac{1}{3} & 0 & 0 \\ \dfrac{2}{3} & -2 & 1 \end{bmatrix} = \begin{bmatrix} 0 & 2 & -1 \\ 1 & -2 & 1 \end{bmatrix} = B$$

3-23 **Example** A summer camp has two types of housing: cabins and bunkhouses. A cabin holds one counselor and eight campers, while a bunkhouse accommodates two counselors and twenty campers. If the camp has fifteen counselors and 132 campers and each cabin and bunkhouse is full, how many cabins and how many bunkhouses are there in the camp? [*Hint:* The inverse of the matrix

$$\begin{bmatrix} 1 & 2 \\ 8 & 20 \end{bmatrix} \quad \text{is} \quad \begin{bmatrix} 5 & -\dfrac{1}{2} \\ -2 & \dfrac{1}{4} \end{bmatrix}$$

Solution Let c be the number of cabins in the camp and b the number of bunkhouses, since these are unknown. There is one counselor in each cabin and two in each bunkhouse, so there are $(1)c + (2)b = c + 2b$ counselors in all. We are told that there are fifteen counselors in the camp, so it must be that

$$c + 2b = 15$$

Similarly, in each cabin there are eight campers and in each bunkhouse twenty campers, so there are $8c + 20b$ campers in all. By the condition of the problem, we then know that

$$8c + 20b = 132$$

Write the unknowns c and b as a matrix thus:

$$X = \begin{bmatrix} c \\ b \end{bmatrix}$$

and let

$$A = \begin{bmatrix} 1 & 2 \\ 8 & 20 \end{bmatrix}$$

Observe that

$$AX = \begin{bmatrix} 1 & 2 \\ 8 & 20 \end{bmatrix} \begin{bmatrix} c \\ b \end{bmatrix} = \begin{bmatrix} c + 2b \\ 8c + 20b \end{bmatrix}$$

the left-hand sides of the two equations above (compare Section 3-3). Since

$$c + 2b = 15$$

$$8c + 20b = 132$$

if we set

$$b = \begin{bmatrix} 15 \\ 132 \end{bmatrix}$$

then the two equations above become the single matrix equation

$$AX = \begin{bmatrix} 1 & 2 \\ 8 & 20 \end{bmatrix} \begin{bmatrix} c \\ b \end{bmatrix} = \begin{bmatrix} 15 \\ 132 \end{bmatrix} = B$$

The problem is then to compute

$$X = \begin{bmatrix} c \\ b \end{bmatrix}$$

so we use the solution $X = (A^{-1})B$ that we found to the matrix equation $AX = B$ above. The statement of the problem informs us that

$$A^{-1} = \begin{bmatrix} 5 & -\dfrac{1}{2} \\ -2 & \dfrac{1}{4} \end{bmatrix}$$

and therefore

$$\begin{bmatrix} c \\ b \end{bmatrix} = X = A^{-1}B = \begin{bmatrix} 5 & -\dfrac{1}{2} \\ -2 & \dfrac{1}{4} \end{bmatrix} \begin{bmatrix} 15 \\ 132 \end{bmatrix} = \begin{bmatrix} 75 - 66 \\ -30 + 33 \end{bmatrix} = \begin{bmatrix} 9 \\ 3 \end{bmatrix}$$

By the definition of matrix equality, this means that there are nine cabins and three bunk-houses in the camp.

Summary

The **inverse** A^{-1} of a **nonsingular** matrix A is a matrix such that

$$A(A^{-1}) = (A^{-1})A = I$$

If A is a nonsingular matrix, then the equation $AX = B$ can be solved for the unknown matrix X; the solution is $X = (A^{-1})B$.

Exercises

● In Exercises 3-83 and 3-84, verify that $(A^{-1})A = A(A^{-1}) = I$.

3-83 $\quad A = \begin{bmatrix} -1 & 2 \\ 0 & 4 \end{bmatrix} \qquad A^{-1} = \begin{bmatrix} -1 & \dfrac{1}{2} \\ 0 & \dfrac{1}{4} \end{bmatrix}$

3-84 $\quad A = \begin{bmatrix} 2 & 1 & 3 \\ 0 & 1 & 2 \\ 1 & 0 & 0 \end{bmatrix} \qquad A^{-1} = \begin{bmatrix} 0 & 0 & 1 \\ -2 & 3 & 4 \\ 1 & -1 & -2 \end{bmatrix}$

● Solve for x in Exercises 3-85 and 3-86.

3-85 $\quad A = \begin{bmatrix} -1 & -2 \\ -3 & 2 \end{bmatrix} \qquad A^{-1} = \begin{bmatrix} -\dfrac{1}{4} & -\dfrac{1}{4} \\ -\dfrac{3}{8} & x \end{bmatrix}$

3-86 $\quad A = \begin{bmatrix} 0 & 0 & 1 \\ 2 & 1 & 2 \\ 1 & -2 & -2 \end{bmatrix} \qquad A^{-1} = \begin{bmatrix} x & \dfrac{2}{5} & \dfrac{1}{5} \\ -\dfrac{6}{5} & \dfrac{1}{5} & x \\ 1 & 0 & 0 \end{bmatrix}$

3-87 Solve $AX = B$ when

$$A = \begin{bmatrix} 2 & 0 \\ 1 & 1 \end{bmatrix} \qquad B = \begin{bmatrix} 1 & 2 \\ 3 & 4 \end{bmatrix} \qquad A^{-1} = \begin{bmatrix} \dfrac{1}{2} & 0 \\ -\dfrac{1}{2} & 1 \end{bmatrix}$$

3-88 Solve $XA = B$ when

$$A = \begin{bmatrix} -1 & 2 & 0 \\ 2 & 1 & 1 \\ 1 & 0 & 1 \end{bmatrix} \qquad B = \begin{bmatrix} -2 & 1 & 0 \\ 3 & 2 & 0 \end{bmatrix} \qquad A^{-1} = \begin{bmatrix} -\dfrac{1}{3} & \dfrac{2}{3} & -\dfrac{2}{3} \\ \dfrac{1}{3} & \dfrac{1}{3} & -\dfrac{1}{3} \\ \dfrac{1}{3} & -\dfrac{2}{3} & \dfrac{5}{3} \end{bmatrix}$$

3-89 Solve $AX = B$ when

$$A = \begin{bmatrix} -1 & 0 & 2 \\ 0 & 0 & -1 \\ 2 & 1 & 0 \end{bmatrix} \qquad B = \begin{bmatrix} -2 \\ 1 \\ 2 \end{bmatrix} \qquad A^{-1} = \begin{bmatrix} -1 & -2 & 0 \\ 2 & 4 & 1 \\ 0 & -1 & 0 \end{bmatrix}$$

3-90 Solve $AX = B + C$ when

$$A = \begin{bmatrix} 2 & 4 \\ 3 & 1 \end{bmatrix} \quad B = \begin{bmatrix} 1 & 2 & 3 \\ -1 & -2 & -4 \end{bmatrix} \quad C = \begin{bmatrix} -2 & -1 & -3 \\ 2 & 1 & 2 \end{bmatrix} \quad A^{-1} = \begin{bmatrix} -\dfrac{1}{10} & \dfrac{2}{5} \\ \dfrac{3}{10} & -\dfrac{1}{5} \end{bmatrix}$$

3-91 Solve $AX + B = C$ when

$$A = \begin{bmatrix} 2 & 0 & 1 \\ -2 & 2 & 0 \\ 2 & 1 & 0 \end{bmatrix} \quad B = \begin{bmatrix} -1 \\ -2 \\ 0 \end{bmatrix} \quad C = \begin{bmatrix} 4 \\ 0 \\ 1 \end{bmatrix} \quad A^{-1} = \begin{bmatrix} 0 & -\dfrac{1}{6} & \dfrac{1}{3} \\ 0 & \dfrac{1}{3} & \dfrac{1}{3} \\ 1 & \dfrac{1}{3} & -\dfrac{2}{3} \end{bmatrix}$$

3-92 Solve $B - AX = C$ when

$$A = \begin{bmatrix} -1 & -1 & -1 \\ 0 & 2 & 1 \\ 1 & 0 & 1 \end{bmatrix} \qquad B = \begin{bmatrix} 1 & 2 & 1 & 0 \\ 3 & 1 & 2 & -1 \\ -1 & -1 & 2 & 1 \end{bmatrix}$$

$$C = \begin{bmatrix} 0 & 0 & 0 & 1 \\ 2 & 0 & 0 & 2 \\ 1 & 1 & 0 & 0 \end{bmatrix} \qquad A^{-1} = \begin{bmatrix} -2 & -1 & -1 \\ -1 & 0 & -1 \\ 2 & 1 & 2 \end{bmatrix}$$

3-93   A factory for the construction of quality furniture has two divisions: a machine shop where the parts of the furniture are fabricated and an assembly and finishing division where the parts are put together into the finished product. Suppose there are twelve employees in the machine shop and twenty in the assembly and finishing division and that each employee works an 8 hour day. Suppose further that the factory produces only two products: chairs and tables. A chair requires 384/17 man-hours of machine shop time and 480/17 man-hours of assembly and finishing time. A table requires 240/17 man-hours of machine shop time and 640/17 man-hours of assembly and finishing time. Assuming that there is unlimited demand for these products and that the manufacturer wishes to keep all employees busy, how many chairs and how many tables can this factory produce each day? [Hint: The inverse of

$$\begin{bmatrix} \dfrac{384}{17} & \dfrac{240}{17} \\[2ex] \dfrac{480}{17} & \dfrac{640}{17} \end{bmatrix} \quad \text{is} \quad \begin{bmatrix} \dfrac{1}{12} & -\dfrac{1}{32} \\[2ex] -\dfrac{1}{16} & \dfrac{1}{20} \end{bmatrix}$$

3-94   A farmer feeds his cattle on a mixture of three standard feeds which we will call type A, type B, and type C. Suppose that a standard unit of type A feed supplies a steer with 10% of the calories, 10% of the protein, and 5% of the carbohydrates it needs each day. Similarly, type B supplies 10% of the calories and 5% of the protein but no carbohydrates, and type C has 5% of the calories, 5% of the protein, and 10% of the carbohydrates. How many units of each type of feed should the farmer give a steer each day so that it gets 100% of the amount of calories, protein, and carbohydrates it requires? [Hint: The inverse of

$$\begin{bmatrix} 10 & 10 & 5 \\ 10 & 5 & 5 \\ 5 & 0 & 10 \end{bmatrix} \quad \text{is} \quad \begin{bmatrix} -\dfrac{2}{15} & \dfrac{4}{15} & -\dfrac{1}{15} \\[2ex] \dfrac{1}{5} & -\dfrac{1}{5} & 0 \\[2ex] \dfrac{1}{15} & -\dfrac{2}{15} & \dfrac{2}{15} \end{bmatrix}$$

3-95   A witch's magic cupboard contains 10 ounces of ground four-leaf clovers and 14 ounces of powdered mandrake root. The cupboard will replenish itself automatically provided she uses up exactly all her supplies. A batch of love potion requires $3\frac{1}{13}$ ounces of ground four-leaf clovers and $2\frac{2}{13}$ ounces of powdered mandrake root. One recipe of a well-known (to witches) cure for the common cold requires $5\frac{5}{13}$ ounces of four-leaf clovers and $10\frac{10}{13}$ ounces of mandrake root. How much of the love potion and the cold remedy should the witch make in order to use up the supply in the cupboard exactly? [Hint: The inverse of

$$\begin{bmatrix} 3\frac{1}{13} & 5\frac{5}{13} \\ 2\frac{2}{13} & 10\frac{10}{13} \end{bmatrix} \quad \text{is} \quad \begin{bmatrix} \dfrac{1}{2} & -\dfrac{1}{4} \\ -\dfrac{1}{10} & \dfrac{1}{7} \end{bmatrix}$$

3-96   A company employs clerks, secretaries, and receptionists. According to the standards for each job, a clerk should be able to file 30 papers and type 4 pages in 1 hour; a secretary should file 10 papers, type 10 pages, and process 6 telephone calls in 1 hour; and a receptionist, who neither files nor types, should handle 20 telephone calls in 1 hour. The company has 560 papers to file, 248 pages of typing to be done, and 140 telephone calls in 1 hour. How many clerks, secretaries, and receptionists does it need to transact its business? [*Hint:* The inverse of

$$\begin{bmatrix} 30 & 10 & 0 \\ 4 & 10 & 0 \\ 0 & 6 & 20 \end{bmatrix} \quad \text{is approximately} \quad \begin{bmatrix} 0.0385 & -0.0385 & 0 \\ -0.0154 & 0.1154 & 0 \\ 0.0046 & -0.0346 & 0.05 \end{bmatrix}.$$

3-5   Computing the Inverse

Because the definition of matrix multiplication is so complicated, the relationship between the numbers in a nonsingular matrix A and those in its inverse A^{-1} is even more complicated; so complicated, in fact, that it is not practical to write down a formula for finding A^{-1}. Instead, A^{-1} is computed from A by means of a procedure which may be thought of as a rather tedious game.

The first step consists of adjoining to the right of the nonsingular, and thus square, matrix A the identity matrix of the same size to form what is called the **augmented matrix.**

3-24   **Example**   Write down the augmented matrices corresponding to

(a) $\begin{bmatrix} -1 & 2 \\ 0 & 4 \end{bmatrix}$

(b) $\begin{bmatrix} -1 & -1 & 2 \\ 0 & 2 & -2 \\ 1 & 2 & 0 \end{bmatrix}$

Solution

(a) $\begin{bmatrix} -1 & 2 & | & 1 & 0 \\ 0 & 4 & | & 0 & 1 \end{bmatrix}$

(b) $\begin{bmatrix} -1 & -1 & 2 & | & 1 & 0 & 0 \\ 0 & 2 & -2 & | & 0 & 1 & 0 \\ 1 & 2 & 0 & | & 0 & 0 & 1 \end{bmatrix}$

The object of the game is to transform the augmented matrix, by means of certain moves we will explain below, into a matrix of the same size in which the identity matrix appears on the *left*. Thus, the augmented matrix of Example 3-24, part (a), is transformed into the matrix

$$\begin{bmatrix} 1 & 0 & | & -1 & \dfrac{1}{2} \\ 0 & 1 & | & 0 & \dfrac{1}{4} \end{bmatrix}$$

and the augmented matrix of part (b) into

$$\begin{bmatrix} 1 & 0 & 0 & | & -\dfrac{2}{3} & -\dfrac{2}{3} & \dfrac{1}{3} \\ 0 & 1 & 0 & | & \dfrac{1}{3} & \dfrac{1}{3} & \dfrac{1}{3} \\ 0 & 0 & 1 & | & \dfrac{1}{3} & -\dfrac{1}{6} & \dfrac{1}{3} \end{bmatrix}$$

The matrix that appears on the right after the moves is the inverse of the original matrix. Schematically, we may write that the object of our game is, given a nonsingular matrix A, to transform the augmented matrix

$$[A \quad | \quad I]$$

into the matrix

$$[I \quad | \quad A^{-1}]$$

There are three kinds of moves which are permitted in the game. The first two types are quite simple, while the last is more complex. The first kind of move we are permitted is to multiply or divide all the numbers in a row of the augmented matrix by the same number, provided the number is not zero. For example, we can multiply the first row of

$$\left[\begin{array}{cc|cc} -1 & 2 & 1 & 0 \\ 0 & 4 & 0 & 1 \end{array}\right]$$

by the number -1 to produce

$$\left[\begin{array}{cc|cc} 1 & -2 & -1 & 0 \\ 0 & 4 & 0 & 1 \end{array}\right]$$

When we determine the inverse of Example 3-24, part (b), at one stage we have the matrix

$$\left[\begin{array}{ccc|ccc} 1 & 1 & -2 & -1 & 0 & 0 \\ 0 & 2 & -2 & 0 & 1 & 0 \\ 0 & 1 & 2 & 1 & 0 & 1 \end{array}\right]$$

and then we divide through the second row by 2 to get

$$\left[\begin{array}{ccc|ccc} 1 & 1 & -2 & -1 & 0 & 0 \\ 0 & 1 & -1 & 0 & \frac{1}{2} & 0 \\ 0 & 1 & 2 & 1 & 0 & 1 \end{array}\right]$$

The second type of allowable move consists of interchanging two rows of the matrix. Thus, interchanging the top two rows of

$$\left[\begin{array}{ccc|ccc} 0 & 2 & -1 & 1 & 0 & 0 \\ 1 & -1 & -1 & 0 & 1 & 0 \\ 2 & 1 & 1 & 0 & 0 & 1 \end{array}\right]$$

we get

$$\left[\begin{array}{ccc|ccc} 1 & -1 & -1 & 0 & 1 & 0 \\ 0 & 2 & -1 & 1 & 0 & 0 \\ 2 & 1 & 1 & 0 & 0 & 1 \end{array}\right]$$

For the third type of move, let us think of each row of the augmented matrix as a matrix with 1 row and many columns. By a **multiple** of a row we mean the matrix obtained by multiplying every number in the row by the same number (other than zero). Then we can add a multiple of a row of the augmented matrix to another row (where addition is performed by the rules of matrix addition) and replace the latter row by the sum. Note that the former row, the one whose multiple we use, is not in any way affected by this operation.

As an example, consider the augmented matrix

$$\left[\begin{array}{ccc|ccc} 1 & 1 & -2 & -1 & 0 & 0 \\ 0 & 1 & -1 & 0 & \dfrac{1}{2} & 0 \\ 0 & 1 & 2 & 1 & 0 & 1 \end{array}\right]$$

We wish to add -1 times the second row to the first row and replace the first row with the sum. Now, -1 times the second row is

$$\left[\begin{array}{ccc|ccc} 0 & -1 & 1 & 0 & -\dfrac{1}{2} & 0 \end{array}\right]$$

We add it to the first row

$$\begin{array}{ll} \begin{bmatrix} 1 & 1 & -2 & \bigm| & -1 & 0 & 0 \end{bmatrix} & \text{(First row)} \\[1em] +\begin{bmatrix} 0 & -1 & 1 & \bigm| & 0 & -\dfrac{1}{2} & 0 \end{bmatrix} & (-1 \text{ times second row}) \\[1.5em] \hline \begin{bmatrix} 1 & 0 & -1 & \bigm| & -1 & -\dfrac{1}{2} & 0 \end{bmatrix} & \end{array}$$

and replace the first row by the sum

$$\left[\begin{array}{ccc|ccc} 1 & 0 & -1 & -1 & -\dfrac{1}{2} & 0 \\ 0 & 1 & -1 & 0 & \dfrac{1}{2} & 0 \\ 0 & 1 & 2 & 1 & 0 & 1 \end{array}\right]$$

Notice that the second row looks just as it did before. We may also add -1 times the second row to the third row

$$\begin{array}{ll} \begin{bmatrix} 0 & 1 & 2 & \bigm| & 1 & 0 & 1 \end{bmatrix} & \text{(Third row)} \\[1em] +\begin{bmatrix} 0 & -1 & 1 & \bigm| & 0 & -\dfrac{1}{2} & 0 \end{bmatrix} & (-1 \text{ times second row}) \\[1.5em] \hline \begin{bmatrix} 0 & 0 & 3 & \bigm| & 1 & -\dfrac{1}{2} & 1 \end{bmatrix} & \end{array}$$

and replace the third row by this new row

$$
\left[\begin{array}{ccc|ccc}
1 & 0 & -1 & -1 & -\dfrac{1}{2} & 0 \\[2mm]
0 & 1 & -1 & 0 & \dfrac{1}{2} & 0 \\[2mm]
0 & 0 & 3 & 1 & -\dfrac{1}{2} & 1
\end{array}\right]
$$

3-25   **Example**   Add 2 times the second row of the following matrix to the first row and replace the first row with the sum:

$$
\left[\begin{array}{cc|cc}
1 & -2 & -1 & 0 \\[2mm]
0 & 1 & 0 & \dfrac{1}{4}
\end{array}\right]
$$

Solution

$$
\begin{array}{l}
\begin{bmatrix} 1 & -2 & | -1 & 0 \end{bmatrix} \quad \text{(First row)} \\[2mm]
+ \begin{bmatrix} 0 & 2 & | \ 0 & \dfrac{1}{2} \end{bmatrix} \quad \text{(2 times second row)} \\[2mm]
\hline
\begin{bmatrix} 1 & 0 & | -1 & \dfrac{1}{2} \end{bmatrix}
\end{array}
$$

and we get

$$
\left[\begin{array}{cc|cc}
1 & 0 & -1 & \dfrac{1}{2} \\[2mm]
0 & 1 & 0 & \dfrac{1}{4}
\end{array}\right]
$$

Now we come to the rules of play by which the allowable moves are used to compute the inverse of a matrix. Recall that a column of an identity matrix, say,

$$
\begin{bmatrix}
1 & 0 & 0 \\
0 & 1 & 0 \\
0 & 0 & 1
\end{bmatrix}
$$

consists of a 1 in the position where the number of the row is the same as the number of the column (which we will call the **diagonal position** in the column) and a 0 in every other position. We wish to transform the left-hand columns of the augmented matrix into columns of an identity matrix, so the rules of play for each column are:

1. Change the number in the diagonal position of the column to a 1.
2. Then change the number in any other position of the column to 0.

These rules of play are applied column-by-column from left to right. It is crucial that the left-to-right rule be obeyed, since otherwise we may undo at some stage the progress we made earlier in turning columns into columns of the identity matrix.

To execute Step 1 of the rules of play, first note whether the number in the diagonal position of the column is 0. If it is 0, interchange the row containing that element with some row below it which contains a nonzero number in the column we are working on. (If all numbers in the column from the diagonal position down are 0, so that it is not possible to carry out this step, the original matrix does not have an inverse—see p. 183.) Thus, the first step in finding the inverse of

$$A = \begin{bmatrix} 0 & 2 \\ -1 & 1 \end{bmatrix}$$

after forming the augmented matrix

$$\left[\begin{array}{cc|cc} 0 & 2 & 1 & 0 \\ -1 & 1 & 0 & 1 \end{array} \right]$$

is to interchange the rows

$$\left[\begin{array}{cc|cc} -1 & 1 & 0 & 1 \\ 0 & 2 & 1 & 0 \end{array} \right]$$

because the number in the diagonal position of the first column in the original matrix was 0. Once there is a nonzero number in the diagonal position, divide the row containing that number by the number itself. This step will change the number in the diagonal position to a 1 and so complete Step 1. Thus, dividing the first row by -1, the matrix above becomes

$$\left[\begin{array}{cc|cc} 1 & -1 & 0 & -1 \\ 0 & 2 & 1 & 0 \end{array} \right]$$

To carry out Step 2 of the rules of play, consider any nonzero number in the column other than the number (now 1) in the diagonal position. Let us call that number a. Add $-a$ times the row containing the diagonal position to the row containing a and replace the row containing a with the sum. The effect of this operation is to change a to 0. Repeat this procedure, in any order, to complete Step 2. In the example of

$$A = \begin{bmatrix} 0 & 2 \\ -1 & 1 \end{bmatrix}$$

we have reached the stage

$$\left[\begin{array}{cc|cc} 1 & -1 & 0 & -1 \\ 0 & 2 & 1 & 0 \end{array}\right]$$

Since the first column already looks like the first column of the 2×2 identity matrix, we turn to the second column. We change the number 2 in the diagonal position of the second column to a 1 by dividing through the second row by 2:

$$\left[\begin{array}{cc|cc} 1 & -1 & 0 & -1 \\ 0 & 1 & \dfrac{1}{2} & 0 \end{array}\right]$$

The only nondiagonal number in the second column is $a = -1$, so $-a = 1$, and we just add the second row to the first

$$\begin{array}{cl} \left[\begin{array}{cc|cc} 1 & -1 & 0 & -1 \end{array}\right] & \text{(First row)} \\[2mm] +\left[\begin{array}{cc|cc} 0 & 1 & \dfrac{1}{2} & 0 \end{array}\right] & \text{(Second row)} \\[2mm] \hline \left[\begin{array}{cc|cc} 1 & 0 & \dfrac{1}{2} & -1 \end{array}\right] & \end{array}$$

and replace the first row by the sum:

$$\left[\begin{array}{cc|cc} 1 & 0 & \dfrac{1}{2} & -1 \\ 0 & 1 & \dfrac{1}{2} & 0 \end{array}\right]$$

to conclude that

$$A^{-1} = \begin{bmatrix} \dfrac{1}{2} & -1 \\ \dfrac{1}{2} & 0 \end{bmatrix}$$

As a check of our work, we should multiply

$$A(A^{-1}) = \begin{bmatrix} 0 & 2 \\ -1 & 1 \end{bmatrix} \begin{bmatrix} \dfrac{1}{2} & -1 \\ \dfrac{1}{2} & 0 \end{bmatrix}$$

to be sure the product is I.

3-26 **Example** Find the inverse of

$$\begin{bmatrix} 2 & -1 \\ 1 & 1 \end{bmatrix}$$

Solution The augmented matrix is

$$\left[\begin{array}{cc|cc} 2 & -1 & 1 & 0 \\ 1 & 1 & 0 & 1 \end{array}\right]$$

and since the number 2 in the diagonal position of the first column is nonzero, we turn it into a 1 by dividing through the first row by 2:

$$\left[\begin{array}{cc|cc} 1 & -\dfrac{1}{2} & \dfrac{1}{2} & 0 \\ 1 & 1 & 0 & 1 \end{array}\right]$$

To make the 1 in the lower left position a 0, we add -1 times the first row to the second row:

$$\begin{array}{ll} \left[\begin{array}{cc|cc} 1 & 1 & 0 & 1 \end{array}\right] & \text{(Second row)} \\[2ex] +\left[\begin{array}{cc|cc} -1 & \dfrac{1}{2} & -\dfrac{1}{2} & 0 \end{array}\right] & (-1 \text{ times first row}) \\[3ex] \hline \left[\begin{array}{cc|cc} 0 & \dfrac{3}{2} & -\dfrac{1}{2} & 1 \end{array}\right] \end{array}$$

and replace the second row by this sum

$$\left[\begin{array}{cc|cc} 1 & -\dfrac{1}{2} & \dfrac{1}{2} & 0 \\ 0 & \dfrac{3}{2} & -\dfrac{1}{2} & 1 \end{array}\right]$$

The number in the diagonal position of the second row is now 3/2, so we divide the second row by 3/2 (which is the same as multiplying by 2/3) in order to change that number to 1:

$$\begin{bmatrix} 1 & -\dfrac{1}{2} & \bigm| & \dfrac{1}{2} & 0 \\[2ex] 0 & 1 & \bigm| & -\dfrac{1}{3} & \dfrac{2}{3} \end{bmatrix}$$

The nondiagonal number in the second column is $a = -1/2$ and since $-a = -(-1/2) = 1/2$, the rules of play direct us to add $1/2$ times the second row to the first row:

$$\begin{aligned}
&\begin{bmatrix} 1 & -\dfrac{1}{2} & \bigm| & \dfrac{1}{2} & 0 \end{bmatrix} && \text{(First row)} \\[2ex]
+\ &\begin{bmatrix} 0 & \dfrac{1}{2} & \bigm| & -\dfrac{1}{6} & \dfrac{1}{3} \end{bmatrix} && \left(\dfrac{1}{2}\text{ times second row}\right) \\[2ex]
&\begin{bmatrix} 1 & 0 & \bigm| & \dfrac{1}{3} & \dfrac{1}{3} \end{bmatrix}
\end{aligned}$$

Replacing the first row by this sum completes the computation of the inverse:

$$\begin{bmatrix} 1 & 0 & \bigm| & \dfrac{1}{3} & \dfrac{1}{3} \\[2ex] 0 & 1 & \bigm| & -\dfrac{1}{3} & \dfrac{2}{3} \end{bmatrix}$$

We can now conclude that the inverse is

$$\begin{bmatrix} \dfrac{1}{3} & \dfrac{1}{3} \\[2ex] -\dfrac{1}{3} & \dfrac{2}{3} \end{bmatrix}$$

Let us see what happens when we apply the rules of play to a matrix which has no inverse. Recall our claim in Section 3-5 that the matrix

$$\begin{bmatrix} 1 & 2 \\ 3 & 6 \end{bmatrix}$$

has no inverse. In attempting to find the inverse, we form the augmented matrix

$$\begin{bmatrix} 1 & 2 & \bigm| & 1 & 0 \\ 3 & 6 & \bigm| & 0 & 1 \end{bmatrix}$$

Since the number in the diagonal position of the first column is already 1, we proceed immediately to the next stage, which is to add -3 times the first row to the second:

$$\begin{array}{r} [3 \quad 6 \mid 0 \quad 1] \quad \text{(Second row)} \\ +[-3 \quad -6 \mid -3 \quad 0] \quad (-3 \text{ times first row}) \\ \hline [0 \quad 0 \mid -3 \quad 1] \end{array}$$

We substitute this for the second row:

$$\left[\begin{array}{cc|cc} 1 & 2 & 1 & 0 \\ 0 & 0 & -3 & 1 \end{array}\right]$$

The next step should be to turn the number in the diagonal position of the second column into a 1. But a 0 occupies that position and there is no number below it in the column to interchange with it, so we cannot continue. This outcome will always occur if the matrix we started with was singular; that is, the rules of play will force us to a point from which it is impossible to go on. The correct conclusion, then, is that the original matrix has no inverse.

3-27 **Example** Compute the inverse of the matrix

$$\left[\begin{array}{ccc} -1 & 0 & 2 \\ 0 & 0 & -1 \\ 2 & 1 & 0 \end{array}\right]$$

Solution We will just list the steps of the process and let the reader supply the reasoning.

$$\left[\begin{array}{ccc|ccc} -1 & 0 & 2 & 1 & 0 & 0 \\ 0 & 0 & -1 & 0 & 1 & 0 \\ 2 & 1 & 0 & 0 & 0 & 1 \end{array}\right]$$

$$\left[\begin{array}{ccc|ccc} 1 & 0 & -2 & -1 & 0 & 0 \\ 0 & 0 & -1 & 0 & 1 & 0 \\ 2 & 1 & 0 & 0 & 0 & 1 \end{array}\right]$$

$$\begin{array}{r} [2 \quad 1 \quad 0 \mid 0 \quad 0 \quad 1] \\ +[-2 \quad 0 \quad 4 \mid 2 \quad 0 \quad 0] \\ \hline [0 \quad 1 \quad 4 \mid 2 \quad 0 \quad 1] \end{array}$$

$$\begin{bmatrix} 1 & 0 & -2 & | & -1 & 0 & 0 \\ 0 & 0 & -1 & | & 0 & 1 & 0 \\ 0 & 1 & 4 & | & 2 & 0 & 1 \end{bmatrix}$$

$$\begin{bmatrix} 1 & 0 & -2 & | & -1 & 0 & 0 \\ 0 & 1 & 4 & | & 2 & 0 & 1 \\ 0 & 0 & -1 & | & 0 & 1 & 0 \end{bmatrix}$$

$$\begin{bmatrix} 1 & 0 & -2 & | & -1 & 0 & 0 \\ 0 & 1 & 4 & | & 2 & 0 & 1 \\ 0 & 0 & 1 & | & 0 & -1 & 0 \end{bmatrix}$$

$$\begin{array}{r} [\; 1 \quad 0 \quad -2 \;|\; -1 \quad 0 \quad 0] \\ +[\; 0 \quad 0 \quad 2 \;|\; 0 \quad -2 \quad 0] \\ \hline [\; 1 \quad 0 \quad 0 \;|\; -1 \quad -2 \quad 0] \end{array}$$

$$\begin{bmatrix} 1 & 0 & 0 & | & -1 & -2 & 0 \\ 0 & 1 & 4 & | & 2 & 0 & 1 \\ 0 & 0 & 1 & | & 0 & -1 & 0 \end{bmatrix}$$

$$\begin{array}{r} [\; 0 \quad 1 \quad 4 \;|\; 2 \quad 0 \quad 1] \\ +[\; 0 \quad 0 \quad -4 \;|\; 0 \quad 4 \quad 0] \\ \hline [\; 0 \quad 1 \quad 0 \;|\; 2 \quad 4 \quad 1] \end{array}$$

$$\begin{bmatrix} 1 & 0 & 0 & | & -1 & -2 & 0 \\ 0 & 1 & 0 & | & 2 & 4 & 1 \\ 0 & 0 & 1 & | & 0 & -1 & 0 \end{bmatrix}$$

So the inverse is

$$\begin{bmatrix} -1 & -2 & 0 \\ 2 & 4 & 1 \\ 0 & -1 & 0 \end{bmatrix}$$

If you compare Examples 3-26 and 3-27, you will observe that the number of individual steps increases substantially when we go from a 2×2 to a 3×3 matrix. As the size of the matrix increases further, the number of individual steps required to find the inverse increases in a spectacular manner. As we will see in Section 3-7, input–output economics requires the inversion of huge matrices, a process that involves an almost unimaginably large number of steps. On the other hand, the rules of play for performing the inversion are quite straightforward, so the reader should have little difficulty believing that computer programs for finding the inverse of a matrix have been written. Thus, a computer with sufficiently fast operating speed and a large enough storage capacity is perfectly capable of computing the inverses of the very large matrices used by economists.

Summary

Given a nonsingular matrix A, its inverse A^{-1} is obtained by transforming the **augmented matrix**

$$[A \mid I]$$

to

$$[I \mid A^{-1}]$$

by means of the following procedures: Work column-by-column and left-to-right. If the number in the **diagonal position** of the column (where the row number equals the column number) is 0, change it to a nonzero number by interchanging the row containing the diagonal position with a lower row. Once a nonzero number occupies this diagonal position, divide through the row containing the diagonal position by the same nonzero number. Then change all the other numbers in the column to zeros as follows: For any other nonzero number a in the column, add $-a$ times the row containing the diagonal position to the row containing a and substitute the sum for the row containing a.

Exercises

● *In Exercises 3-97 through 3-100, write the corresponding augmented matrix.*

3-97 $\begin{bmatrix} 2 & \dfrac{1}{2} \\ 1 & -1 \end{bmatrix}$

3-98 $\begin{bmatrix} -1 & 2 & 0 \\ \dfrac{1}{2} & \dfrac{1}{3} & 2 \\ -1 & 1 & \dfrac{1}{2} \end{bmatrix}$

3-99 $\begin{bmatrix} 1 & -2 & 0 & 1 \\ 0 & 0 & 1 & 2 \\ -3 & 1 & -1 & -1 \\ \dfrac{1}{2} & 2 & 0 & 1 \end{bmatrix}$

3-100 $\begin{bmatrix} 0 & 0 & 1 & 0 & 1 \\ 2 & 1 & 2 & -1 & 0 \\ 3 & 0 & 0 & 0 & 1 \\ -1 & -1 & 2 & 0 & 1 \\ 1 & 0 & 1 & 0 & 0 \end{bmatrix}$

3-101   Interchange the first and third rows.

$$\left[\begin{array}{ccc|ccc} 0 & 1 & -1 & 1 & 0 & 0 \\ 0 & 2 & 0 & 0 & 1 & 0 \\ 2 & 3 & 0 & 0 & 0 & 1 \end{array}\right]$$

3-102   Multiply the second row by -2.

$$\left[\begin{array}{ccc|ccc} 1 & 1 & 3 & 2 & 0 & 0 \\ 0 & -\dfrac{1}{2} & 2 & -1 & 1 & 0 \\ 0 & 2 & -1 & \dfrac{1}{2} & 0 & 1 \end{array}\right]$$

3-103   Interchange the second and fourth rows.

$$\left[\begin{array}{cccc|cccc} 1 & 2 & 1 & 2 & \dfrac{1}{2} & 0 & 0 & 0 \\ 0 & 0 & 2 & 1 & -1 & 2 & 0 & 0 \\ 0 & 0 & 3 & 4 & 3 & 0 & 1 & 0 \\ 0 & 3 & -1 & 2 & 1 & 0 & 0 & 1 \end{array}\right]$$

3-104   Add -2 times the first row to the second row and substitute the sum for the second row.

$$\left[\begin{array}{cc|cc} 1 & 2 & 1 & 0 \\ 2 & 1 & 0 & 1 \end{array}\right]$$

3-105   Add $1/2$ times the second row to the first row and substitute the sum for the first row.

$$\left[\begin{array}{ccc|ccc} 1 & -\dfrac{1}{2} & 3 & -\dfrac{1}{4} & 0 & 0 \\ 0 & 1 & 1 & -\dfrac{1}{4} & 1 & 0 \\ 0 & 2 & -1 & 1 & 0 & 1 \end{array}\right]$$

3-106   Add -3 times the third row to the second row and substitute the sum for the second row.

$$\left[\begin{array}{cccc|cccc} 1 & 0 & 0 & 1 & \dfrac{1}{4} & -\dfrac{1}{2} & 2 & 0 \\ 0 & 1 & 3 & 0 & \dfrac{1}{2} & 2 & 0 & 0 \\ 0 & 0 & 1 & 1 & \dfrac{1}{8} & -\dfrac{1}{2} & 1 & 0 \\ 0 & 0 & 1 & 2 & 1 & -1 & 0 & 1 \end{array}\right]$$

● In Exercises 3-107 through 3-114, a stage of the computation of the inverse of a matrix is shown. State in words what the next step should be, but do not actually carry out the arithmetic.

3-107
$$\left[\begin{array}{ccc|ccc} 1 & 2 & -1 & 1 & 0 & 0 \\ 0 & -1 & -2 & -1 & 1 & 0 \\ 0 & 0 & -1 & 0 & 0 & 1 \end{array}\right]$$

3-108
$$\left[\begin{array}{ccc|ccc} 1 & 1 & 0 & \dfrac{1}{2} & 0 & 0 \\ 0 & 0 & -1 & \dfrac{1}{2} & 1 & 0 \\ 0 & -1 & 0 & 1 & 0 & 1 \end{array}\right]$$

3-109
$$\left[\begin{array}{cc|cc} 1 & -\dfrac{1}{4} & \dfrac{1}{4} & 0 \\ 0 & 1 & -1 & 2 \end{array}\right]$$

3-110
$$\left[\begin{array}{ccc|ccc} 1 & -\dfrac{1}{2} & -\dfrac{1}{2} & \dfrac{1}{2} & 0 & 0 \\ 0 & 1 & \dfrac{1}{2} & 0 & \dfrac{1}{2} & 0 \\ 0 & \dfrac{1}{2} & \dfrac{1}{2} & -\dfrac{1}{2} & 0 & 1 \end{array}\right]$$

3-111
$$\left[\begin{array}{cc|cc} 1 & \dfrac{3}{2} & \dfrac{1}{2} & 0 \\ 0 & -\dfrac{17}{2} & -\dfrac{5}{2} & 0 \end{array}\right]$$

3-112
$$\left[\begin{array}{cccc|cccc} 1 & 2 & -1 & 0 & 1 & 0 & 0 & 0 \\ 0 & \dfrac{1}{2} & 2 & -1 & 0 & 1 & 0 & 0 \\ -\dfrac{1}{3} & 2 & 0 & 0 & 0 & 0 & 1 & 0 \\ 0 & 1 & 1 & 1 & 0 & 0 & 0 & 1 \end{array}\right]$$

3-113
$$\left[\begin{array}{cc|cc} 1 & -\dfrac{2}{5} & \dfrac{1}{5} & 0 \\ 3 & 4 & 0 & 1 \end{array}\right]$$

3-114
$$\left[\begin{array}{cccc|cccc} 1 & 0 & -1 & 0 & 1 & 0 & 0 & 0 \\ 0 & 1 & -1 & 1 & 1 & 1 & 0 & 0 \\ 0 & 0 & 0 & -1 & -1 & -1 & 1 & 0 \\ 0 & 0 & 1 & 2 & -1 & 0 & 0 & 1 \end{array}\right]$$

● *Find the inverses of the matrices in Exercises 3-115 through 3-122.*

3-115
$$\begin{bmatrix} 0 & -2 \\ 1 & 4 \end{bmatrix}$$

3-116
$$\begin{bmatrix} 2 & 3 \\ -1 & -2 \end{bmatrix}$$

3-117
$$\begin{bmatrix} 2 & 0 & 0 \\ 2 & -1 & 0 \\ 0 & -1 & 1 \end{bmatrix}$$

3-118
$$\begin{bmatrix} 2 & -1 & 0 \\ 1 & -1 & -2 \\ 1 & 0 & 1 \end{bmatrix}$$

3-119
$$\begin{bmatrix} 0 & 1 & -1 \\ 2 & -1 & 1 \\ 3 & -2 & -1 \end{bmatrix}$$

3-120
$$\begin{bmatrix} 2 & -1 & 0 \\ 0 & 3 & 1 \\ 2 & 1 & 2 \end{bmatrix}$$

3-121
$$\begin{bmatrix} 1 & -1 & \frac{1}{2} & -3 \\ 0 & 1 & 0 & 3 \\ -1 & 0 & \frac{1}{2} & 0 \\ 0 & 0 & 0 & 3 \end{bmatrix}$$

3-122
$$\begin{bmatrix} 0 & 0 & -2 & 0 \\ -1 & 0 & 3 & -1 \\ 2 & 4 & 0 & 0 \\ 1 & 3 & -3 & 1 \end{bmatrix}$$

3-123   An ice cream shop sells only ice cream sodas and milk shakes. It puts 1 ounce of syrup
and 4 ounces of ice cream in an ice cream soda and 1 ounce of syrup and 3 ounces of ice
cream in a milk shake. If the store used 4 gallons of ice cream and 5 quarts of syrup in
a day, how many ice cream sodas and how many milk shakes did it sell? [*Hint:* 1 quart
= 32 ounces; 1 gallon = 128 ounces.]

3-124   A farmer feeds her cattle a mixture of two types of feed. One standard unit of type *A* feed
supplies a steer with 10% of its minimum daily requirement of protein and 15% of its
requirement of carbohydrates. Type *B* feed contains 12% of the requirement of protein
and 8% of the requirement of carbohydrates in a standard unit. If the farmer wishes to
feed her cattle exactly 100% of their minimum daily requirement of protein and carbohydrates, how many units of each type of feed should she give a steer each day?

3-125   A large corporation pays its vice-presidents $100,000 a year in salary, 100 shares of stock,
and an entertainment allowance of $20,000. A division manager receives $70,000 in salary,
50 shares of stock, and $5000 for official entertainment. The assistant manager of a divi-
sion receives $40,000 in salary, but neither stock nor entertainment allowance. If the
corporation pays out $1,600,000 in salaries, 1000 shares of stock, and $150,000 in expense
allowances to its vice-presidents, division managers, and assistant division managers in
a year, how many vice-presidents, division managers, and assistant division managers
does the company have?

3-126   An automobile service station employs mechanics and station attendants. Each works
8 hours a day. An attendant only pumps gas, while mechanics are expected to spend
3/4 of their time repairing automobiles and 1/4 of their time pumping gas. Suppose it
takes 1/10 man-hour to service an automobile that comes in for gas. If the service station
owner wants to be able to sell gas to 320 cars a day and have 24 man-hours of mechanics'
time available for repair work, how many attendants and how many mechanics should
be hired?

3-6   Matrix Equations

A historian, studying an important voyage of exploration, is unable to find any record
of the number of seamen and the number of officers the ship carried on that voyage.
However, the historian does discover that religious medals, one for each participant in
the voyage, were donated by a religious charity and that 40 such medals were donated.
Further, the quartermaster on the voyage requested that enough wine be carried on board
to allow 348 ounces per day. The historian estimates that, at that time, seamen were allowed a ration of 8 ounces of wine a day and officers were allowed 12 ounces. A final
piece of information uncovered by the historian is that the owner of the ship had been
required to put on deposit 2 shillings for each seaman and 5 shillings for each officer as
a fund to help the dependents of any men that died on the voyage and that the amount
put on deposit totalled 101 shillings.

We will now see how the historian can use this information together with some matrix techniques to calculate the number of seamen and the number of officers who made the voyage. The historian wishes to know x, the number of seamen, and y, the number of officers. We thus have a matrix of unknowns,

$$X = \begin{bmatrix} x \\ y \end{bmatrix}$$

Since 40 medals were donated for the voyage, one for each participant, we conclude that 40 people in all were on board; that is,

$$x + y = 40$$

Each seaman received 8 ounces of wine a day which required $8x$ ounces for all the seamen and, similarly, $12y$ ounces a day were required for the officers' wine allowance. The daily wine consumption of the ship was 348 ounces, so we conclude that

$$8x + 12y = 348$$

Applying the same reasoning to the information about the amount of money on deposit produces the equation

$$2x + 5y = 101$$

If we let

$$A = \begin{bmatrix} 1 & 1 \\ 8 & 12 \\ 2 & 5 \end{bmatrix} \quad \text{and} \quad B = \begin{bmatrix} 40 \\ 348 \\ 101 \end{bmatrix}$$

then

$$AX = \begin{bmatrix} 1 & 1 \\ 8 & 12 \\ 2 & 5 \end{bmatrix} \begin{bmatrix} x \\ y \end{bmatrix} = \begin{bmatrix} x + y \\ 8x + 12y \\ 2x + 5y \end{bmatrix}$$

Just as in Section 3-4, the three equations

$$x + y = 40$$
$$8x + 12y = 348$$
$$2x + 5y = 101$$

are therefore equivalent to the single matrix equation $AX = B$. Thus, the solution to the historian's problem amounts to solving the matrix equation $AX = B$ for X.

In Section 3-4, we solved the matrix equation $AX = B$ for X when A was nonsingular by finding the inverse A^{-1} of A and then letting $X = (A^{-1})B$. This solution cannot work for the historian's problem, because the matrix A is not square—it is 3×2—and thus has no inverse. However, we shall see that a modification of the method we described in Section 3-5 for finding the inverse of a nonsingular matrix can be used to analyze all matrix equations of the form $AX = B$ and, in particular, solve the problem posed by the historian.

We can think of finding an inverse A^{-1} as the solution to the matrix equation

$$A(A^{-1}) = I$$

for the unknown matrix A^{-1}. The procedure of Section 3-5 was to use the augmented matrix

$$[A \mid I]$$

and then convert the left-hand side to the identity matrix to obtain

$$[I \mid A^{-1}]$$

Now, in order to solve the matrix equation

$$AX = B$$

we begin with the corresponding augmented matrix

$$[A \mid B]$$

So the historian should begin with the augmented matrix

$$\begin{bmatrix} 1 & 1 & \vline & 40 \\ 8 & 12 & \vline & 348 \\ 2 & 5 & \vline & 101 \end{bmatrix}$$

Of course, it will be impossible to turn the part of the matrix to the left of the line into an identity matrix, because it is the wrong size, but let us begin as if we were setting out to do just this. The first goal will be to use the admissible moves we learned in Section 3-5 to convert the first column to the form

$$\begin{bmatrix} 1 \\ 0 \\ 0 \end{bmatrix}$$

Adding -8 times the first row of the augmented matrix to the second row of the matrix gives us

$$\left[\begin{array}{cc|c} 1 & 1 & 40 \\ 0 & 4 & 28 \\ 2 & 5 & 101 \end{array}\right]$$

Similarly, we add -2 times the first row to the third row:

$$\left[\begin{array}{cc|c} 1 & 1 & 40 \\ 0 & 4 & 28 \\ 0 & 3 & 21 \end{array}\right]$$

Next, we go through the steps that will turn the second column into the second column of a 3×3 identity matrix. We divide the second row of the augmented matrix by 4 to produce

$$\left[\begin{array}{cc|c} 1 & 1 & 40 \\ 0 & 1 & 7 \\ 0 & 3 & 21 \end{array}\right]$$

and then add appropriate multiples of the second row to the other two rows and end up with

$$\left[\begin{array}{cc|c} 1 & 0 & 33 \\ 0 & 1 & 7 \\ 0 & 0 & 0 \end{array}\right]$$

We started with an augmented matrix of the form

$$[A \mid B]$$

and by means of the moves described in Section 3-5, produced an augmented matrix

$$[A' \mid B']$$

where

$$A' = \begin{bmatrix} 1 & 0 \\ 0 & 1 \\ 0 & 0 \end{bmatrix} \qquad B' = \begin{bmatrix} 33 \\ 7 \\ 0 \end{bmatrix}$$

The relationship between the two augmented matrices is described by the following fact from the theory of matrices: If an augmented matrix

$$[A \mid B]$$

is transformed into an augmented matrix

$$[A' \mid B']$$

by means of interchanging rows, multiplying rows by constants, and adding multiples of one row to another row, then the solution X to the matrix equation

$$AX = B$$

and the solution X' to the matrix equation

$$A'X' = B'$$

are the same matrix, that is, $X = X'$.

Let us apply this fact to the historian's problem. Starting with the augmented matrix

$$[A \mid B] = \begin{bmatrix} 1 & 1 & 40 \\ 8 & 12 & 348 \\ 2 & 5 & 101 \end{bmatrix}$$

we ended up with the augmented matrix

$$[A' \mid B'] = \begin{bmatrix} 1 & 0 & 33 \\ 0 & 1 & 7 \\ 0 & 0 & 0 \end{bmatrix}$$

Then letting

$$X' = \begin{bmatrix} x' \\ y' \end{bmatrix}$$

the matrix equation $A'X' = B'$ is

$$\begin{bmatrix} 1 & 0 \\ 0 & 1 \\ 0 & 0 \end{bmatrix} \begin{bmatrix} x' \\ y' \end{bmatrix} = \begin{bmatrix} 33 \\ 7 \\ 0 \end{bmatrix}$$

Multiplying out the two matrices on the left and then replacing the matrix equation with the corresponding ordinary equations, gives us

$$(1)x' + (0)y' = 33$$
$$(0)x' + (1)y' = 7$$
$$(0)x' + (0)y' = 0$$

which is the same as

$$x' = 33$$
$$y' = 7$$
$$0 = 0$$

Since we know that $X = X'$, we can conclude that $x = 33$ and $y = 7$; the answer to the historian's problem is that there were 33 seamen and 7 officers on the ship. As a check of the answer, we can substitute back into the original equations and find that

$$33 + 7 = 40$$
$$8(33) + 12(7) = 348$$
$$2(33) + 5(7) = 101$$

Now, let us turn to the general situation. We are interested in matrix equations of the form $AX = B$, where A is an $n \times m$ matrix, B is an $n \times 1$ matrix, and we are trying to solve for an $m \times 1$ matrix X of unknowns. The analysis of such an equation is based on the techniques of Section 3-5 just as in the example above. We form the augmented matrix

$$[A \mid B]$$

and use the moves from Section 3-5 to turn the columns of A into the corresponding columns of an $n \times n$ identity matrix to the extent possible. We thus produce a new augmented matrix

$$[A' \mid B']$$

According to the fact from matrix theory stated above, the solution X' to the matrix equation $A'X' = B'$ is the same as the solution X to the original equation $AX = B$. The advantage of this is that the form of A' is so simple that the nature of the solution X is evident.

3-28   **Example**   Solve the equation $AX = B$ where

$$X = \begin{bmatrix} x \\ y \\ z \end{bmatrix} \qquad A = \begin{bmatrix} 1 & 2 & -1 \\ 2 & 3 & 0 \\ 4 & 1 & 2 \\ 2 & -1 & 1 \\ -2 & -1 & 0 \end{bmatrix} \qquad B = \begin{bmatrix} 3 \\ -1 \\ 3 \\ 6 \\ -5 \end{bmatrix}$$

Solution   We form the matrix

$$[A \mid B] = \begin{bmatrix} 1 & 2 & -1 & 3 \\ 2 & 3 & 0 & -1 \\ 4 & 1 & 2 & 3 \\ 2 & -1 & 1 & 6 \\ -2 & -1 & 0 & -5 \end{bmatrix}$$

Since the number in the diagonal position of the first column is already 1, we go right to the next step and add -2 times the first row to the second row:

$$\begin{bmatrix} 1 & 2 & -1 & 3 \\ 0 & -1 & 2 & -7 \\ 4 & 1 & 2 & 3 \\ 2 & -1 & 1 & 6 \\ -2 & -1 & 0 & -5 \end{bmatrix}$$

Adding appropriate multiples of the first row to the remaining three rows converts the first column to the desired form:

$$\begin{bmatrix} 1 & 2 & -1 & 3 \\ 0 & -1 & 2 & -7 \\ 0 & -7 & 6 & -9 \\ 0 & -5 & 3 & 0 \\ 0 & 3 & -2 & 1 \end{bmatrix}$$

Since the number in the second row and second column is -1, we divide through the second row by -1:

$$\begin{bmatrix} 1 & 2 & -1 & 3 \\ 0 & 1 & -2 & 7 \\ 0 & -7 & 6 & -9 \\ 0 & -5 & 3 & 0 \\ 0 & 3 & -2 & 1 \end{bmatrix}$$

Now, adding various multiples of the second row to the other rows, we have

$$\begin{bmatrix} 1 & 0 & 3 & -11 \\ 0 & 1 & -2 & 7 \\ 0 & 0 & -8 & 40 \\ 0 & 0 & -7 & 35 \\ 0 & 0 & 4 & -20 \end{bmatrix}$$

We divide the third row by -8 to put a 1 in the diagonal position of the third column:

$$\begin{bmatrix} 1 & 0 & 3 & -11 \\ 0 & 1 & -2 & 7 \\ 0 & 0 & 1 & -5 \\ 0 & 0 & -7 & 35 \\ 0 & 0 & 4 & -20 \end{bmatrix}$$

The final steps adjust the third column:

$$\begin{bmatrix} 1 & 0 & 0 & 4 \\ 0 & 1 & 0 & -3 \\ 0 & 0 & 1 & -5 \\ 0 & 0 & 0 & 0 \\ 0 & 0 & 0 & 0 \end{bmatrix}$$

This is the required form for the augmented matrix $[A' \mid B']$. The matrix equation $A'X = B'$ is

$$\begin{bmatrix} 1 & 0 & 0 \\ 0 & 1 & 0 \\ 0 & 0 & 1 \\ 0 & 0 & 0 \\ 0 & 0 & 0 \end{bmatrix} \begin{bmatrix} x \\ y \\ z \end{bmatrix} = \begin{bmatrix} 4 \\ -3 \\ -5 \\ 0 \\ 0 \end{bmatrix}$$

This produces the ordinary equations

$$x = 4$$

$$y = -3$$

$$z = -5$$

$$0 = 0$$

$$0 = 0$$

and the solution X to the original equation $AX = B$ is

$$X = \begin{bmatrix} 4 \\ -3 \\ -5 \end{bmatrix}$$

You may have noticed that the last 2 rows of A and B in Example 3-28 and the last row of the A and B matrices of the historian's problem seem not to have made much of a contribution to the solution of the problem, because they just reduce to the obvious statement $0 = 0$. You might wonder whether we could replace the given equation $AX = B$ of Example 3-28 by the simpler matrix equation

$$\begin{bmatrix} 1 & 2 & -1 \\ 2 & 3 & 0 \\ 4 & 1 & 2 \end{bmatrix} \begin{bmatrix} x \\ y \\ z \end{bmatrix} = \begin{bmatrix} 3 \\ -1 \\ 3 \end{bmatrix}$$

that results from simply deleting the last 2 rows of A and B. We could then solve this new equation $AX = B$ by calculating $X = (A^{-1})B$ as in Section 3-5. This procedure would indeed produce the correct answer. Similarly, if the historian had ignored the information concerning the number of shillings put on deposit by the owner of the ship, the matrix equation that resulted would have been of the form $AX = B$, with A nonsingular, and, again, the solution from Section 3-5 would give the correct number of seamen and officers on the ship. Nevertheless, it is a poor idea to throw away information in order to turn the problem into one with a convenient form, as the next paragraphs demonstrate.

Let us now suppose that the historian attempts to apply the same technique to determine the number of seamen and officers on another voyage where this information cannot be obtained from contemporary records. The total number of persons on the voyage is known to be 29, the daily wine allowance for all aboard the ship was 252 ounces, and the monthly payroll was 30,000 maravedis. The historian still assumes the daily wine allowance per man was 8 ounces for seamen and 12 ounces for officers. Contemporary accounts indicate that the usual salary for a seaman was 1000 maravedis a month and that officers received twice that amount. Therefore, the equations for this voyage, with x the number of seamen and y the number of officers, are

$$x + y = 29$$

$$8x + 12y = 252$$

$$1000x + 2000y = 30{,}000$$

In matrix form, we have the equation $AX = B$, where

$$A = \begin{bmatrix} 1 & 1 \\ 8 & 12 \\ 1000 & 2000 \end{bmatrix} \qquad B = \begin{bmatrix} 29 \\ 252 \\ 30{,}000 \end{bmatrix}$$

We form the augmented matrix,

$$\begin{bmatrix} 1 & 1 & 29 \\ 8 & 12 & 252 \\ 1000 & 2000 & 30{,}000 \end{bmatrix}$$

and proceed as before:

$$\begin{bmatrix} 1 & 1 & 29 \\ 0 & 4 & 20 \\ 0 & 1000 & 1000 \end{bmatrix}$$

$$\begin{bmatrix} 1 & 0 & 24 \\ 0 & 1 & 5 \\ 0 & 0 & -4000 \end{bmatrix}$$

This time,

$$A' = \begin{bmatrix} 1 & 0 \\ 0 & 1 \\ 0 & 0 \end{bmatrix} \qquad B' = \begin{bmatrix} 24 \\ 5 \\ -4000 \end{bmatrix}$$

and the ordinary equations corresponding to the matrix equation $A'X = B'$ are

$$x = 24$$

$$y = 5$$

$$0 = -4000$$

What is the significance of the obviously false equation $0 = -4000$? One might be tempted to ignore that equation and just use the results $x = 24$ and $y = 5$ from the first two equations. If there were 24 seamen and 5 officers on the ship, then certainly there were 29 men in all. Furthermore, at 8 ounces a day of wine for each seaman and 12 for each officer, the total daily consumption would be 252 ounces, because

$$(8)(24) + (12)(5) = 252$$

But at 1000 maravedis a month for a seaman and 2000 for an officer, the total monthly payroll would be

$$(1000)(24) + (2000)(5) = 34{,}000$$

which is 4000 maravedis more than the records indicate it was. Thus, the import of the last, false equation $0 = -4000$ is that there are no numbers x and y for which

$$x + y = 29$$
$$8x + 12y = 252$$
$$1000x + 2000y = 30{,}000$$

are all true. In this case, the corresponding matrix equation $AX = B$ has no solution.

In general, suppose $AX = B$ is transformed into the equation $A'X = B'$ by the moves of Section 3-5 and the latter matrix equation corresponds to ordinary equations of which one is in the form

$$0 = a$$

where a is not zero. Then the conclusion is that the equation $AX = B$ has no solution, that is, for the given matrices A and B there is no matrix X with the property that the product AX is equal to B.

Why was the historian unable to determine the number of seamen and officers on this particular voyage? The three pieces of information about the voyage (total number of men aboard, the assumed daily wine allowance, and monthly payroll) are not consistent. Consequently, no solution is possible. The inconsistency might have arisen because the historian's assumptions concerning the wine allowances or the salaries of the time are incorrect. Or it might be that the historian's assumptions are right, but the contemporary records contain mistakes. The mathematics cannot identify the cause of the trouble. It can only indicate that something is wrong. The historian had best recheck the records and look for additional information that might serve to locate the error.

3-29 **Example** Solve the equation $AX = B$ if possible, where

$$X = \begin{bmatrix} w \\ x \\ y \\ z \end{bmatrix} \qquad A = \begin{bmatrix} 0 & 2 & 3 & 0 \\ 3 & 2 & -1 & 1 \\ 0 & 0 & 1 & 2 \\ 2 & -1 & -1 & 0 \\ 3 & 0 & 2 & 0 \end{bmatrix} \qquad B = \begin{bmatrix} 0 \\ 2 \\ -1 \\ -1 \\ -2 \end{bmatrix}$$

Solution Since the number in the first row and first column of

$$\left[\begin{array}{cccc|c} 0 & 2 & 3 & 0 & 0 \\ 3 & 2 & -1 & 1 & 2 \\ 0 & 0 & 1 & 2 & -1 \\ 2 & -1 & -1 & 0 & -1 \\ 3 & 0 & 2 & 0 & -2 \end{array}\right]$$

is 0, we interchange the first 2 rows:

$$\left[\begin{array}{cccc|c} 3 & 2 & -1 & 1 & 2 \\ 0 & 2 & 3 & 0 & 0 \\ 0 & 0 & 1 & 2 & -1 \\ 2 & -1 & -1 & 0 & -1 \\ 3 & 0 & 2 & 0 & -2 \end{array}\right]$$

We combine a number of steps below as we change the matrix to the required form:

$$\left[\begin{array}{cccc|c} 1 & \dfrac{2}{3} & -\dfrac{1}{3} & \dfrac{1}{3} & \dfrac{2}{3} \\[6pt] 0 & 2 & 3 & 0 & 0 \\ 0 & 0 & 1 & 2 & -1 \\[6pt] 0 & -\dfrac{7}{3} & -\dfrac{1}{3} & -\dfrac{2}{3} & -\dfrac{7}{3} \\[6pt] 0 & -2 & 3 & -1 & -4 \end{array}\right]$$

$$\left[\begin{array}{cccc|c} 1 & 0 & -\dfrac{4}{3} & \dfrac{1}{3} & \dfrac{2}{3} \\[6pt] 0 & 1 & \dfrac{3}{2} & 0 & 0 \\[6pt] 0 & 0 & 1 & 2 & -1 \\[6pt] 0 & 0 & \dfrac{19}{6} & -\dfrac{2}{3} & -\dfrac{7}{3} \\[6pt] 0 & 0 & 6 & -1 & -4 \end{array}\right]$$

$$\begin{bmatrix} 1 & 0 & 0 & 3 & -\dfrac{2}{3} \\[2mm] 0 & 1 & 0 & -3 & \dfrac{3}{2} \\[2mm] 0 & 0 & 1 & 2 & -1 \\[2mm] 0 & 0 & 0 & -7 & \dfrac{5}{6} \\[2mm] 0 & 0 & 0 & -13 & 2 \end{bmatrix}$$

$$\begin{bmatrix} 1 & 0 & 0 & 0 & -\dfrac{13}{42} \\[2mm] 0 & 1 & 0 & 0 & \dfrac{8}{7} \\[2mm] 0 & 0 & 1 & 0 & -\dfrac{16}{21} \\[2mm] 0 & 0 & 0 & 1 & -\dfrac{5}{42} \\[2mm] 0 & 0 & 0 & 0 & \dfrac{19}{42} \end{bmatrix}$$

The last row of the matrix above corresponds to the ordinary equation

$$0 = \frac{19}{42}$$

which is false, so we conclude that the original matrix equation $AX = B$ has no solution.

Suppose we are given a matrix equation $AX = B$, where, as in Section 3-5, the matrix A is nonsingular. If we apply the method of this section to the augmented matrix

$$[A|B]$$

then we will end up with

$$[A'|B'] = [I|(A^{-1})B]$$

The method of this section is more efficient for solving the matrix equation $AX = B$ when A is nonsingular than is the method of Sections 3-4 and 3-5. It often happens, however, that the actual problem consists not of solving a single matrix equation, but rather several equations of the form $AX = B$, where the A matrix remains the same and the B matrix changes. An example of just this situation is furnished by input–output economics, because the input–output matrix is quite constant over time, but the demand matrix changes frequently, as we shall see in Section 3-7. In such cases, it is more efficient to calculate

A^{-1} and then just use matrix multiplication to find $(A^{-1})B$ for the various matrices B. Furthermore, there are other uses of the inverse matrix besides solving matrix equations of the form $AX = B$, so our efforts in the previous two sections were by no means a waste of time.

We have been considering matrix equations $AX = B$, where A has at least as many rows as columns. The method, however, requires no such restriction on the matrix A.

3-30   **Example**   Solve the equation $AX = B$ if possible, where

$$A = \begin{bmatrix} 2 & 1 & -2 \\ 1 & -1 & 1 \end{bmatrix} \qquad X = \begin{bmatrix} x \\ y \\ z \end{bmatrix} \qquad B = \begin{bmatrix} 3 \\ 4 \end{bmatrix}$$

Solution   Form the augmented matrix

$$\begin{bmatrix} 2 & 1 & -2 & 3 \\ 1 & -1 & 1 & 4 \end{bmatrix}$$

and perform the usual operations

$$\begin{bmatrix} 1 & \dfrac{1}{2} & -1 & \dfrac{3}{2} \\ 0 & -\dfrac{3}{2} & 2 & \dfrac{5}{2} \end{bmatrix}$$

$$\begin{bmatrix} 1 & 0 & -\dfrac{1}{3} & \dfrac{7}{3} \\ 0 & 1 & -\dfrac{4}{3} & -\dfrac{5}{3} \end{bmatrix}$$

At this point, we have obtained a 2×2 identity matrix for the first 2 columns and any operations on the third column will just complicate the first 2 columns, so we view this last matrix as

$$[A'|B']$$

and write out $A'X = B'$, which is

$$\begin{bmatrix} 1 & 0 & -\dfrac{1}{3} \\ 0 & 1 & -\dfrac{4}{3} \end{bmatrix} \begin{bmatrix} x \\ y \\ z \end{bmatrix} = \begin{bmatrix} \dfrac{7}{3} \\ -\dfrac{5}{3} \end{bmatrix}$$

or, as ordinary equations,

$$x - \left(\frac{1}{3}\right)z = \frac{7}{3}$$

$$y - \left(\frac{4}{3}\right)z = -\frac{5}{3}$$

The values of x and y will depend on what z is. For example, if $z = 0$, then $x = 7/3$ and $y = -5/3$, while if $z = 1$, then

$$x - \left(\frac{1}{3}\right)(1) = \frac{7}{3} \qquad \text{or} \qquad x = \frac{8}{3}$$

$$y - \left(\frac{4}{3}\right)(1) = -\frac{5}{3} \qquad \text{or} \qquad y = -\frac{1}{3}$$

That is to say,

$$X = \begin{bmatrix} \dfrac{7}{3} \\ -\dfrac{5}{3} \\ 0 \end{bmatrix} \quad \text{and} \quad X = \begin{bmatrix} \dfrac{8}{3} \\ -\dfrac{1}{3} \\ 1 \end{bmatrix}$$

are both solutions to $AX = B$ in this case. Since there is a solution X for every possible value for z, there are actually an infinite number of solutions to this matrix equation.

The fact that A has more columns than rows does not, however, guarantee that $AX = B$ will have an infinite number of solutions, as a final example demonstrates.

3-31   **Example**   Solve the equation $AX = B$ if possible, where

$$A = \begin{bmatrix} 1 & 1 & -2 \\ 3 & 3 & -6 \end{bmatrix} \qquad X = \begin{bmatrix} x \\ y \\ z \end{bmatrix} \qquad B = \begin{bmatrix} -1 \\ 1 \end{bmatrix}$$

Solution   We form the augmented matrix

$$\left[\begin{array}{ccc|c} 1 & 1 & -2 & -1 \\ 3 & 3 & -6 & 1 \end{array} \right]$$

and convert the first column to the required form by adding -3 times the first row to the second row. The result is the matrix

$$\left[\begin{array}{ccc|c} 1 & 1 & -2 & -1 \\ 0 & 0 & 0 & 4 \end{array}\right]$$

But the last row corresponds to the ordinary equation

$$0 = 4$$

which is false, so $AX = B$ has no solution.

Summary

To solve a matrix equation of the form $AX = B$, where A is a given $n \times m$ matrix, B is a given $n \times 1$ matrix, and X is an $m \times 1$ matrix of unknowns, first form the $n \times (m + 1)$ augmented matrix

$$[A|B]$$

To the extent possible, transform columns of A, working from left to right, to the columns of an identity matrix by means of the techniques of Section 3-5. The allowable moves are interchange of rows, multiplication and division through a row by a nonzero constant, and addition of a nonzero multiple of one row to another row. After these moves have been used to produce a matrix

$$[A'|B']$$

the matrix equation $A'X = B'$ has the property that its solution X is also the solution to the original equation $AX = B$. If one of the ordinary equations corresponding to the matrix equation $A'X = B'$ is of the type $0 = a$, where a is not zero, then neither $A'X = B'$ nor $AX = B$ has a solution. Otherwise, $AX = B$ has either a single solution or an infinite number of solutions.

Exercises

● *In Exercises 3-127 through 3-140, analyze the matrix equation $AX = B$ for the given matrices. Indicate if there is no solution to the equation. If there is one solution, find out what it is. If there are an infinite number of solutions, write down any two of them.*

3-127 $X = \begin{bmatrix} x \\ y \end{bmatrix}$ $A = \begin{bmatrix} 1 & 1 \\ 4 & 1 \\ 1 & 4 \end{bmatrix}$ $B = \begin{bmatrix} -1 \\ -3 \\ -2 \end{bmatrix}$

3-128 $X = \begin{bmatrix} x \\ y \end{bmatrix}$ $A = \begin{bmatrix} 2 & -1 \\ 3 & -1 \end{bmatrix}$ $B = \begin{bmatrix} 0 \\ 4 \end{bmatrix}$

3-129 $X = \begin{bmatrix} x \\ y \\ z \end{bmatrix}$ $A = \begin{bmatrix} 0 & 2 & 1 \\ -1 & 1 & 0 \\ 2 & -2 & 1 \\ 1 & 1 & 2 \end{bmatrix}$ $B = \begin{bmatrix} 2 \\ 0 \\ -1 \\ 1 \end{bmatrix}$

3-130 $X = \begin{bmatrix} x \\ y \\ z \end{bmatrix}$ $A = \begin{bmatrix} 1 & -1 & 0 \\ 1 & 2 & 2 \end{bmatrix}$ $B = \begin{bmatrix} 2 \\ 5 \end{bmatrix}$

3-131 $X = \begin{bmatrix} x \\ y \\ z \end{bmatrix}$ $A = \begin{bmatrix} -1 & -1 & 0 \\ 2 & 1 & 3 \\ 0 & 1 & 3 \\ 1 & 1 & 0 \\ 0 & 0 & 1 \end{bmatrix}$ $B = \begin{bmatrix} 1 \\ -1 \\ 2 \\ -2 \\ 0 \end{bmatrix}$

3-132 $X = \begin{bmatrix} x \\ y \\ z \end{bmatrix}$ $A = \begin{bmatrix} 1 & 2 & 0 \\ 0 & -1 & -2 \\ 2 & 0 & 2 \end{bmatrix}$ $B = \begin{bmatrix} -1 \\ -2 \\ -3 \end{bmatrix}$

3-133 $X = \begin{bmatrix} x \\ y \\ z \end{bmatrix}$ $A = \begin{bmatrix} 2 & -1 & 0 \\ -1 & 3 & 2 \\ 1 & 2 & 2 \end{bmatrix}$ $B = \begin{bmatrix} -2 \\ 1 \\ -1 \end{bmatrix}$

3-134 $X = \begin{bmatrix} x \\ y \\ z \end{bmatrix}$ $A = \begin{bmatrix} 0 & 1 & -1 \\ 0 & 1 & 1 \\ 1 & 2 & 1 \\ -2 & 0 & 1 \\ 2 & 1 & 0 \end{bmatrix}$ $B = \begin{bmatrix} 1 \\ 1 \\ 0 \\ 4 \\ -1 \end{bmatrix}$

3-135 $X = \begin{bmatrix} w \\ x \\ y \\ z \end{bmatrix}$ $A = \begin{bmatrix} 1 & 2 & -1 & -2 \\ 1 & -1 & 0 & 0 \\ -1 & 2 & 0 & 1 \end{bmatrix}$ $B = \begin{bmatrix} -1 \\ -1 \\ -1 \end{bmatrix}$

3-136 $X = \begin{bmatrix} w \\ x \\ y \\ z \end{bmatrix}$ $A = \begin{bmatrix} 0 & 1 & 0 & 2 \\ 1 & -1 & 2 & 1 \\ -1 & 2 & 0 & 0 \\ 0 & 1 & 2 & 0 \\ 0 & 0 & 1 & 1 \end{bmatrix}$ $B = \begin{bmatrix} -1 \\ 2 \\ 0 \\ 1 \\ 3 \end{bmatrix}$

3-137 $X = \begin{bmatrix} v \\ w \\ x \\ y \\ z \end{bmatrix}$ $A = \begin{bmatrix} 0 & 1 & 2 & 3 & 1 \\ 1 & 3 & 2 & 0 & 2 \\ 1 & 2 & 0 & -3 & 1 \end{bmatrix}$ $B = \begin{bmatrix} 0 \\ 1 \\ -1 \end{bmatrix}$

3-138 $X = \begin{bmatrix} x \\ y \\ z \end{bmatrix}$ $A = \begin{bmatrix} 2 & -1 & 0 \\ -2 & 2 & 1 \\ 2 & 0 & 1 \end{bmatrix}$ $B = \begin{bmatrix} 2 \\ -1 \\ 1 \end{bmatrix}$

3-139 $X = \begin{bmatrix} x \\ y \\ z \end{bmatrix}$ $A = \begin{bmatrix} 1 & 0 & 1 \\ 0 & -1 & 0 \\ 2 & 0 & 0 \\ 1 & 2 & 0 \\ 0 & 1 & 0 \\ 1 & 0 & 0 \end{bmatrix}$ $B = \begin{bmatrix} 2 \\ 1 \\ 0 \\ -1 \\ -2 \\ 0 \end{bmatrix}$

3-140 $X = \begin{bmatrix} w \\ x \\ y \\ z \end{bmatrix}$ $A = \begin{bmatrix} 1 & 1 & 0 & 2 \\ 0 & 0 & 1 & 0 \\ 1 & 1 & 2 & -1 \end{bmatrix}$ $B = \begin{bmatrix} 2 \\ 1 \\ 0 \end{bmatrix}$

3-141   A traveler just returned from Europe spent $15 a day for housing in England, $10 a day in France, and $10 a day in Spain. For food the traveler spent $10 a day in England, $15 a day in France, and $10 a day in Spain. The traveler spent $5 a day in each country for incidental expenses. The traveler's records of the trip indicate a total of $170 spent for housing, $160 for food, and $70 for incidental expenses while traveling in these countries. Calculate the number of days the traveler spent in each of the countries or show that the records must be incorrect, because the amounts spent are incompatible with each other.

3-142   An intelligence agent knows that 60 aircraft, consisting of fighter planes and bombers, are stationed at a certain secret airfield. The agent wishes to determine how many of the 60 are fighter planes and how many are bombers. There is a type of rocket carried by both sorts of planes; the fighter carries six of these rockets, the bomber only two. The agent learns that 250 rockets are required to arm every plane at this airfield. Furthermore, the agent overhears a remark that there are twice as many fighter planes as bombers at the base (that is, the number of fighter planes minus twice the number of bombers equals zero). Calculate the number of fighter planes and bombers at the airfield or show that the agent's information must be incorrect, because it is inconsistent.

3-143   A card player holds a thirteen card bridge hand. She says that she has three more red cards than black cards (the number of red cards minus the number of black cards equals three), that twice the number of red cards in the hand plus three times the number of black cards adds up to 31, and that three times the number of red cards plus the number of black cards equals 29. Calculate the number of red and of black cards in the hand, if possible.

3-144   A man refuses to tell anyone his age, but he likes to drop hints about it. For example, he once remarked that twice his own age plus his mother's age added up to 140, and on another occasion he mentioned that his age plus his father's age added up to 105. Another time, he said that the sum of his age and his mother's age is 30 years more than his father's age (that is, his age plus his mother's age minus his father's age equals 30). Calculate the man's age or show that his hints contradict one another.

3-145   Messages intercepted from the headquarters of an army division give the total strength of the division to be 10,000 soldiers. Of this number, some are combat soldiers, some are male soldiers assigned to support duties, and the rest are female soldiers with support duties. Another message indicates that there are three times as many male as female support soldiers (that is, the number of male support soldiers minus three times the number of female support soldiers equals zero). A third message implies that the number of combat soldiers plus four times the number of female support soldiers equals the total division size, that is, 10,000 soldiers. Are the three messages consistent with one another? If they are, determine whether there is enough information given to calculate the number of combat soldiers in the division and, if there is, calculate this number.

3-146   An investor remarks to a stockbroker that all her stock holdings are in three companies, Eastern Airlines, Hilton Hotels, and McDonald's, and that 2 days ago the value of her stocks went down $350 but yesterday the value increased by $600. The broker recalls that 2 days ago the price of Eastern Airlines stock dropped by $1 a share, Hilton Hotels dropped $1.50, but the price of McDonald's stock rose by $0.50. The broker also remembers that yesterday the price of Eastern Airlines stock rose $1.50, there was a further drop of $0.50 a share in Hilton Hotels stock, and McDonald's stock rose $1. Show that the broker does not have enough information to calculate the number of shares the investor owns of each company's stock, but that when the investor says that she owns 200 shares of McDonald's stock, the broker can calculate the number of shares of Eastern Airlines and Hilton Hotels.

3-7 The Leontief Model

In the example of Section 3-1, the 1958 American economy was divided into six sectors: final nonmetal (FN), final metal (FM), basic metal (BM), basic nonmetal (BN), energy (E), and services (S). The input–output table (repeated here as Table 3-3) meant that it required, for example, 0.348 unit of basic nonmetal production to produce 1 unit of final nonmetal output and 0.025 unit of energy to produce 1 unit of services. In addition, a bill of demands (in millions) was given:

FN	$ 99,640
FM	$ 75,548
BM	$ 14,444
BN	$ 33,501
E	$ 23,527
S	$263,985

We turn now to the problem posed in Section 3-1: How many units of output from each sector are needed in order to run the economy and fill the bill of demands?

Table 3-3

	FN	FM	BM	BN	E	S
FN	0.170	0.004	0	0.029	0	0.008
FM	0.003	0.295	0.018	0.002	0.004	0.016
BM	0.025	0.173	0.460	0.007	0.011	0.007
BN	0.348	0.037	0.021	0.403	0.011	0.048
E	0.007	0.001	0.039	0.025	0.358	0.025
S	0.120	0.074	0.104	0.123	0.173	0.234

The units of output from each sector required to run the economy and fill the bill of demands is unknown, so we follow tradition and call these unknowns x. In our example, there are six quantities which are, at the moment, unknown. The number of units of final nonmetal production required to solve the problem will be our first unknown, because this sector is represented by the first row of the input–output matrix. The unknown quantity of final nonmetal units will be represented by the symbol x_1. Following the same pattern, we represent the unknown quantities in the following manner:

x_1 = units of final nonmetal production required

x_2 = units of final metal production required

x_3 = units of basic metal production required

x_4 = units of basic nonmetal production required

x_5 = units of energy required

x_6 = units of services required

These six numbers are the quantities we are attempting to calculate.

Let the symbol s_{11} represent the number of units of final nonmetal production required to produce x_1 units of final nonmetal output. We will not know how large s_{11} is until we determine x_1 but, for the present, we need the symbol. The symbol s_{12} will stand for the number of units of final nonmetal output required to produce x_2 units of final metal output. Likewise, s_{13}, s_{14}, s_{15}, and s_{16} represent final nonmetal units used by each of the remaining sectors to produce x_3, x_4, x_5, and x_6 units of output, respectively. The total need x_1 for final nonmetal units is the demand for 99,640 units plus the sum of the needs for final nonmetal units from the six sectors. The notation we have established permits us to write

$$x_1 = s_{11} + s_{12} + s_{13} + s_{14} + s_{15} + s_{16} + 99{,}640$$

In just the same way, we obtain the equations

$$x_2 = s_{21} + s_{22} + s_{23} + s_{24} + s_{25} + s_{26} + 75{,}548$$

$$x_3 = s_{31} + s_{32} + s_{33} + s_{34} + s_{35} + s_{36} + 14{,}444$$

$$x_4 = s_{41} + s_{42} + s_{43} + s_{44} + s_{45} + s_{46} + 33{,}501$$

$$x_5 = s_{51} + s_{52} + s_{53} + s_{54} + s_{55} + s_{56} + 23{,}527$$

$$x_6 = s_{61} + s_{62} + s_{63} + s_{64} + s_{65} + s_{66} + 263{,}985$$

where, throughout, the symbol s_{ij} represents the number of units of output from the sector represented by the ith row of the input–output matrix required to produce x_j units of output from the sector represented by the jth column.

Now according to the input–output table, it takes 0.170 unit of final nonmetal input to produce 1 unit of final nonmetal output. By the assumption underlying this kind of economic theory, it then takes $0.170x_1$ units of final nonmetal input to produce x_1 final nonmetal units. For example, if x_1 were 120,000, then it would require $(0.170)(120,000) = 20,400$ final nonmetal units to produce the 120,000 units. Since s_{11} was defined to be the number of units of final nonmetal input required to produce x_1 final nonmetal units, we obtain the equation

$$s_{11} = 0.170x_1$$

Similarly, from the table, it takes 0.004 unit of final nonmetal production to make 1 final metal unit, so it takes $0.004x_2$ to make x_2 units. Thus, we conclude that

$$s_{12} = 0.004x_2$$

In the same way, we have

$$s_{13} = 0x_3 = 0$$
$$s_{14} = 0.029x_4$$
$$s_{15} = 0x_5 = 0$$
$$s_{16} = 0.008x_6$$

Substituting into the equation

$$x_1 = s_{11} + s_{12} + s_{13} + s_{14} + s_{15} + s_{16} + 99{,}640$$

provides us with

$$x_1 = 0.170x_1 + 0.004x_2 + 0 + 0.029x_4 + 0 + 0.008x_6 + 99{,}640$$

In the same way, we know that

$$s_{21} = 0.003x_1$$

that is, in order to produce x_1 units of final nonmetal output, it will take $0.003x_1$ units of final metal input to do the job. Substituting for s_{21} and the rest in the equations for x_2, x_3, and so on above, we now have the following equations:

$$x_1 = 0.170x_1 + 0.004x_2 + 0 \qquad + 0.029x_4 + 0 \qquad + 0.008x_6 + 99{,}640$$

$$x_2 = 0.003x_1 + 0.295x_2 + 0.018x_3 + 0.002x_4 + 0.004x_5 + 0.016x_6 + 75{,}548$$

$$x_3 = 0.025x_1 + 0.173x_2 + 0.460x_3 + 0.007x_4 + 0.011x_5 + 0.007x_6 + 14{,}444$$

$$x_4 = 0.348x_1 + 0.037x_2 + 0.021x_3 + 0.403x_4 + 0.011x_5 + 0.048x_6 + 33{,}501$$

$$x_5 = 0.007x_1 + 0.001x_2 + 0.039x_3 + 0.025x_4 + 0.358x_5 + 0.025x_6 + 23{,}527$$

$$x_6 = 0.120x_1 + 0.074x_2 + 0.104x_3 + 0.123x_4 + 0.173x_5 + 0.234x_6 + 263{,}985$$

It might be possible to solve these equations by techniques of elementary algebra for the unknowns x_1, x_2, x_3, x_4, x_5, and x_6, given enough time and effort. Keep in mind, however, that we want a procedure that will work for an 81 sector economic model like the one Leontief actually used for the 1958 American economy and not just for simplified six sector examples. Thus, the next step, in which we turn the problem of computing the numbers x_1, x_2, x_3, x_4, x_5, and x_6 into a matrix arithmetic problem, is of fundamental importance in obtaining a practical, and not just theoretical, solution to the problem.

Let A be the 6×6 matrix corresponding to the input–output table.

$$A = \begin{bmatrix} 0.170 & 0.004 & 0 & 0.029 & 0 & 0.008 \\ 0.003 & 0.295 & 0.018 & 0.002 & 0.004 & 0.016 \\ 0.025 & 0.173 & 0.460 & 0.007 & 0.011 & 0.007 \\ 0.348 & 0.037 & 0.021 & 0.403 & 0.011 & 0.048 \\ 0.007 & 0.001 & 0.039 & 0.025 & 0.358 & 0.025 \\ 0.120 & 0.074 & 0.104 & 0.123 & 0.173 & 0.234 \end{bmatrix}$$

Call A the input–output matrix. The bill of demands gives rise to a 6×1 demand matrix:

$$D = \begin{bmatrix} 99{,}640 \\ 75{,}548 \\ 14{,}444 \\ 33{,}501 \\ 23{,}527 \\ 263{,}985 \end{bmatrix}$$

In addition, we have a 6×1 matrix of unknowns,

$$X = \begin{bmatrix} x_1 \\ x_2 \\ x_3 \\ x_4 \\ x_5 \\ x_6 \end{bmatrix}$$

called the **intensity matrix.** We claim that the six ordinary equations above are the same as the single matrix equation

$$X = AX + D$$

If you check that

$$AX = \begin{bmatrix} 0.170 & 0.004 & 0 & 0.029 & 0 & 0.008 \\ 0.003 & 0.295 & 0.018 & 0.002 & 0.004 & 0.016 \\ 0.025 & 0.173 & 0.460 & 0.007 & 0.011 & 0.007 \\ 0.348 & 0.037 & 0.021 & 0.403 & 0.011 & 0.048 \\ 0.007 & 0.001 & 0.039 & 0.025 & 0.358 & 0.025 \\ 0.120 & 0.074 & 0.104 & 0.123 & 0.173 & 0.234 \end{bmatrix} \begin{bmatrix} x_1 \\ x_2 \\ x_3 \\ x_4 \\ x_5 \\ x_6 \end{bmatrix}$$

$$= \begin{bmatrix} 0.170x_1 + 0.004x_2 + 0 & + 0.029x_4 + 0 & + 0.008x_6 \\ 0.003x_1 + 0.295x_2 + 0.018x_3 + 0.002x_4 + 0.004x_5 + 0.016x_6 \\ 0.025x_1 + 0.173x_2 + 0.460x_3 + 0.007x_4 + 0.011x_5 + 0.007x_6 \\ 0.348x_1 + 0.037x_2 + 0.021x_3 + 0.403x_4 + 0.011x_5 + 0.048x_6 \\ 0.007x_1 + 0.001x_2 + 0.039x_3 + 0.025x_4 + 0.358x_5 + 0.025x_6 \\ 0.120x_1 + 0.074x_2 + 0.104x_3 + 0.123x_4 + 0.173x_5 + 0.234x_6 \end{bmatrix}$$

you should have no difficulty, using the definitions of matrix addition and equality of matrices, in verifying that our claim is correct.

In the general input–output model, Leontief assumes that the coefficients of the input–output table for an n-sector economy have been determined on the basis of economic data, producing an $n \times n$ input–output matrix A. The bill of demands is computed from a knowledge of the economic situation, and it becomes the $n \times 1$ matrix D. The number of units required for each of the n sectors in order to run the economy and fill the bill of demands is calculated by solving the matrix equation

$$X = AX + D$$

where X is the $n \times 1$ matrix of unknowns, in the same order of sectors as the rows of A and D.

We know enough about matrices from the previous sections to solve the equation

$$X = AX + D$$

First, subtracting the $n \times 1$ matrix AX from both sides gives us

$$X - AX = D$$

Letting I represent the $n \times n$ identity matrix, we know that $IX = X$, so if we substitute into the equation, it becomes

$$IX - AX = D$$

Applying the appropriate form of the distributive law:

$$(I - A)X = D$$

Let $B = I - A$. Then the equation becomes

$$BX = D$$

As in Section 3-6, if $B = I - A$ is a nonsingular $n \times n$ matrix, we can solve for the matrix of unknowns (the intensity matrix)

$$X = B^{-1}D$$

or, in terms of the original notation,

$$X = (I - A)^{-1}D$$

We have worked out the method by which the computer solves the problem of determining the required total output for each sector. The input–output matrix A is subtracted from the identity matrix of the same size. The matrix $I - A$ is usually nonsingular. In case it is not, a very small numerical ajdustment to A, which has no economic significance, will make $I - A$ nonsingular. Next, the inverse of $I - A$ is calculated. This is a very formidable task in the case of a large matrix, but one within the capability of present-day computers. Finally, the matrix $(I - A)^{-1}$ is multiplied by the $n \times 1$ demand matrix D to give the $n \times 1$ matrix X of answers.

Let us apply the method to the example above, where

$$A = \begin{bmatrix} 0.170 & 0.004 & 0 & 0.029 & 0 & 0.008 \\ 0.003 & 0.295 & 0.018 & 0.002 & 0.004 & 0.016 \\ 0.025 & 0.173 & 0.460 & 0.007 & 0.011 & 0.007 \\ 0.348 & 0.037 & 0.021 & 0.403 & 0.011 & 0.048 \\ 0.007 & 0.001 & 0.039 & 0.025 & 0.358 & 0.025 \\ 0.120 & 0.074 & 0.104 & 0.123 & 0.173 & 0.234 \end{bmatrix}$$

and

$$D = \begin{bmatrix} 99,640 \\ 75,548 \\ 14,444 \\ 33,501 \\ 23,527 \\ 263,985 \end{bmatrix}$$

We subtract

$$I - A = \begin{bmatrix} 1 & 0 & 0 & 0 & 0 & 0 \\ 0 & 1 & 0 & 0 & 0 & 0 \\ 0 & 0 & 1 & 0 & 0 & 0 \\ 0 & 0 & 0 & 1 & 0 & 0 \\ 0 & 0 & 0 & 0 & 1 & 0 \\ 0 & 0 & 0 & 0 & 0 & 1 \end{bmatrix} - \begin{bmatrix} 0.170 & 0.004 & 0 & 0.029 & 0 & 0.008 \\ 0.003 & 0.295 & 0.018 & 0.002 & 0.004 & 0.016 \\ 0.025 & 0.173 & 0.460 & 0.007 & 0.011 & 0.007 \\ 0.348 & 0.037 & 0.021 & 0.403 & 0.011 & 0.048 \\ 0.007 & 0.001 & 0.039 & 0.025 & 0.358 & 0.025 \\ 0.120 & 0.074 & 0.104 & 0.123 & 0.173 & 0.234 \end{bmatrix}$$

$$= \begin{bmatrix} 0.830 & -0.004 & 0 & -0.029 & 0 & -0.008 \\ -0.003 & 0.705 & -0.018 & -0.002 & -0.004 & -0.016 \\ -0.025 & -0.173 & 0.540 & -0.007 & -0.011 & -0.007 \\ -0.348 & -0.037 & -0.021 & 0.597 & -0.011 & -0.048 \\ -0.007 & -0.001 & -0.039 & -0.025 & 0.642 & -0.025 \\ -0.120 & -0.074 & -0.104 & -0.123 & -0.173 & 0.766 \end{bmatrix}$$

By the method of Section 3-5, we find the inverse (actually an approximation),

$$(I - A)^{-1} = \begin{bmatrix} 1.234 & 0.014 & 0.006 & 0.064 & 0.007 & 0.018 \\ 0.017 & 1.436 & 0.057 & 0.012 & 0.020 & 0.032 \\ 0.071 & 0.465 & 1.877 & 0.019 & 0.045 & 0.031 \\ 0.751 & 0.134 & 0.100 & 1.740 & 0.066 & 0.124 \\ 0.060 & 0.045 & 0.130 & 0.082 & 1.578 & 0.059 \\ 0.339 & 0.236 & 0.307 & 0.312 & 0.376 & 1.349 \end{bmatrix}$$

Therefore,

$$X = (I - A)^{-1}D$$

$$= \begin{bmatrix} 1.234 & 0.014 & 0.006 & 0.064 & 0.007 & 0.018 \\ 0.017 & 1.436 & 0.057 & 0.012 & 0.020 & 0.032 \\ 0.071 & 0.465 & 1.877 & 0.019 & 0.045 & 0.031 \\ 0.751 & 0.134 & 0.100 & 1.740 & 0.066 & 0.124 \\ 0.060 & 0.045 & 0.130 & 0.082 & 1.578 & 0.059 \\ 0.339 & 0.236 & 0.307 & 0.312 & 0.376 & 1.349 \end{bmatrix} \begin{bmatrix} 99,640 \\ 75,548 \\ 14,444 \\ 33,501 \\ 23,527 \\ 263,985 \end{bmatrix} = \begin{bmatrix} 131,161 \\ 120,324 \\ 79,194 \\ 178,936 \\ 66,703 \\ 426,542 \end{bmatrix}$$

Since, from the definition,

$$X = \begin{bmatrix} x_1 \\ x_2 \\ x_3 \\ x_4 \\ x_5 \\ x_6 \end{bmatrix}$$

we have the solutions

$$x_1 = 131{,}161$$
$$x_2 = 120{,}324$$
$$x_3 = 79{,}194$$
$$x_4 = 178{,}936$$
$$x_5 = 66{,}703$$
$$x_6 = 426{,}542$$

In other words, it would require 131,161 units ($131,161 million worth) of final nonmetal production, 120,324 units of final metal output, 79,194 units of basic metal products, and so on to run the 1958 American economy and completely fill the stated bill of demands.

If a new bill of demands D' is calculated for a subsequent time period, the required intensities of the various sectors of the economy are calculated readily by the computer; just by performing the matrix multiplication $(I - A)^{-1}D'$, using the same inverse matrix $(I - A)^{-1}$. However, if new technology or new economic circumstances change the relationship between two sectors, and so change the input–output matrix A to a new matrix A', then an inverse $(I - A')^{-1}$ must be calculated all over again, and that is a considerably more difficult task.

Summary

Let A be the matrix corresponding to the input–output table and let D be the matrix corresponding to the bill of demands. Define an $n \times 1$ matrix X, the **intensity matrix,** by letting each number in X represent the number of units of production required from some sector to fill the final demand for that sector and to produce required input to all sectors. The same order of sectors is used for the rows of $A, D,$ and X. The problem of determining the production required from each sector to run the economy and fill the bill of demands is solved by computing

$$X = (I - A)^{-1}D$$

Exercises

3-147 A much simplified version of Leontief's 42 sector analysis of the 1947 American economy divides the economy into just three sectors: agriculture, manufacturing, and the household (i.e., the sector of the economy which produces labor). It consists of the following input–output table:

	Agriculture	Manufacturing	Household
Agriculture	0.245	0.102	0.051
Manufacturing	0.099	0.291	0.279
Household	0.433	0.372	0.011

The bill of demands (in billions of dollars) is

Agriculture	2.88
Manufacturing	31.45
Household	30.91

(*a*) Write the input–output matrix A, the demand matrix D, and (in symbols) the intensity matrix X. (*b*) Compute $I - A$. (*c*) Check that

$$(I - A)^{-1} = \begin{bmatrix} 1.454 & 0.291 & 0.157 \\ 0.533 & 1.763 & 0.525 \\ 0.837 & 0.791 & 1.278 \end{bmatrix}$$

is an approximation to the inverse of $I - A$ by calculating $(I - A)^{-1}(I - A)$. (*d*) Use the matrix of part (*c*) to compute X. (*e*) Explain the meaning of the numbers in X in dollars.

3-148 An analysis of the 1958 Israeli economy* is here simplified by grouping the economy into three sectors: agriculture, manufacturing, and energy. The input–output table is

	Agriculture	Manufacturing	Energy
Agriculture	0.293	0	0
Manufacturing	0.014	0.207	0.017
Energy	0.044	0.010	0.216

*Wassily Leontief, *Input–Output Economics* (New York: Oxford University Press, 1966), pp. 54–57.

Exports (in thousands of Israeli pounds) were

Agriculture	138,213
Manufacturing	17,597
Energy	1,786

(a) Write the input–output matrix A and the demand (export) matrix D. (b) Compute $I - A$. (c) Check that

$$(I - A)^{-1} = \begin{bmatrix} 1.414 & 0 & 0 \\ 0.027 & 1.261 & 0.027 \\ 0.080 & 0.016 & 1.276 \end{bmatrix}$$

is an approximation to the inverse of $I - A$ by calculating $(I - A)^{-1}(I - A)$. (d) Use the matrix of part (c) to determine the number of Israeli pounds worth of agricultural products, manufactured goods, and energy required to run this model of the Israeli economy and export the stated value of products.

3-149   An input–output analysis of the 1963 British economy[*] is simplified below in terms of four sectors: nonmetals (N), metals (M), energy (E), and services (S).

	N	M	E	S
N	0.184	0.101	0.355	0.059
M	0.062	0.199	0.075	0.031
E	0.029	0.023	0.150	0.015
S	0.104	0.112	0.075	0.076

The bill of demands (in millions of pounds) is

N	10,271
M	5,987
E	1,161
S	13,780

(a) Write the input-output matrix A and the demand matrix D. (b) Compute $I - A$.

[*]L. S. Berman, "Development of Input–Output Statistics," in W. F. Grossling, ed., *Input–Output in the United Kingdom* (*Proc. 1968 Manchester Conf.*) (London: Frank Cass, 1970), pp. 34–35).

(c) Check that

$$(I - A)^{-1} = \begin{bmatrix} 1.272 & 0.191 & 0.557 & 0.097 \\ 0.110 & 1.276 & 0.164 & 0.056 \\ 0.050 & 0.044 & 1.203 & 0.024 \\ 0.162 & 0.194 & 0.152 & 1.102 \end{bmatrix}$$

is an approximation to the inverse of $I - A$ by calculating $(I - A)^{-1}(I - A)$. (d) Use the matrix of part (c) to compute the number of pounds worth of nonmetal and metal products, energy, and services required to run the British economy and fill the bill of demands.

3-150 Given a square matrix A, let $A^2 = AA$, $A^3 = AAA$, and so on. If the sums of the numbers in each column of A are positive but add up to less than one, then, according to matrix theory,

$$(I - A)^{-1} = I + A + A^2 + A^3 + \cdots$$

where the dots indicate that the pattern continues indefinitely. Compute $I + A + A^2$ for the matrix A of Exercise 3-148 and compare it to the matrix for $(I - A)^{-1}$ given in that exercise.

Concluding Essay

At this point it may seem that input–output economics should be solving the economic problems of the world. But, while input–output economics is a very useful tool, it is not a universal panacea—nor is it as easy as it looks. In *Business Week* (see References), Leontief is quoted as saying, "Some people got their hands burned, they thought it [input–output analysis] was simple. It's not." Indeed, the input–output concept seems almost trivial, and the mechanical manipulation of data in the table is not at all difficult for a modern computer. On the other hand, the formation of the input–output table requires considerable research and the table must be used with discretion.

The construction of a table is difficult for several reasons. One of the main problems in constructing the table for the economy of a country is that authentic industry totals are not up-to-date. For example, an input–output table issued by the United States government in the spring of 1968 was based on figures for 1963. Another difficulty is the question of how far to break down an industry. For instance, should energy be a single sector or should it be divided into petroleum, electricity, and coal? The trend in input–output analysis has been to increase the number of subdivisions. In 1958 the United States government table contained 52 industries, while by 1963 the table had expanded to 250 industries. When a company is setting up an input–output model, one of the main difficulties is the classification of industries. The government table uses the Census Bureau's Standard Industrial Classification (SIC) numbers. Very few companies code their suppliers and customers by SIC number, so the process of checking invoices, totaling sales or purchases, and then assigning the totals to an SIC group is tedious. However, some businesses consider the trouble and expense worthwhile. For example, the Celanese Corporation utilizes a service run by IBM and Dun and Bradstreet that relates customers to SIC numbers in order to acquire classifications for enough of its customers and vendors, so that Celanese is able to construct a useful input–output model of its operations.

The noted mathematician E. T. Bell once wrote, "the invention of matrices illustrates . . . the power and suggestiveness of a well-devised notation; it also exemplifies the fact . . . that a trivial notational device may be the germ of a vast theory having innumerable applications."* The prominence of matrices in modern mathematics can hardly be over-emphasized. The fact that matrices play a crucial part in each of the next three chapters of this book merely reflects the situation in mathematics as a whole. Indeed, the use of matrices in the applications of mathematics is by no means limited to input–output economics. The succeeding chapters illustrate the employment of matrices in business and psychology, but the social sciences are not the only beneficiaries of this subject. Modern physics is so dependent on the matrix concept, and on more advanced facts from the theory of matrices, that it is difficult to imagine what twentieth century theoretical physics would have been like without matrices. The use of matrices in such fields as ecology and population genetics is rather less sophisticated than in physics, but it is nevertheless of great importance. Thus, throughout much of mathematics, and in the applications of mathematics to a wide range of human concerns, matrices have a most significant role.

*E. T. Bell, *The Development of Mathematics*, (New York: McGraw-Hill, 1945), p. 205.

References

A fine article on Leontief and his work is "The 1973 Nobel Prize for Economic Science" by Walter Isard and Phyllis Kaniss in *Science* (Nov. 9, 1973). More articles about input–output economics may be found in *Scientific American*. First, "Input Output Economics" by Wassily Leontief appeared in the October 1951 issue. Then, "The Economic Effects of Disarmament" by Wassily Leontief and Marvin Hoffenberg was published in April 1961. The application of input–output analysis to the economies of developing nations was described in "The Structure of Development" by Leontief in September 1963. Leontief also wrote "The Structure of the U.S. Economy," which appeared in April 1965. The standard reference for Leontief's work is his book *The Structure of the American Economy— 1919–1939*, rev. ed. (New York: Oxford University Press, 1951).

An article in *Business Week* (Sept. 23, 1967), entitled "Planners Put Big Picture on Grid," tells how large corporations are making use of Leontief's work.

More information on the mathematics of matrices can be found in texts with titles including the words "matrix theory," "linear algebra," and "modern algebra." Among the many good texts of this type are the following: Hugh Campbell, *An Introduction to Matrices, Vectors and Linear Programming* (New York: Appleton-Century-Crofts, 1965). Marvin Marcus and Henryk Minc, *Introduction to Linear Algebra* (New York: Macmillan, 1965). Neal McCoy, *Introduction to Modern Algebra* (Boston: Allyn and Bacon, 1960). Lowell Paige and Dean Swift, *Elements of Linear Algebra* (Boston: Ginn and Co., 1961).

Chapter 4
Predicting
Brand-Share
by Markov Chain
Methods

Introductory Essay

When a company develops a new product, it is usually test-marketed in an area of the country before it is distributed nationally. The company must determine as quickly as possible whether the product is successful, so that if it is, the company can give it nationwide distribution before a rival firm brings out a similar product. Or, if the product is a failure, the company can stop production and hold its losses to a minimum.

An example of this kind of test-marketing is illustrated by a study written by Benjamin Lipstein for the *Harvard Business Review* (see References). This article shows how predictions on one product were made from a study in the Chicago area. Fictitious names are used, but the information

223

is based on an actual marketing situation.

When a product Lipstein calls "Electra margarine" was test-marketed from November 1958 to May 1959, its brand share (the percentage of the available market that the new brand took over) rose quickly in December, dropped in January, and rose again in February. During the same test, the percentage of margarine buyers who were new triers of Electra rose to 15% in mid-December but then slipped in mid-January to 4%, a percentage that remained relatively stable through the 6 month trial period. At the same time, the percentage of repeat buyers of Electra showed a steady increase, while the average repeat rate for competitive brands dropped.

To understand what this all means in terms of Electra's success or failure in the marketplace, it helps to look at what was happening to all margarine sales. Before Electra was introduced, brand-loyal margarine buyers (those who devoted 3/4 or more of their purchases of margarine in any 6 week period to one brand) made up 70% of the market. During the first 6 weeks of Electra's introduction, brand-loyal buyers declined to 65%, primarily because some of them were trying

Electra. The number of buyers of competitive brands decreased further (to 58%) with the advent of the Electra advertising campaign and coupon distribution. Then, the percentage of repeat buyers of all margarines increased, because repeat buyers of Electra margarine were included in the total. The repeat rate for Electra at the end of 4 months was 2% above the market average, but this did not necessarily guarantee a satisfactory brand-share for Electra, because the whole market was in turmoil.

As the market began to stabilize toward the end of the 6 month trial period, Electra's total share of the market was still dwindling, the new trier rate had leveled off, but the repeat buyer rate continued to climb. In other words, Electra seemed to have claimed a share of the market and acquired some loyal customers. However, were there enough of these repeat buyers and were their ranks going to continue to increase?

One way to make a prediction is to use a concept from applied probability called a "Markov chain process." As we shall see, this technique is a combination of probability theory and matrix arithmetic which can be used to predict the probability of a future action.

4-1 Matrices and Probability

A simplified version of the brand-share problem described in the introductory essay is the following: Suppose that at a certain time 14% of all margarine buyers have purchased

Electra brand margarine. Studies of buyer behavior indicate that 23% of Electra purchasers will choose the same brand the next time they buy margarine. Ten percent of the people who bought another brand the previous time will switch to Electra. What percentage of all margarine buyers will select Electra for their margarine in the next time period? We will assume that each buyer makes one margarine purchase in each time period.

We may solve this problem by the partition technique of probability theory that was the subject of Section 2-10. Let B_1 be the set of all purchasers of Electra margarine in the given time period and let B_2 be the set of margarine buyers who chose another brand at that time. If a margarine purchaser is chosen at random, the probability is $p(B_1) = .14$ that this person bought Electra brand and therefore $p(B_2) = 1 - .14 = .86$ that the choice was some other brand. Letting $p(A)$ represent the probability that a margarine purchaser chosen at random will select Electra in the next time period, then according to Chapter 2, that probability can be calculated by means of the formula

$$p(A) = p(A \mid B_1) p(B_1) + p(A \mid B_2) p(B_2)$$

The conditional probability $p(A \mid B_1)$ is the probability that an Electra brand buyer in the original time period will buy Electra again in the next period, so we have been told that $p(A \mid B_1) = .23$. Similarly, $p(A \mid B_2)$ represents the probability that a purchaser of a brand other than Electra will switch to Electra in the next time period, and thus we know that $p(A \mid B_2) = .10$. Substituting into the formula tells us that

$$p(A) = (.23)(.14) + (.10)(.86) = .12 \text{ (approximately)}$$

Thus, Electra's share of the margarine market will drop from 14% to 12% from the given time period to the next.

Thus far, the technique of Chapter 2 seems quite adequate for brand-share problems, but let us see what happens when we try to analyze not only how Electra margarine will perform from one buying period to the next, but also how this behavior will compare to that of its major competitor, "B-R Stores" brand margarine. This comparison is important in understanding the marketing situation, because the entrance of Electra brand into the margarine market should affect the sales of other brands.

Suppose that 20% of all margarine buyers in the given time period chose B-R margarine. The results of a study of buyer behavior, the same one that said that 23% of Electra buyers would repeat their choice, can be conveniently summarized in Table 4-1.

The meaning of the first column is that there is a probability of .23 that a purchaser of Electra margarine in the first time period will choose the same brand in the next, a probability of .04 that the buyer will switch to B-R brand, and therefore a probability of $1 - (.23 + .04) = .73$ that some brand other than Electra or B-R will be purchased. Similarly, the second column describes the expected behavior of B-R brand purchasers: 5% will switch to Electra,

Table 4-1

Original period / Next period	Electra	B-R	Others
Electra	.23	.05	.12
B-R	.04	.25	.15
Others	.73	.70	.73

70% to some other brand, and 25% will repeat their purchase of B-R brand margarine. The final column predicts the behavior of purchasers of the other brands of margarine.

Now, in the language of Chapter 2, we partition the margarine buyers according to their purchase in the given time period into three classes: Electra buyers B_1, B-R buyers B_2, and purchasers of other brands B_3. We will suppose that $p(B_1) = .14$ and $p(B_2) = .20$, so $p(B_3) = .66$. We would like to know the proportion of Electra buyers, B-R buyers, and purchasers of other brands in the next buying period. Consequently, it is not enough to specify a single type of success as we did above, where $p(A)$ represented the probability that a purchaser of margarine would choose Electra in the next period. The only alternative (failure) in that case was the choice of a brand other than Electra. Now we are interested in three different types of outcomes, not just two.

For a useful mathematical description of what it is we want to find out, note that the partition B_1, B_2, B_3 of the sample space U of all margarine buyers was not in any sense arbitrary. It classified the members of U according to a fixed criterion, namely, margarine purchased in a given time period. In other words, we are imagining that each member of U has been examined with respect to brand purchase and placed in an appropriate subset of the partition as a result. Formally, we say that each outcome can be in exactly one of three possible **states,** Electra purchaser, B-R purchaser, and purchaser of another brand, and that the partition B_1, B_2, B_3 is determined by which outcomes (purchasers) are in which state (which brand they choose). We continue to assume that each buyer makes one margarine choice in each time period.

We would like to know the probability that in the next time period a margarine purchaser will buy Electra, B-R, or some other brand. That is, we wish to know the probability that an outcome will be in each one of the same three possible states in the next time period. Just as in the original time period, we can examine each outcome to determine which state it is in during the next time period. This examination produces a new partition A_1, A_2, A_3 of the sample space U of all purchasers, where A_1 denotes those who choose Electra brand during the next time period, A_2 those who choose B-R brand, and A_3 those who select one of the others. Notice that A_1 refers to the same state as subset B_1 (Electra margarine), although as a subset of U it may have quite different members. Also, A_2 and B_2 refer to the same state (B-R brand), while A_3 and B_3 refer to the same state (other brand). The problem, now, is to calculate $p(A_1)$, $p(A_2)$, and $p(A_3)$ from the available information.

With respect to the partitions B_1, B_2, B_3 and A_1, A_2, A_3, the numbers in Table 4-1 represent all possible conditional probabilities. For example, the given probability of .23 that an Electra purchaser in the given time period will again buy Electra in the next period is the conditional probability that an outcome will be in subset A_1 in the next time period, under the condition that the outcome was in subset B_1 in the original time period. In symbols, the table informs us that $p(A_1 \mid B_1) = .23$. The probability that a B-R purchaser will switch to Electra is the probability that an outcome will be in subset A_1, given that the outcome was originally in B_2. Thus, the table states that $p(A_1 \mid B_2) = .05$. In the same way, we read from the table that the probability a purchaser of a brand other than Electra or B-R will choose B-R brand in the next time period is $p(A_2 \mid B_3) = .15$.

We will view Table 4-1 as we did the input–output tables of Chapter 3, as a matrix:

$$\begin{bmatrix} .23 & .05 & .12 \\ .04 & .25 & .15 \\ .73 & .70 & .73 \end{bmatrix}$$

However, there is more to it than that, since the numbers in the matrix represent probabilities. Specifically,

$$\begin{bmatrix} .23 & .05 & .12 \\ .04 & .25 & .15 \\ .73 & .70 & .73 \end{bmatrix} = \begin{bmatrix} p(A_1 \mid B_1) & p(A_1 \mid B_2) & p(A_1 \mid B_3) \\ p(A_2 \mid B_1) & p(A_2 \mid B_2) & p(A_2 \mid B_3) \\ p(A_3 \mid B_1) & p(A_3 \mid B_2) & p(A_3 \mid B_3) \end{bmatrix}$$

The matrix is denoted by P and called a **transition matrix,** because the conditional probability $p(A_i \mid B_j)$ represents the probability of a transition from state j in the original time period to state i in the succeeding period. Thus, for example, $p(A_3 \mid B_2)$ is the probability that a margarine purchaser in state 2 (B-R brand) in the given time period will make a transition to state 3 (other brand) in the next period. Even a probability like $p(A_1 \mid B_1)$, that an Electra purchaser will stay with Electra, is a transition probability in the sense that it tells us the probability that no transition from state 1 will take place. The numbers in each column of the transition matrix will always add up to one, because, for instance, a buyer of Electra brand margarine in the previous period must either repeat that choice, change to B-R brand, or change to some other brand. So

$$p(A_1 \mid B_1) + p(A_2 \mid B_1) + p(A_3 \mid B_1) = 1$$

The formula from Section 2-10 for $p(A_1)$, the probability that a margarine buyer will select Electra brand in the next time period, is

$$p(A_1) = p(A_1 \mid B_1)p(B_1) + p(A_1 \mid B_2)p(B_2) + p(A_1 \mid B_3)p(B_3)$$

(We are thinking of A_1 as success and everything else as failure.) Similarly, for B-R brand,

$$p(A_2) = p(A_2 \mid B_1)p(B_1) + p(A_2 \mid B_2)p(B_2) + p(A_2 \mid B_3)p(B_3)$$

There is a corresponding formula for $p(A_3)$. Thinking in terms of matrices, we observe that the three equations for calculating $p(A_1)$, $p(A_2)$, and $p(A_3)$ can be replaced by the single matrix equation

$$\begin{bmatrix} p(A_1) \\ p(A_2) \\ p(A_3) \end{bmatrix} = \begin{bmatrix} p(A_1 \mid B_1) & p(A_1 \mid B_2) & p(A_1 \mid B_3) \\ p(A_2 \mid B_1) & p(A_2 \mid B_2) & p(A_2 \mid B_3) \\ p(A_3 \mid B_1) & p(A_3 \mid B_2) & p(A_3 \mid B_3) \end{bmatrix} \begin{bmatrix} p(B_1) \\ p(B_2) \\ p(B_3) \end{bmatrix}$$

In more compact notation, if we let

$$S^{(0)} = \begin{bmatrix} p(B_1) \\ p(B_2) \\ p(B_3) \end{bmatrix} \qquad S^{(1)} = \begin{bmatrix} p(A_1) \\ p(A_2) \\ p(A_3) \end{bmatrix}$$

and, as before,

$$P = \begin{bmatrix} p(A_1 \mid B_1) & p(A_1 \mid B_2) & p(A_1 \mid B_3) \\ p(A_2 \mid B_1) & p(A_2 \mid B_2) & p(A_2 \mid B_3) \\ p(A_3 \mid B_1) & p(A_3 \mid B_2) & p(A_3 \mid B_3) \end{bmatrix}$$

then we have obtained the formula

$$S^{(1)} = PS^{(0)}$$

Specifically, in the margarine marketing problem, we were given that

$$P = \begin{bmatrix} .23 & .05 & .12 \\ .04 & .25 & .15 \\ .73 & .70 & .73 \end{bmatrix}$$

and that

$$S^{(0)} = \begin{bmatrix} .14 \\ .20 \\ .66 \end{bmatrix}$$

Therefore, multiplying the matrices and rounding off the answers, we have

$$S^{(1)} = PS^{(0)} = \begin{bmatrix} .23 & .05 & .12 \\ .04 & .25 & .15 \\ .73 & .70 & .73 \end{bmatrix} \begin{bmatrix} .14 \\ .20 \\ .66 \end{bmatrix} = \begin{bmatrix} .121 \\ .155 \\ .724 \end{bmatrix} \text{(approximately)}$$

Thus, the calculation predicts that Electra's share of the market will drop from 14% to 12% in the next purchasing period and that, at the same time, B-R Stores' share will drop from 20% to about $15\frac{1}{2}$%.

The 3 × 1 matrix $S^{(0)}$ represents the probability that an outcome is in each of the possible states determining the partition B_1, B_2, B_3 at the original time period (that is, the 0 time period). Since A_1, A_2, A_3 is a partition of outcomes for the next time period with respect to the same states, the notation $S^{(1)}$ indicates the probability that an outcome will be in each state in that period (the 1 period). The similarity of notation, $S^{(0)}$ and $S^{(1)}$, is chosen to emphasize the fact that the probabilities in the two matrices are calculated with respect to the same states.

We now turn to a general class of probability problems of which the brand-share problem we have been studying is a special case. We imagine a sample space U and a fixed set of states $s_1, s_2, \ldots, s_k$ initially determining a partition of U. That is, we assume we have a way of deciding which state each element of U is in and each element is in exactly one state. We suppose we are to perform an experiment that will result in placing each member of U in one of the states $s_1, s_2, \ldots, s_k$—possibly the state it is in already or perhaps another one. We are given a $k \times k$ transition matrix P of numbers p_{ij}, where p_{ij} is the probability that an element of U in state s_j when the experiment begins will be in state s_i at the conclusion of the experiment. Thus, if we let $B_1, B_2, \ldots, B_k$ be the partition of U determined by the states when the experiment begins and $A_1, A_2, \ldots, A_k$ be the corresponding partition afterward, then the p_{ij} are the conditional probabilities

$$p_{ij} = p(A_i \mid B_j)$$

The numbers in each column of the transition matrix will always add up to one, because an element of U in state s_j before the experiment must end up in one of the states $s_1, s_2, \ldots, s_k$ after the experiment, which implies that

$$p_{1j} + p_{2j} + \cdots + p_{kj} = 1$$

for each choice of j.

We let

$$
S^{(0)} = \begin{bmatrix} p(s_1)^{(0)} \\ p(s_2)^{(0)} \\ \vdots \\ p(s_k)^{(0)} \end{bmatrix} \qquad
S^{(1)} = \begin{bmatrix} p(s_1)^{(1)} \\ p(s_2)^{(1)} \\ \vdots \\ p(s_k)^{(1)} \end{bmatrix}
$$

where $p(s_i)^{(0)}$ is the probability that an element of U is in state s_i before the experiment begins and $p(s_i)^{(1)}$ is the probability that an element is in state s_i after the conclusion of the experiment. Thus, in terms of the partition, $p(s_i)^{(0)} = p(B_i)$, while $p(s_i)^{(1)} = p(A_i)$. As a consequence of Section 2-10, we know that

$$
S^{(1)} = PS^{(0)}
$$

4-1 **Example**   A sampling of opinion with regard to a coming election indicates that 60% of the voters who voted for the Democratic candidate in the previous election will again vote for the Democrat, while 40% will vote for the Republican opponent. On the other hand, 80% of those who voted for the Republican in the last election will again support the Republican, while 20% will vote for the Democrat. If the Democrat won the previous election with 55% of the vote against 45% for the Republican, who would we expect to win the present election? What percentage of the vote will the winner receive?

Solution   We describe the possible states as s_1, voting for the Democrat, and s_2, voting for the Republican. We have been given data represented by Table 4-2.

Table 4-2

Present election \ Previous election	s_1	s_2
s_1	.60	.20
s_2	.40	.80

Thus, the transition matrix is

$$
P = \begin{bmatrix} .60 & .20 \\ .40 & .80 \end{bmatrix}
$$

Since, in the last election, 55% of the voters chose the Democrat, that is, were in state s_1, and 45% chose the Republican, then

$$S^{(0)} = \begin{bmatrix} .55 \\ .45 \end{bmatrix}$$

The prediction for the election is that

$$S^{(1)} = PS^{(0)} = \begin{bmatrix} .60 & .20 \\ .40 & .80 \end{bmatrix} \begin{bmatrix} .55 \\ .45 \end{bmatrix} = \begin{bmatrix} .42 \\ .58 \end{bmatrix}$$

The Republican will win the election with 58% of the vote.

4-2   **Example**   Suppose that in a study of size of a certain species of animals, it is found that, of the male offspring of large males, 50% are also large, 40% are of medium size, and 10% are small. If the father is of medium size, then 20% of the male offspring are large, 60% medium, and 20% small. If the father is small, 5% of his male offspring are large, 70% medium, and 25% small. If in the present adult male population 30% are large, 60% medium, and 10% small, what will the distribution of sizes be among males in the next generation?

Solution   Let the possible states (sizes) s_1, s_2, s_3 be large, medium, and small, respectively. The sample space U consists of the young males and $p(s_1)^{(0)}$, for example, is the probability that the father of a member of U is large, while $p(s_1)^{(1)}$ represents the probability that the young animal himself will be large. From the data of the problem, the transition matrix is

$$P = \begin{bmatrix} .50 & .20 & .05 \\ .40 & .60 & .70 \\ .10 & .20 & .25 \end{bmatrix}$$

The distribution by size of the present male adult population means that

$$S^{(0)} = \begin{bmatrix} .30 \\ .60 \\ .10 \end{bmatrix}$$

and we calculate

$$S^{(1)} = PS^{(0)} = \begin{bmatrix} .275 \\ .55 \\ .175 \end{bmatrix}$$

So in the next generation, $27\frac{1}{2}\%$ of the males will be large, 55% medium, and $17\frac{1}{2}\%$ small.

Summary

Suppose that each element in a sample space U can be in exactly one of a set of **states** $s_1, s_2, \ldots, s_k$. An experiment is performed, and for each i and j between 1 and k, there is a given probability p_{ij} that an element of U in state j before the experiment begins will be transformed to state i as a result of the experiment. The $k \times k$ matrix P with p_{ij} in the ith row and the jth column is the **transition matrix.** The numbers in each column of P add up to one. Let $p(s_i)^{(0)}$ be the probability that an element of U is in state s_i before the experiment and let $p(s_i)^{(1)}$ be the probability that an element of U is in state s_i after the experiment. Define $k \times 1$ matrices $S^{(0)}$ and $S^{(1)}$ by letting the number in the ith row and $S^{(0)}$ be $p(s_i)^{(0)}$ and of $S^{(1)}$ be $p(s_i)^{(1)}$. Then

$$S^{(1)} = PS^{(0)}$$

Exercises

4-1   In an automobile leasing company's study of its regular customers who lease a new car every year, it is found that 30% of those who leased sedans the previous year change to station wagons, while 40% of those who had leased wagons change to sedans. The rest of the customers repeat the type of vehicle they had before. If 20% of the cars leased to these customers last year were wagons, what percentage of these customers will lease wagons this year?

4-2   The weather in a certain city is classified as fair, cloudy (without rain), and rainy. An investigation of past records indicates the probability that a fair day will be followed by another fair day is .60, by a cloudy day .25, and by a rainy day .15. A cloudy day has a .30 probability of being followed by a fair day, .40 by a cloudy day, and .30 by a rainy day. A rainy day has a .50 chance of being followed by a fair day, .20 by a cloudy day, and .30 by another day of rain. If the forecast for tomorrow's weather is cloudy with a 50% chance of rain (and thus a 50% chance of cloudiness without rain), what does this information suggest the forecast for the day after tomorrow should be?

4-3   A door-to-door salesperson finds that, while regular customers will buy something several times a year, they seldom make a purchase at both of two successive visits. Specifically, only 35% of the regular customers who bought something the last time he called will buy something the next time. On the other hand, if a regular customer did not buy anything during the previous visit, there is a 55% probability that he will make a sale. If during the last round of visits to regular customers, the salesperson sold something to 40% of them, what proportion of visits does he expect will produce sales in the present round of visits?

4-4   A chess player notices that her performance in the second game of a tournament is very much influenced by the outcome of the first game. The probabilities that she will win, lose, or draw the second game, given the outcome of the first, are presented in the table:

Second game \ First game	Win	Lose	Draw
Win	.50	.20	.40
Lose	.10	.30	.20
Draw	.40	.50	.40

In the first games of previous tournaments she has won 60% of the time, lost 10%, and drawn 30%. Estimate the probability that in the next tournament she enters she will win the second game she plays.

4-5 Suppose the probability that a person will be divorced is .35 if his or her parents were divorced and .20 otherwise. If 25% of the people in the previous generation were divorced, what proportion of the present generation will be divorced?

4-6 In a certain farming area, wheat farms are classified as poor, fair, good, or excellent in yield per acre. A new variety of wheat is introduced and used in all the farms in this area. As a result of using the new variety, it is expected that a farm classified as poor has a 50% chance of remaining in the poor classification, a 40% chance of improving its yield to fair, and a 10% chance of improving sufficiently so that it is classified as good. For a farm rated fair, there is a 5% risk that the new variety will cause the farm to drop to poor because of a lower yield per acre, a 45% chance that it will retain its fair classification, a 40% chance that it will advance to good, and a 10% chance that its yield will move it up to the excellent classification. If a farm is classified as good, there is a 5% chance that the introduction of the new variety of wheat will cause a drop in yield to fair, a 55% chance that it will continue to have a good yield, and a 40% chance that its yield will improve so that it is classified as excellent. A farm rated as excellent runs a 5% risk of dropping to good and there is a 95% probability that it will still have an excellent yield with the new variety of wheat. If at present 15% of all farms in the area are rated as poor, 30% fair, 40% good, and 15% excellent, what will the distribution of classifications be after the introduction of the new variety of wheat?

4-7 A study of the records of previous classes indicates the probability of various grades in the second semester of a 1 year calculus course, among the students who received grades of A, B, C, and D or F in the first semester, to be:

Second semester \ First semester	A	B	C	D/F
A	.80	.40	.20	.05
B	.15	.40	.30	.10
C	.05	.15	.30	.40
D/F	0	.05	.20	.45

Among the students who are going on to the second semester this year, 20% received As, 40% Bs, 35% Cs, and 5% Ds in the first semester. What is the expected distribution of grades in the second semester of the course?

4-8   Suppose the probability that a girl whose mother graduated from college will also graduate from college is .70, that she will graduate from high school but not from college is .25, and that she will not finish high school is .05. If the mother finished high school, the probability that the daughter will graduate from college is .50, will graduate from high school but not college is .40, and will not finish high school is .10. If the mother did not graduate from high school, the probabilities are .25 that the daughter will finish college, .55 that she will graduate from high school, and .20 that she will not finish high school. If 30% of the mother's generation graduated from college, 50% graduated from high school but not college, and 20% did not finish high school, what percentage of the daughters will graduate from college? What percentage will graduate from high school only? What percentage will not graduate from high school?

4-9   If an adult of genotype aa mates with one of genotype AA, the offspring will all be of type Aa. If a type aa mates with type Aa, then according to Mendel's law, the probability that the offspring will be of type Aa is .50 and of type aa .50. Finally, the offspring of parents who are both of genotype aa must be of the same genotype. Suppose the female of some animal species is known from her physical characteristics to be of type aa with respect to some gene, but the identity of her mate is unknown. If 30% of the males of this species are of genotype AA, 50% of type Aa, and 20% of type aa, and mating is independent of this gene, what is the probability the first offspring of this female will be of genotype AA? What is the probability the first offspring will be of genotype Aa? What is the probability the first offspring will be of genotype aa?

4-10   A college offers majors in four groups: arts, humanities, natural sciences, and social sciences. In addition to choosing a major, each student at this college must select a minor subject for graduation. Suppose a study of major and minor choices by students at this college shows the probability that an arts major will choose a minor also in the arts (for example, a music major chooses an art history minor) is .50, that an arts major will choose a humanities minor is .30, a natural sciences minor .05, and a social sciences minor .15. The same study indicates the probabilities that a humanities major will choose an arts minor is .20, .55 for another humanities subject, .05 for natural sciences, and .20 for social sciences. For a natural sciences major, the chances are .10 for an arts minor, .25 for the humanities, .45 for a natural sciences minor other than the major, and .20 for the social sciences. Finally, the probability that a social sciences major will select an arts minor is .15, a humanities minor .35, a natural sciences minor .20, and another social science .30. If the present distribution of students at the college by major is 20% in the arts, 25% humanities, 20% natural sciences, and 35% social sciences, what should we expect the distribution of students according to minors to be at this college?

4-2 Markov
Chain Processes

A company that markets a new product, such as the Electra brand margarine we have been using as an example, is not primarily interested in short-term behavior, i.e., the product's immediate popularity among potential buyers. Rather, the company is concerned with its product's long-term prospects—whether the product will continue to command a significant share of the market.

In Section 4-1, we saw that if the company could collect data on the likelihood that a margarine buyer would repeat the brand purchased or switch to another brand, then it was possible, given the product's share of the market in a base period, to predict the brand-share in the next buying period. Now, if the company can assume the market has settled down sufficiently after the introduction of the new brand so that the table of buyers' tendencies to repeat or switch brands describes a stable phenomenon, then it is possible to use this information to make long-range predictions of brand-share. We continue to assume that each purchaser makes one purchase of margarine in each time period.

To illustrate the meaning of the assumption of stability we repeat the table we used earlier as Table 4-3. The probability is .23 that an Electra purchaser will stay with Electra brand in the next period, .04 of a switch from Electra to B-R, .05 of a switch in the opposite direction, and so on. The assumption we must make to produce long-term predictions is that the table remains valid over the period of time we wish to predict. That is, in each succeeding time period, Electra brand will retain 23% of the people who bought that brand, while 4% change to B-R, 25% of B-R buyers repeat the brand purchased, and so on. This behavior is assumed to be independent of the number of buyers who chose each brand. For example, whether Electra brand commands a large share of the market or a small share, 23% of its buyers in one period will repeat their purchase in the next.

Table 4-3

Next period \ Original period	Electra	B-R	Others
Electra	.23	.05	.12
B-R	.04	.25	.15
Others	.73	.70	.73

So now we assume the table of margarine buyer behavior is valid for each succeeding time period and use this information and the techniques of Section 4-1 to study the market shares of Electra and B-R brand margarine over several time periods. Recall from the brand-share example that in the original time period, Electra had been the choice of 14% of all margarine buyers surveyed and B-R had been the choice of 20%. In the notation of that section, we write

$$S^{(0)} = \begin{bmatrix} .14 \\ .20 \\ .66 \end{bmatrix}$$

The table of buyer behavior becomes the transition matrix

$$P = \begin{bmatrix} .23 & .05 & .12 \\ .04 & .25 & .15 \\ .73 & .70 & .73 \end{bmatrix}$$

and we calculated in Section 4-1 that the market shares in the next time period would be

$$S^{(1)} = PS^{(0)} = \begin{bmatrix} .121 \\ .155 \\ .724 \end{bmatrix} \quad \text{(approximately)}$$

If we think of the 1 time period as a new base for prediction, then the next time period should be numbered 2, and the market shares for the various brands in that period should be represented by a matrix

$$S^{(2)} = \begin{bmatrix} p(s_1)^{(2)} \\ p(s_2)^{(2)} \\ p(s_3)^{(2)} \end{bmatrix}$$

where $p(s_1)^{(2)}$ is the percentage of the margarine buyers who choose Electra brand in the 2 time period, $p(s_2)^{(2)}$ is the same for B-R brand, and $p(s_3)^{(2)}$ represents the rest of the buyers in that time period. The assumption is that buyer behavior does not change over time. This means that we should use the same transition matrix P to calculate market shares in the period 2 as we used for period 1. The point is, then, that the situation is exactly the same as before, except for the number of the base time period (1 instead of 0). Consequently, it must be that

$$S^{(2)} = PS^{(1)}$$

For this particular example, we calculate that, to four decimal places,

$$S^{(2)} = \begin{bmatrix} .23 & .05 & .12 \\ .04 & .25 & .15 \\ .73 & .70 & .73 \end{bmatrix} \begin{bmatrix} .121 \\ .155 \\ .724 \end{bmatrix} = \begin{bmatrix} .1225 \\ .1522 \\ .7253 \end{bmatrix}$$

Therefore, Electra will continue to hold about 12% of the market in period 2, but B-R Stores brand will drop approximately another .003 from period 1 to period 2.

We could have found the figures in the matrix $S^{(2)}$ another way—without first calculating $S^{(1)}$. Assuming the transition matrix is the same in each period, we have the two equations

$$S^{(1)} = PS^{(0)} \qquad S^{(2)} = PS^{(1)}$$

If we substitute the right-hand side of the first equation for $S^{(1)}$ in the second, we have

$$S^{(2)} = P(PS^{(0)}) = P^2 S^{(0)} \qquad \text{(where } P^2 = PP\text{)}$$

Thus, we could first have calculated

$$P^2 = PP = \begin{bmatrix} .23 & .05 & .12 \\ .04 & .25 & .15 \\ .73 & .70 & .73 \end{bmatrix} \begin{bmatrix} .23 & .05 & .12 \\ .04 & .25 & .15 \\ .73 & .70 & .73 \end{bmatrix} = \begin{bmatrix} .1425 & .1080 & .1227 \\ .1287 & .1695 & .1518 \\ .7288 & .7225 & .7255 \end{bmatrix}$$

and, rounding off the answer to four decimal places,

$$S^{(2)} = \begin{bmatrix} .1425 & .1080 & .1227 \\ .1287 & .1695 & .1518 \\ .7288 & .7225 & .7255 \end{bmatrix} \begin{bmatrix} .14 \\ .20 \\ .66 \end{bmatrix} = \begin{bmatrix} .1257 \\ .1521 \\ .7222 \end{bmatrix}$$

This calculation is somewhat more accurate than before because we avoid the error caused by rounding off $S^{(1)}$ to three decimal places.

Similarly, if we want to calculate the brand-shares $S^{(3)}$ in the next time period without determining the intervening brand-share figures, we can see that our assumption about the continuing validity of the transition matrix implies that

$$S^{(3)} = PS^{(2)}$$

and combine this with the equation above to obtain

$$S^{(3)} = P(P^2 S^{(0)}) = P^3 S^{(0)} \qquad \text{(where } P^3 = PP^2 = PPP\text{)}$$

Now let us consider the general type of probability problem of which the long-range prediction of brand-share is a particular example. There is a sample space U and a fixed set of states $s_1, s_2, \ldots, s_k$, just as there were in Section 4-1. Also, we still imagine an experiment that put each element of U into one of the states $s_1, s_2, \ldots, s_k$. The new aspect of the situation is that we now suppose that the experiment is repeated several times; each time under identical circumstances. The probability p_{ij} that an element of U will change from state s_j to state s_i as a result of a performance of the experiment is thus assumed to be the same each time the experiment takes place. The transition matrix P of the probabilities p_{ij} therefore remains the same each time the experiment is repeated. This type of repeated experiment probability procedure is called a **Markov chain process.** We still write $p(s_i)^{(0)}$ for the probability that an element of U is in state s_i at the beginning of the process and $S^{(0)}$ for the corresponding $k \times 1$ matrix. Let $p(s_i)^{(m)}$ stand for the probability that a member of U is in state s_i after the experiment has been repeated m times under identical circumstances described by the transition matrix P. For $S^{(m)}$ the $k \times 1$ matrix with $p(s_i)^{(m)}$ in the ith row, we have an equation

$$S^{(m)} = P^m S^{(0)}$$

where P^m is the product

$$P^m = P \cdot P \cdots P \qquad (m \text{ copies of } P)$$

4-3 **Example** In a certain species of animals, the probability that the male offspring of a large male will also be large is .50, that it will be of medium size is .40, and that it will be small is .10. For the male offspring of a male of medium size the probabilities are .20 that they will be large, .60 medium, and .20 small. If the father is small, the corresponding probabilities are .05 for large, .70 for medium, and .25 for small. If in the present population, 30% of the males are large, 60% are medium, and 10% are small, and if the probabilities of male offspring of various sizes from fathers of different sizes do not change from generation to generation, what will be the distribution of sizes of males in four generations?

Solution We observed in Example 4-2 that in this situation

$$P = \begin{bmatrix} .50 & .20 & .05 \\ .40 & .60 & .70 \\ .10 & .20 & .25 \end{bmatrix} \quad \text{and} \quad S^{(0)} = \begin{bmatrix} .30 \\ .60 \\ .10 \end{bmatrix}$$

We wish to know $S^{(4)}$, which, according to the general theory, is calculated by the formula

$$S^{(4)} = P^4 S^{(0)}$$

We can compute that

$$P^4 = \begin{bmatrix} .257 & .244 & .238 \\ .562 & .571 & .575 \\ .181 & .185 & .187 \end{bmatrix} \quad \text{(approximately)}$$

and therefore

$$S^{(4)} = \begin{bmatrix} .257 & .244 & .238 \\ .562 & .571 & .575 \\ .181 & .185 & .187 \end{bmatrix} \begin{bmatrix} .30 \\ .60 \\ .10 \end{bmatrix} = \begin{bmatrix} .25 \\ .57 \\ .18 \end{bmatrix} \quad \text{(approximately)}$$

so 25% of the males will be large in four generations, 57% of medium size, and 18% small.

Next, let us go back and carry out the calculation of margarine market shares one more period into the future than we did before, still using the same transition matrix. The next time period is number 3, and since we have shown that, to two decimal places of accuracy,

$$S^{(2)} = \begin{bmatrix} .12 \\ .15 \\ .73 \end{bmatrix}$$

we readily calculate that, if we again round off to two decimal places,

$$S^{(3)} = PS^{(2)} = \begin{bmatrix} .12 \\ .15 \\ .73 \end{bmatrix}$$

In other words, to two decimal places of accuracy there is no change in market penetration for Electra and B-R brand margarines from period 2 to period 3. We might therefore suspect that the margarine market has settled down to the extent that Electra margarine can expect to continue to receive 12% of all margarine sales and B-R brand 15%. We can show this is indeed the case by supposing that at some future time period m it is true that

$$S^{(m)} = \begin{bmatrix} .12 \\ .15 \\ .73 \end{bmatrix}$$

Then matrix multiplication shows us, when we round off after two places,

$$S^{(m+1)} = PS^{(m)} = \begin{bmatrix} .23 & .05 & .12 \\ .04 & .25 & .15 \\ .73 & .70 & .73 \end{bmatrix} \begin{bmatrix} .12 \\ .15 \\ .73 \end{bmatrix} = \begin{bmatrix} .12 \\ .15 \\ .73 \end{bmatrix}$$

This implies the market shares for Electra and B-R will remain constant. The conclusion, of course, assumes the customers' behavior with regard to brand loyalty and switching does not change.

Thus, it may happen in a Markov chain process that after a certain number of repetitions of the experiment, there is no further change in the probability that an element of U will be in each of the possible states. In symbols, we may say, for some m,

$$S^{(m)} = S^{(m+1)} = S^{(m+2)} = \cdots$$

in which case we say the Markov chain process has reached **equilibrium.** (Strictly speaking, $S^{(m)}$, $S^{(m+1)}$, and the rest will usually be changing very slightly as they approach equilibrium, and they never quite reach it. However, as we calculate the numbers in these matrices only to a few decimal places of accuracy, they will appear to be equal from some point on.)

Since Markov chain processes are used most often to find out how something will turn out in the long run, e.g., what share of the market a product can hope to command on a (more-or-less) permanent basis, the question of whether the process reaches equilibrium is a central one in the subject. Unfortunately, not all Markov chain processes can reach an equilibrium position, as the following example demonstrates. Let

$$S^{(0)} = \begin{bmatrix} \dfrac{1}{4} \\ \dfrac{3}{4} \end{bmatrix} \qquad P = \begin{bmatrix} 0 & 1 \\ 1 & 0 \end{bmatrix}$$

then

$$S^{(1)} = \begin{bmatrix} \dfrac{3}{4} \\ \dfrac{1}{4} \end{bmatrix} \qquad S^{(2)} = \begin{bmatrix} \dfrac{1}{4} \\ \dfrac{3}{4} \end{bmatrix} \qquad S^{(3)} = \begin{bmatrix} \dfrac{3}{4} \\ \dfrac{1}{4} \end{bmatrix}$$

and so on. Thus, the process forever jumps back and forth as $S^{(m)}$ changes, depending on whether m is even or odd.

There is, however, a large class of Markov chain processes we know reach equilibrium, and these include many that arise in practical problems. A transition matrix P is said to be **regular** if for some number r the matrix P^r contains no zeros. A Markov chain process in which the transition matrix is regular always reaches equilibrium. The transition matrix

$$P = \begin{bmatrix} .23 & .05 & .12 \\ .04 & .25 & .15 \\ .73 & .70 & .73 \end{bmatrix}$$

of the margarine problem is already without zeros, so we know it is regular and therefore this process must reach equilibrium.

4-4   **Example**   Show that the transition matrix below is regular.

$$P = \begin{bmatrix} .9 & .3 & .5 \\ .1 & .6 & 0 \\ 0 & .1 & .5 \end{bmatrix}$$

Solution   Since P itself contains zeros, we calculate

$$P^2 = \begin{bmatrix} .84 & .50 & .70 \\ .15 & .39 & .05 \\ .01 & .11 & .25 \end{bmatrix}$$

which has no zeros and shows that P is regular.

Once we know the Markov chain process reaches equilibrium because the transition matrix is regular, we still must find out what $S^{(m)}$ will be when equilibrium is reached. One procedure is simply to calculate $S^{(1)}$, $S^{(2)}$, and so on until, at last, we come to a number m for which $S^{(m)} = S^{(m+1)}$. Then

$$S^{(m+2)} = PS^{(m+1)} = PS^{(m)} = S^{(m+1)}$$

and so on, which tells us equilibrium has been reached. Another way is to calculate P^2, P^3, P^4, and so on. If we continue, then according to the mathematical theory, when P is regular we will reach a number m for which all the columns of P^m are identical and the columns of P^m are equal to $S^{(m)}$ at equilibrium. In Example 4-3, we found that

$$P^4 = \begin{bmatrix} .257 & .244 & .238 \\ .562 & .571 & .575 \\ .181 & .185 & .187 \end{bmatrix}$$

Continued calculation will establish that, rounding off to two decimal places,

$$P^6 = \begin{bmatrix} .25 & .25 & .25 \\ .57 & .57 & .57 \\ .18 & .18 & .18 \end{bmatrix}$$

Eventually, 25% of the males will be large, 57% will be medium, and 18% will be small; and those percentages will remain constant thereafter.

Both of these procedures for finding the equilibrium values (calculating $S^{(1)}$, $S^{(2)}$, and so on, or calculating P^2, P^3, and the rest) are very unsatisfactory because they involve numerous matrix multiplications and there is no way of telling how many such multiplications will be required for a particular problem. There is, however, an efficient method for determining $S^{(m)}$ when equilibrium is reached. This involves just a fixed number of steps, where that number depends only on the number of states of the Markov chain process. We will describe this technique in Section 4-3.

Summary

A **Markov chain process** is a sequence of identical experiments of the following type: There is a sample space U and a set of states $s_1, s_2, \ldots, s_k$. The effect of the experiment is to place each element of U in one of the states. The probability p_{ij} that an element in state s_j at the beginning of the experiment will be in state s_i at the conclusion is the same each time the experiment is performed. The $k \times k$ matrix P of the p_{ij} is the **transition matrix** of the Markov chain process.

Let $p(s_i)^{(0)}$ be the probability that an element of U is in state s_i before the first performance of the experiment in a Markov chain process and let $p(s_i)^{(m)}$ be the probability that an element of U is in state s_i after the experiment has taken place m times. Define $S^{(0)}$ to be the $k \times 1$ matrix of $p(s_i)^{(0)}$ and $S^{(m)}$ the $k \times 1$ matrix of $p(s_i)^{(m)}$. Then $S^{(m)} = P^m S^{(0)}$, where P^m is the product of m copies of the transition matrix P.

A Markov chain process reaches **equilibrium** if $S^{(m)} = S^{(m+1)}$ for some m.

The transition matrix P of a Markov chain process is **regular** if P^r contains no zeros for some positive whole number r. A Markov chain process whose transition matrix is regular always reaches equilibrium.

If all the columns of P^m are identical for some m, then the Markov chain process has reached equilibrium after m repetitions of the experiment and the columns of P^m are equal to $S^{(m)}$.

Exercises

● *In Exercises 4-11 through 4-14, calculate $S^{(1)}$, $S^{(2)}$, and $S^{(3)}$ for the Markov chain process with the $S^{(0)}$ and P given.*

4-11 $S^{(0)} = \begin{bmatrix} .30 \\ .70 \end{bmatrix}$ $P = \begin{bmatrix} .25 & .05 \\ .75 & .95 \end{bmatrix}$

4-12 $S^{(0)} = \begin{bmatrix} .1 \\ .8 \\ .1 \end{bmatrix}$ $P = \begin{bmatrix} 0 & .1 & .6 \\ .2 & 0 & 0 \\ .8 & .9 & .4 \end{bmatrix}$

4-13 $S^{(0)} = \begin{bmatrix} 0 \\ 1 \\ 0 \end{bmatrix}$ $P = \begin{bmatrix} .2 & 0 & .4 \\ .2 & .4 & .2 \\ .6 & .6 & .4 \end{bmatrix}$

4-14 $S^{(0)} = \begin{bmatrix} .3 \\ .3 \\ .4 \end{bmatrix}$ $P = \begin{bmatrix} 1 & .7 & 0 \\ 0 & .1 & 0 \\ 0 & .2 & 1 \end{bmatrix}$

4-15 Suppose in some population of animals there is a genetic defect carried only by males. The probability is .80 that the male offspring of a male with the defect will also have it and a male whose male parent does not have the defect still has a probability of .10 of possessing the defect through mutation. If, in some generation, 25% of the males possess the defect, what proportion of males in the next three generations will have it?

4-16 The weather in a certain city is classified as fair, cloudy (without rain), and rainy. An investigation of weather records indicates the probability that a fair day will be followed by another fair day is .50, by a cloudy day .30, and by a rainy day .20. A cloudy day has a .40 probability of being followed by a fair day, .40 by a cloudy day, and .20 by a rainy day. A rainy day has a .45 chance of being followed by a fair day, .25 by a cloudy day, and .30 by another day of rain. If it is raining today, that is,

$$S^{(0)} = \begin{bmatrix} 0 \\ 0 \\ 1 \end{bmatrix}$$

what is the probability of fair weather, cloudy weather, and rain in each of the next 3 days, based just on this information?

- In Exercises 4-17 and 4-18, calculate $S^{(3)}$ for the given $S^{(0)}$ and P, without also calculating $S^{(1)}$ and $S^{(2)}$, but rather by calculating P^3.

4-17 $S^{(0)} = \begin{bmatrix} .22 \\ .78 \end{bmatrix}$ $P = \begin{bmatrix} .2 & .4 \\ .8 & .6 \end{bmatrix}$

4-18 $S^{(0)} = \begin{bmatrix} .1 \\ .2 \\ .7 \end{bmatrix}$ $P = \begin{bmatrix} 0 & 0 & .3 \\ .5 & .8 & .7 \\ .5 & .2 & 0 \end{bmatrix}$

4-19   A study of coffee buying habits indicates that about 70% of the people who bought Jones brand coffee in a given buying period would again choose Jones brand in the next period, while 20% of those who chose another brand in the given period would switch to Jones brand. If, at the time of the study, 34% of the coffee buyers in a certain area chose Jones brand, calculate P^3 to find out what Jones brand's share of the market will be in that area three time periods later.

4-20   A college's records indicate that each year 10% of its humanities majors change to a major in the social sciences, while the rest remain in the humanities. Records also show that 10% of its social science majors change to majors in the humanities, 10% change to the natural sciences, and the rest remain in the social sciences. Natural science majors, however, do not change out of that area of studies. If at that college 50% of the freshmen chose majors in the humanities, 30% in the social sciences, and 20% in the natural sciences, calculate the proportion of majors in each area for these same students in their senior year without doing the same for the intervening years (that is, by calculating P^3).

- In Exercises 4-21 and 4-22, verify that the given transition matrix is regular.

4-21 $\begin{bmatrix} 0 & .2 & 0 \\ .1 & 0 & .3 \\ .9 & .8 & .7 \end{bmatrix}$

4-22 $\begin{bmatrix} .5 & .1 & 0 & 0 \\ 0 & 0 & .2 & 0 \\ 0 & .9 & 0 & .1 \\ .5 & 0 & .8 & .9 \end{bmatrix}$

- In Exercises 4-23 and 4-24, verify that the given Markov chain process appears to reach equilibrium at the indicated values if numbers are rounded off to two decimal places of accuracy.

4-23 $S^{(0)} = \begin{bmatrix} 1 \\ 0 \end{bmatrix}$ $P = \begin{bmatrix} .15 & .20 \\ .85 & .80 \end{bmatrix}$ Equilibrium $= \begin{bmatrix} .19 \\ .81 \end{bmatrix}$

4-24 $S^{(0)} = \begin{bmatrix} .33 \\ .34 \\ .33 \end{bmatrix}$ $P = \begin{bmatrix} .1 & .3 & .7 \\ 0 & .2 & 0 \\ .9 & .5 & .3 \end{bmatrix}$ Equilibrium $= \begin{bmatrix} .44 \\ 0 \\ .56 \end{bmatrix}$

4-25   Suppose there is a society in which all women are classified in the opinion of the community as either noble or common on the basis both of birth and accomplishments. Suppose also that the probability that the daughter of a noble woman will be noble is .70, but the probability of attaining nobility drops to just .10 for the daughter of a woman who is not noble. Show that, in the long run, the society will stabilize with 25% of all women considered noble.

4-3 Equilibrium

A Markov chain process reaches equilibrium after m repetitions of the experiment if the probability that an element of the sample space U will be in each of the states s_1, s_2, . . . , s_k does not change as a result of further repetitions of the experiment. In symbols,

$$p(s_i)^{(m)} = p(s_i)^{(m+1)} = p(s_i)^{(m+2)} = \cdots$$

for each i from 1 to k. In matrix terms,

$$S^{(m)} = S^{(m+1)} = S^{(m+2)} = \cdots$$

We observed in Section 4-2 that once we reach the stage m where $S^{(m)} = S^{(m+1)}$, we have in fact reached equilibrium and need not check that $S^{(m+1)} = S^{(m+2)}$ and the rest. Now, since it is always true that

$$PS^{(m)} = S^{(m+1)}$$

our job is to find the matrix $S^{(m)}$ for which

$$PS^{(m)} = S^{(m)}$$

For example, suppose a salesperson for a manufacturer of handtools (hammers, screwdrivers, and such) calls on hardware stores in some area at regular intervals. The hardware stores do not generally need to order tools from the salesperson on each visit, because their stock on hand may be sufficient for their expected sales until the next visit. On the other hand, each store will order fairly often, because it does not want to tie up money and display space through a large inventory of handtools. The salesperson observes on a recent round of visits that 20% of the stores that ordered tools in the previous visit ordered again, while 70% of the stores that had not ordered the previous time did place an order. If this is the customary behavior of the hardware stores in a single round of visits, at what percentage of the stores can the salesperson expect to receive an order for tools? We can present the information through Table 4-4.

Table 4-4

Present visit \ Previous visit	Order	No order
Order	.20	.70
No order	.80	.30

The transition matrix for this Markov chain process is

$$P = \begin{bmatrix} .20 & .70 \\ .80 & .30 \end{bmatrix}$$

When the process reaches equilibrium, we will have a matrix

$$S^{(m)} = \begin{bmatrix} p(s_1)^{(m)} \\ p(s_2)^{(m)} \end{bmatrix}$$

where $p(s_1)^{(m)}$ is the probability a store will order and $p(s_2)^{(m)}$ is the probability it will not, such that

$$PS^{(m)} = S^{(m)}$$

Forgetting the context of Markov chain processes for the moment, we are trying to solve a matrix equation for a 2×1 matrix $S^{(m)}$, given a constant matrix P. Therefore, we can simplify the notation by replacing $S^{(m)}$ by a matrix of unknowns

$$X = \begin{bmatrix} x \\ y \end{bmatrix}$$

and attempting to solve the matrix equation

$$PX = X$$

for X. Subtracting X from both sides of the equation gives

$$PX - X = 0$$

where 0 denotes the 2×1 matrix of zeros. As in Section 3-3, we can use the property of the 2×2 identity matrix I, namely $IX = X$, and the distributive law to rewrite the equation $PX - X = 0$ as

$$(P - I)X = 0$$

Now we can see that there is an obvious solution to the equation, namely

$$X = 0 = \begin{bmatrix} 0 \\ 0 \end{bmatrix}$$

But $X = 0$ is a useless solution, because the numbers in $S^{(m)}$, which X replaced in the equation, were $p(s_1)^{(m)}$, the probability a store would place an order, and $p(s_2)^{(m)}$, the probability it would not. Certainly these probabilities must add up to one. Thus, the matrix X must not only be a solution to the equation $PX = X$, or $(P - I)X = 0$, but the numbers in X must also add up to one, that is, $x + y = 1$.

In order to combine the two equations $(P - I)X = 0$ and $x + y = 1$, we write out the matrix equation as

$$(P - I)X = \left(\begin{bmatrix} .20 & .70 \\ .80 & .30 \end{bmatrix} - \begin{bmatrix} 1 & 0 \\ 0 & 1 \end{bmatrix} \right) \begin{bmatrix} x \\ y \end{bmatrix} = \begin{bmatrix} -.80 & .70 \\ .80 & -.70 \end{bmatrix} \begin{bmatrix} x \\ y \end{bmatrix}$$

$$= \begin{bmatrix} (-.80)x + .70y \\ .80x + (-.70)y \end{bmatrix} = \begin{bmatrix} 0 \\ 0 \end{bmatrix}$$

We observe that the matrix equation $(P - I)X = 0$ represents two ordinary equations, which we add to the other condition imposed on X:

$$x + \qquad y = 1$$
$$(-.80)x + \qquad .70y = 0$$
$$.80x + (-.70)y = 0$$

Notice that if we let

$$M = \begin{bmatrix} 1 & 1 \\ -.80 & .70 \\ .80 & -.70 \end{bmatrix}$$

then MX represents the left-hand sides of the equations, that is,

$$MX = \begin{bmatrix} 1 & 1 \\ -.80 & .70 \\ .80 & -.70 \end{bmatrix} \begin{bmatrix} x \\ y \end{bmatrix} = \begin{bmatrix} x + y \\ (-.80)x + .70y \\ .80x + (-.70)y \end{bmatrix}$$

Representing the right-hand sides by

$$B = \begin{bmatrix} 1 \\ 0 \\ 0 \end{bmatrix}$$

the three equations above become the single matrix equation

$$MX = B$$

In conclusion, we have replaced the two conditions on X, namely $PX = X$ and $x + y = 1$, by a single matrix equation $MX = B$, where M is of the form

$$M = \begin{bmatrix} 1 & 1 \\ P & - & I \end{bmatrix}$$

(I is the 2×2 identity matrix) and

$$B = \begin{bmatrix} 1 \\ 0 \\ 0 \end{bmatrix}$$

The general problem of finding equilibrium for a Markov chain process is that of finding the matrix of probabilities

$$S^{(m)} = \begin{bmatrix} p(s_1)^{(m)} \\ \vdots \\ p(s_k)^{(m)} \end{bmatrix}$$

for which $PS^{(m)} = S^{(m)}$. If we replace $S^{(m)}$ by a matrix

$$X = \begin{bmatrix} x_1 \\ \vdots \\ x_k \end{bmatrix}$$

of unknowns, it is not enough to try to solve $PX = X$, because the $p(s_i)^{(m)}$ represent probabilities and this implies

$$p(s_1)^{(m)} + \cdots + p(s_k)^{(m)} = 1$$

Therefore, we must also require

$$x_1 + \cdots + x_k = 1$$

The conditions can be represented by the single matrix equation $MX = B$, where

$$M = \begin{bmatrix} 1 \cdots 1 \\ P - I \end{bmatrix}$$

(I is the $k \times k$ identity matrix) and

$$B = \begin{bmatrix} 1 \\ 0 \\ \vdots \\ 0 \end{bmatrix}$$

The equilibrium problem, then, has been reduced to the problem of solving a matrix equation of the form $MX = B$, where X is a $k \times 1$ matrix of unknowns. The matrix M is a $(k + 1) \times k$ matrix, since P is $k \times k$ and we have added 1 row—the row of 1s. Therefore, B is a $(k + 1) \times 1$ matrix.

In Section 3-6, we studied matrix equations of the form $AX = B$, where A is a given $n \times m$ matrix, B is a given $n \times 1$ matrix, and X is an $m \times 1$ matrix of unknowns. The equation $MX = B$ above is of this type, so just as in Chapter 3, we form the augmented matrix

$$[M \,|\, B]$$

and then produce a matrix

$$[M' \,|\, B']$$

where the columns of M' are as much like those of an identity matrix as possible. The form of M' is simple enough so the solution to $M'X = B'$, which is the same as that of $MX = B$, will be evident. In Chapter 3, we saw examples of equations of the form $AX = B$ which had an infinite number of solutions and others with no solution at all. However, when the transition matrix P is regular, the Markov chain process reaches equilibrium. In other words, the equation $MX = B$, where

$$M = \begin{bmatrix} 1 \cdots 1 \\ P - I \end{bmatrix} \qquad B = \begin{bmatrix} 1 \\ 0 \\ \vdots \\ 0 \end{bmatrix}$$

has a single solution which is the equilibrium position for the process.

For example, we apply this procedure to the handtool salesperson's problem we described earlier in this section. In that case,

$$M = \begin{bmatrix} 1 & 1 \\ -.80 & .70 \\ .80 & -.70 \end{bmatrix} \qquad B = \begin{bmatrix} 1 \\ 0 \\ 0 \end{bmatrix}$$

We form the augmented matrix, expressing the numbers in fraction form for greater accuracy in calculation,

$$\begin{bmatrix} 1 & 1 & | & 1 \\ -.80 & .70 & | & 0 \\ .80 & -.70 & | & 0 \end{bmatrix} = \begin{bmatrix} 1 & 1 & | & 1 \\ -\dfrac{8}{10} & \dfrac{7}{10} & | & 0 \\ \dfrac{8}{10} & -\dfrac{7}{10} & | & 0 \end{bmatrix}$$

and go through the steps required for the solution:

$$\begin{bmatrix} 1 & 1 & | & 1 \\ 0 & \dfrac{15}{10} & | & \dfrac{8}{10} \\ 0 & -\dfrac{15}{10} & | & -\dfrac{8}{10} \end{bmatrix}$$

$$\begin{bmatrix} 1 & 1 & | & 1 \\ 0 & 1 & | & \dfrac{8}{15} \\ 0 & -\dfrac{15}{10} & | & -\dfrac{8}{10} \end{bmatrix}$$

$$\begin{bmatrix} 1 & 0 & | & \dfrac{7}{15} \\ 0 & 1 & | & \dfrac{8}{15} \\ 0 & 0 & | & 0 \end{bmatrix}$$

Thus, the solution to $MX = B$ is

$$X = \begin{bmatrix} x \\ y \end{bmatrix} = \begin{bmatrix} \dfrac{7}{15} \\ \dfrac{8}{15} \end{bmatrix} = \begin{bmatrix} .47 \\ .53 \end{bmatrix} \text{ (approximately)}$$

which means the salesperson can expect to receive an order from 47% of the hardware stores visited on each round.

4-5 **Example** An automobile leasing company's study indicates that 60% of its regular customers who lease a two-door sedan in one time period will lease the same type of car the next time, 30% will change to a four-door sedan, and the other 10% will switch to a station wagon. If a customer leases a four-door sedan, the probability is .25 that he will change to a two-door sedan the next time, .55 that a four-door sedan will be chosen again, and .20 that a station wagon will be chosen. Of the customers who leased a station wagon in the previous period, none will switch to a two-door sedan, 50% will change to a four-door sedan, and the other 50% will again choose a station wagon. Suppose these buying patterns remain constant and have done so long enough so that the demand for each type of car is the same from period to period. When the leasing company orders new cars from the manufacturer, what proportion should be two-door sedans, four-door sedans, and station wagons?

Solution The matrix of unknowns is

$$X = \begin{bmatrix} x \\ y \\ z \end{bmatrix}$$

where x refers to two-door sedans, y to four-door sedans, and z to station wagons. The transition matrix is

$$P = \begin{bmatrix} .60 & .25 & 0 \\ .30 & .55 & .50 \\ .10 & .20 & .50 \end{bmatrix}$$

Although we do not need the information to solve the problem, the reader might wish to verify that P is regular, because P^2 contains no zeros. We form

$$M = \begin{bmatrix} 1 & 1 & 1 \\ P - I & & \end{bmatrix} = \begin{bmatrix} 1 & 1 & 1 \\ -.40 & .25 & 0 \\ .30 & -.45 & .50 \\ .10 & .20 & -.50 \end{bmatrix}$$

and

$$B = \begin{bmatrix} 1 \\ 0 \\ 0 \\ 0 \end{bmatrix}$$

Thus, in order to solve the equation $MX = B$, we form the augmented matrix (with entries in fraction form) and calculate, combining a number of steps,

$$
\begin{bmatrix}
1 & 1 & 1 & 1 \\
-\dfrac{2}{5} & \dfrac{1}{4} & 0 & 0 \\
\dfrac{3}{10} & -\dfrac{9}{20} & \dfrac{1}{2} & 0 \\
\dfrac{1}{10} & \dfrac{1}{5} & -\dfrac{1}{2} & 0
\end{bmatrix}
$$

$$
\begin{bmatrix}
1 & 1 & 1 & 1 \\
0 & \dfrac{13}{20} & \dfrac{2}{5} & \dfrac{2}{5} \\
0 & -\dfrac{3}{4} & \dfrac{1}{5} & -\dfrac{3}{10} \\
0 & \dfrac{1}{10} & -\dfrac{3}{5} & -\dfrac{1}{10}
\end{bmatrix}
$$

$$
\begin{bmatrix}
1 & 0 & \dfrac{5}{13} & \dfrac{5}{13} \\
0 & 1 & \dfrac{8}{13} & \dfrac{8}{13} \\
0 & 0 & \dfrac{43}{65} & \dfrac{21}{130} \\
0 & 0 & \dfrac{43}{65} & \dfrac{21}{130}
\end{bmatrix}
$$

$$
\begin{bmatrix}
1 & 0 & 0 & \dfrac{325}{1118} \\
0 & 1 & 0 & \dfrac{260}{559} \\
0 & 0 & 1 & \dfrac{21}{86} \\
0 & 0 & 0 & 0
\end{bmatrix}
$$

So the solution to the equation is

$$
X = \begin{bmatrix} \dfrac{325}{1118} \\ \dfrac{260}{559} \\ \dfrac{21}{86} \end{bmatrix} = \begin{bmatrix} .291 \\ .465 \\ .244 \end{bmatrix} \quad \text{(approximately)}
$$

which means that about 29.1% of the cars the leasing company orders should be two-door sedans, 46.5% should be four-door sedans, and the remaining 24.4% should be station wagons.

Summary

A Markov chain process with $k \times k$ transition matrix P reaches **equilibrium** after m repetitions of the experiment if $PS^{(m)} = S^{(m)}$. The equilibrium matrix $S^{(m)}$ can be calculated by solving a matrix equation $MX = B$ for a $k \times 1$ matrix X of unknowns, where M is the $(k + 1) \times k$ matrix

$$M = \begin{bmatrix} 1 \cdots 1 \\ P - I \end{bmatrix}$$

with I the $k \times k$ identity matrix, and B is the $(k + 1) \times 1$ matrix

$$B = \begin{bmatrix} 1 \\ 0 \\ \vdots \\ 0 \end{bmatrix}$$

The method of solution can be found in Section 3-6. In a Markov chain process in which the transition matrix P is **regular** (that is, P^r contains no zeros, for some r), the corresponding equation $MX = B$ always has a single solution; and that solution is the equilibrium matrix for the process.

Exercises

● *In Exercises 4-26 through 4-29, the matrix given is the transition matrix of a Markov chain process. Find the equilibrium matrix for the process.*

4-26 $\begin{bmatrix} .6 & .4 \\ .4 & .6 \end{bmatrix}$

4-27 $\begin{bmatrix} .2 & .8 & .4 \\ .8 & 0 & .2 \\ 0 & .2 & .4 \end{bmatrix}$

4-28 $\begin{bmatrix} .35 & .95 \\ .65 & .05 \end{bmatrix}$

4-29 $\begin{bmatrix} .1 & .1 & .7 \\ 0 & .5 & .2 \\ .9 & .4 & .1 \end{bmatrix}$

4-30 In a study of voting patterns in some country, it is observed that 9/10 of the daughters of women who vote in most elections also vote regularly, while only 3/4 of the daughters of women who do not usually vote are themselves regular voters. If this pattern were to continue, eventually what proportion of the women in that country would be in the habit of voting in most elections?

4-31 Suppose the occupations available to men in a certain primitive agricultural society are those of farmer, craftsman, and leader (i.e., ruler or priest). Studies indicate the probability that the son of a man who has a particular occupation will follow each of the available occupations is given by the following table:

Son's occupation \ Father's occupation	Farmer	Craftsman	Leader
Farmer	.90	.70	.30
Craftsman	.10	.20	.20
Leader	0	.10	.50

It is observed that the proportion of men in each occupation stays the same from generation to generation. What proportion of men are farmers, craftsmen, and leaders in this society?

4-32 The males in an animal population are classified as aggressive and passive. It is observed that 70% of the male offspring of aggressive males are themselves aggressive, while 50% of the offspring of passive males are aggressive. In the long run, assuming these percentages remain constant, what proportion of the males in this animal population will be aggressive?

4-33 Suppose a study of coffee buying indicates that the probability of brand change in the next time period for Jones brand, Smith brand, and all other brands is given by the following table:

Next period \ Present period	Jones	Smith	Others
Jones	.75	.10	.10
Smith	.05	.65	.05
Others	.20	.25	.85

If this table remains valid in the future so that the market for coffee becomes stable, what percentage of the coffee market will belong to Jones brand, Smith brand, and to all other brands?

4-34   A chess player's probability of winning, drawing, or losing a game is influenced by her performance in the previous game to the extent that, if she wins a game, in the next game the probability is .50 that she will win, .40 that she will draw, and .10 that she will lose. If she draws, the probabilities become .20 for a win, .50 for another draw, and .30 for a loss in the next game. If she loses, the probability is .30 that she will win her next game, .40 that she will draw, and .30 that she will lose again. What will her long-term win-draw-loss proportion be; that is, what will be the probabilities of win, draw, or loss when equilibrium is reached?

Brand-Share of Electra Margarine (Technical Essay)

When Benjamin Lipstein conducted the marketing study described in the introductory essay, he did not restrict his attention to just two brands of margarine. The simplified example we have been using in this chapter divided the margarine buyers into those that purchased Electra brand, those that chose B-R Stores brand, and those that bought some other brand. Lipstein's study took into account three other brands of margarine in addition to Electra and B-R; he called these Gloria, Meadowlark, and Aunt Mary's brands. He grouped the remaining margarines into a class of other brands. In addition, he acknowledged that not all margarine buyers will make a purchase in any given time period. In order to represent these individuals, he had a separate category for consumers who did not buy margarine during the time period.

Thus, in Benjamin Lipstein's study, the possible states a margarine buyer could be in were:

a_1 = Electra brand

a_2 = Gloria brand

a_3 = B-R Stores brand

a_4 = Aunt Mary's brand

a_5 = Meadowlark brand

a_6 = All other brands

a_7 = Did not buy margarine during time period

Lipstein's paper contains a table (Table 4-5) of brand switching and brand loyalty which represents the situation in the Chicago margarine market shortly after the introduction of the new brand, Electra.

Table 4-5

Next period \ Original period	Electra	Gloria	B-R	Aunt Mary's	Meadow-lark	Other	Did not buy
Electra	.12	.05	.03	.02	.04	.03	.05
Gloria	.05	.25	.02	.05	.01	.05	.03
B-R	.07	.03	.21	.01	.03	.03	.04
Aunt Mary's	.04	.02	.05	.23	.02	.04	.01
Meadowlark	.03	.02	.03	.04	.22	.05	.02
Other	.28	.26	.26	.25	.30	.23	.28
Did not buy	.41	.37	.40	.40	.38	.57	.57

Source: Benjamin Lipstein, "Tests for Test Marketing," *Harvard Business Review* 76 (March–April 1961).

The table can be used, as in Section 4-1, to make short-term predictions of buyer behavior.

For example, if the purchasing pattern in a period shortly after Electra margarine's introduction was

Electra	.08
Gloria	.06
B-R Stores	.10
Aunt Mary's	.02
Meadowlark	.04
All other	.26
Did not buy	.44

then, in the notation of Section 4-1, we have

$$S^{(0)} = \begin{bmatrix} .08 \\ .06 \\ .10 \\ .02 \\ .04 \\ .26 \\ .44 \end{bmatrix}$$

The transition matrix, P, is specified by the table. We can estimate consumer behavior in the next buying period by multiplying, and we find that, to two decimal places,

$$S^{(1)} = PS^{(0)} = \begin{bmatrix} .05 \\ .05 \\ .06 \\ .03 \\ .04 \\ .26 \\ .51 \end{bmatrix}$$

In other words, in the next time period, the share of the margarine market for each brand in the study is

Electra	.05
Gloria	.05
B-R Stores	.06
Aunt Mary's	.03
Meadowlark	.04
All other	.26
Did not buy	.51

If the pattern of buyer switching and loyalty exhibited by the table in Lipstein's paper remained constant, then Section 4-3 would give us a procedure for calculating the long-range brand-shares of the various brands in the study, including Electra. We recall that the technique requires us to form the matrix

$$M = \begin{bmatrix} 1 \cdots 1 \\ P - I \end{bmatrix}$$

where P is the transition matrix and I is therefore the 7×7 identity matrix. The problem then amounts to nothing more than solving the matrix equation $MX = B$ with

$$B = \begin{bmatrix} 1 \\ 0 \\ 0 \\ 0 \\ 0 \\ 0 \\ 0 \\ 0 \end{bmatrix}$$

Using the method of Section 3-6, we construct the augmented matrix $[M|B]$ and carry out the steps of the procedure, with the result, rounded off to two decimal places, shown below. The conclusion is that if the

$$\begin{bmatrix} 1 & 0 & 0 & 0 & 0 & 0 & 0 & .04 \\ 0 & 1 & 0 & 0 & 0 & 0 & 0 & .04 \\ 0 & 0 & 1 & 0 & 0 & 0 & 0 & .09 \\ 0 & 0 & 0 & 1 & 0 & 0 & 0 & .02 \\ 0 & 0 & 0 & 0 & 1 & 0 & 0 & .03 \\ 0 & 0 & 0 & 0 & 0 & 1 & 0 & .27 \\ 0 & 0 & 0 & 0 & 0 & 0 & 1 & .51 \\ 0 & 0 & 0 & 0 & 0 & 0 & 0 & 0 \end{bmatrix}$$

margarine market behavior described by the table were to continue, Electra brand would end up with only 4% of the market. From the same matrix, we would also determine that Gloria brand would command 4% of the market, B-R Stores brand 9%, Aunt Mary's brand 2%, and Meadowlark brand 3%. All other brands combined would be the choice of 27% of the market, and in any given time period, 51% of the potential margarine buyers would fail to make a purchase.

As we mentioned in the introductory essay, the brand-share of Electra margarine did drop to about 4% soon after this product was introduced. However, in the 6 months or so after its introduction, Electra succeeded in

$$[M|B] = \begin{bmatrix} 1 & 1 & 1 & 1 & 1 & 1 & 1 & 1 \\ -.88 & .05 & .03 & .02 & .04 & .03 & .05 & 0 \\ .05 & -.75 & .02 & .05 & .01 & .05 & .03 & 0 \\ .07 & .03 & -.79 & .01 & .03 & .03 & .04 & 0 \\ .04 & .02 & .05 & -.77 & .02 & .04 & .01 & 0 \\ .03 & .02 & .03 & .04 & -.78 & .05 & .02 & 0 \\ .28 & .26 & .26 & .25 & .30 & -.77 & .28 & 0 \\ .41 & .37 & .40 & .40 & .38 & .57 & -.43 & 0 \end{bmatrix}$$

building up the probability that a person who purchased Electra brand in one period would repeat that purchase in the next period. The increase was from the .12 recorded in Table 4-5 to the more respectable .23 figure we used earlier in this chapter. Consequently, the transition matrix for margarine purchases that evolved once the market stabilized after the introduction of Electra brand was quite different from the one in Lipstein's paper. In fact, Electra margarine did succeed in capturing a substantial share of the margarine market—about 12%.

Concluding Essay

Markov chain processes are used in many areas other than marketing studies. One often cited example comes from the field of genetics. An article published in the first volume of the *Proceedings of the Royal Society of Medicine* (p. 165) challenged the Mendelian theory of dominant and recessive characteristics transmitted through the genes. The author claimed that since brachydactyly (the condition of having abnormally short fingers) is a dominant characteristic, then the ratio of brachydactylic persons to normal persons should be three to one. Since brachydactyly is a rare condition, the author concluded that Mendel's theory should be changed. When this came to the attention of the great English mathematician G. H. Hardy, he wrote a letter to *Science* magazine (Vol. 28, 1908) in which he used Markov chains to show that the reasoning that led to a three to one ratio was wrong and that the correct mathematical analysis did not conflict with the Mendelian theory.

Psychologists use Markov chain processes to study a subject's response patterns. Usually, the subject is a rat deciding which way to go in a maze in response to certain stimuli. The experimenter assigns a probability to each possible response pattern and these probabilities serve as a tool for analyzing the subject's behavior.

The Markov chain process is just one part of a larger area of probability called "stochastic processes." These are defined by mathematician Mark Kac as being "the probabilistic analysis of phenomena that vary continuously in time." In other words, stochastic processes study things which change, and the answers are expressed in terms of the probability of the change. In addition to the examples above, stochastic processes are a useful tool in such varied subjects as astronomy, physics, sociology, economics, and ecology.

References

The article "Tests for Test Marketing" by Benjamin Lipstein, which appeared in the *Harvard Business Review* of March–April 1961, formed the basis for the major application of this chapter.

Other interesting applications of Markov processes, particularly in sociology, are discussed in O. J. Bartos, *Simple Models of Group Behavior* (New York: Columbia University Press, 1967).

More of the mathematics involved can be found in James E. Freund's *Introduction to Probability* (Encino, Ca.: Dickenson, 1973), or on a more advanced level in Emanuel Parzen's *Modern Probability Theory and Its Applications* (New York: Wiley, 1960).

Chapter 5
Farm Management:
A Showcase for
Linear Programming

Managing a farm in the most efficient way, that is, producing the maximum profit from the available resources while preserving the fertility of the soil, is an extremely complicated business. In the 1950s, farm management specialists discovered that a mathematical technique called "linear programming" could be very useful in their work. A good example of this approach is in James N. Boles' article in the *Journal of Farm Economics* (see References). This article describes how linear programming was used in developing optimum resource allocations for farms in Kern County, California, where water for irrigation was limited and the government imposed cotton acreage restrictions.

Boles considered a typical farm whose resources consisted of 150 acres of

261

cropland, labor, the management ability of the farm operator, a collection of farm machinery and tools, some working capital, and an irrigation system that could supply only a limited amount of water. The problem was to decide what crops to grow during each planning period and how much land to give to each of the crops so the farmer's profit would be as large as possible.

To perform the mathematical analysis needed for the problem, Boles made certain assumptions that might not be entirely correct. However, they were considered close enough to reality to make the solution obtained from the mathematics the best one possible with the information available. One such assumption was that the farmer could accurately predict the profit per acre for each of the five crops (cotton, potatoes, alfalfa, barley, and sugar beets) normally grown in Kern County. Another was that the farm operator had sufficient working capital, labor, and equipment to undertake any crop allocation plan the mathematics proposed. Also, government restrictions did not permit the farmer to devote more than 60 acres of crop-

land to cotton. Boles also assumed that the farmer would plant no more than 50 acres of potatoes because of the extreme variability of potato prices.

Boles divided the growing season into three periods, with varying amounts of water available to the farmer for irrigation during each period. He then had to consider the added complication that each crop requires different amounts of water during each period. For example, each acre of cotton requires 4 inches of water during the first period, 16.6 inches during the second, and 7.8 inches during the third. But potatoes require 13.3 inches of water per acre during the first period and none for the rest of the growing season.

If we look at all the restrictions which must be taken into account in this relatively simple farming situation, we can see that a farmer's management decisions are not made easily. This chapter will show how linear programming provides a system that can take many different restrictions into account in determining the land utilization plan that will give the farmer the largest possible profit.

5-1 Linear Programming Problems

Let us consider a simple problem of the type described in the introduction. A farmer has 150 acres of land on which he wishes to grow cotton and potatoes. Government restrictions prevent the farmer from devoting more than 60 acres of land to cotton, but he can use as much acreage for potatoes as he wishes. It requires 30 inches per acre of water for irrigation to grow cotton and 15 inches per acre to grow potatoes. The farmer has 3000 acre-inches of water available. If the farmer's profit per acre is $207 for cotton and $200 for potatoes, how many acres of cotton and how many acres of potatoes should he grow in order to obtain the largest possible profit from his land?

If there were no limitations, the farmer's best decision would be to devote the entire 150 acres to cotton, since he would then make a profit of $207 an acre—$7 an acre more than he would earn for each acre of potatoes. It might seem that, considering the government-imposed limitation on cotton acreage, the farmer should grow as much cotton as he can; that is, he should devote 60 acres to cotton and use the other 90 acres for potatoes. But he must consider the water limitation also. Now, 60 acres of cotton requires $(60)(30) = 1800$ acre-inches of water, because each acre uses 30 inches of water. On the other hand, 90 acres of potatoes needs $(90)(15) = 1350$ acre-inches of water at 15 inches per acre. Thus, it would require $1800 + 1350 = 3150$ acre-inches of water to grow 60 acres of cotton and 90 acres of potatoes. But the farmer has only 3000 acre-inches of water available for irrigation, so this scheme is impossible.

5-1   **Example**   In the problem just described, is the farmer permitted to plant the following amounts of each crop: (a) 70 acres of cotton and 80 acres of potatoes, (b) 40 acres of cotton and 100 acres of potatoes, (c) 40 acres of cotton and 120 acres of potatoes, (d) no cotton and 150 acres of potatoes, (e) 57 acres of cotton and 88 acres of potatoes?

Solution   (a) No, because the farmer is restricted to at most 60 acres of cotton.

(b) There are fewer than 60 acres of cotton planted and fewer than 150 acres in all. The 40 acres of cotton will require $(40)(30) = 1200$ acre-inches of water and the 100 acres of potatoes will use $(100)(15) = 1500$ acre-inches, so the total requirement would be 2700 acre-inches, which is less than the 3000 acre-inches available. We conclude that such a planting scheme is permissible under the conditions of the problem.

(c) No, because $40 + 120$ is more than 150 acres.

(d) Certainly, fewer than 60 acres of cotton are planted. Exactly 150 acres of land are required. Since $(150)(15) = 2250$ acre-inches is less than 3000, there is enough water available for this program, so it could be accomplished.

(e) Since 57 is less than 60 and $57 + 88 = 145$ is less than 150, it remains to check the water requirement. For cotton, that is $(57)(30) = 1710$ acre-inches and for potatoes it is $(88)(15) = 1320$. Thus, it would require $1710 + 1320 = 3030$ acre-inches of water to carry out this planting program, and since only 3000 acre-inches are available, it cannot be used.

If the farmer devotes 60 acres of his land to cotton, he will use $(60)(30) = 1800$ acre-inches of the 3000 available for this purpose, leaving 1200 acre-inches for potatoes. Since each acre of potatoes requires 15 inches of water, the farmer could plant $1200/15 = 80$ acres of potatoes to use up the remaining water. Thus, it might seem that the farmer's best plan would be to grow 60 acres of cotton, 80 acres of potatoes, and leave the remaining 10 acres unplanted. The farmer's profit is $207 per acre for cotton, so 60 acres of cotton will earn $(207)(60) = \$12,420$. The profit on 80 acres of potatoes at $200 an acre is $(200)(80) = \$16,000$. We calculate that this scheme will give the farmer $\$12,400 + \$16,000 = \$28,400$ profit.

The profit of $28,400 is not, however, the highest profit he could make from his land. For example, suppose the farmer planted 56 acres of cotton and 85 acres of potatoes. This requires a total of 141 acres, so there is enough land. There is also enough water available because 56 acres of cotton uses $(56)(30) = 1680$ acre-inches and 85 acres of potatoes uses $(85)(15) = 1275$ acre-inches, or 2955 acre-inches in all. The profit from the cotton would be $(56)(207) = \$11,592$ and from the potatoes $(85)(200) = \$17,000$, so the total profit would be $28,592, or $192 more than that obtained by planting 60 acres of cotton and 80 acres of potatoes. Clearly, the farmer could do even better than that, because there are 9 acres of land and 45 acre-inches of water unused. He could, for example, plant 3 more acres of potatoes to use up the available water and thus increase his profits by an additional $600.

5-2   **Example** In the problem being described, how many acre-inches of water are required and what is the total profit to the farmer if he uses the following programs, each of which uses no more than 150 acres of land: (*a*) 40 acres of cotton and 100 acres of potatoes, (*b*) no cotton and 150 acres of potatoes, (*c*) 57 acres of cotton and 86 acres of potatoes, (*d*) 51 acres of cotton and 98 acres of potatoes?

Solution (*a*) Since cotton requires 30 inches of water per acre and potatoes 15 inches, this planting scheme needs $(40)(30) + (100)(15) = 2700$ acre-inches of water. The profit from the cotton is $(40)(207) = \$8280$ and from potatoes $(100)(200) = \$20,000$, so the total profit is $28,280.

(*b*) Devoting all the acreage to potatoes requires $(150)(15) = 2250$ acre-inches of water. The profit from the potatoes is $(150)(200) = \$30,000$.

(*c*) The water requirement is $(57)(30) = 1710$ acre-inches for cotton and $(86)(15) = 1290$ acre-inches for potatoes, so all 3000 acre-inches of water will be used. The profit from cotton is $(57)(207) = \$11,799$ and from potatoes is $(86)(200) = \$17,200$ which comes to $28,999 in all.

(*d*) The water needs are $(51)(30) + (98)(15) = 3000$ acre-inches again, but the profit rises to

$$(51)(207) + (98)(200) = \$10,557 + \$19,600 = \$30,157$$

It should be clear by now that trying to guess at the distribution of acreage between cotton and potatoes that will produce the most profit for the farmer is not likely to lead to the correct answer. Furthermore, even if we did happen to guess correctly, we would have no way of being sure it was the best answer. Thus, we require a theory that will tell us how to solve the farmer's problem and others like it, and assure us that the answer we get is really the best possible.

Let us begin by describing the farmer's problem in mathematical terms. The two numbers we are trying to find are the number of acres of cotton the farmer should plant—call this number x—and the number of acres of potatoes he should plant—call that y. Since there are 150 acres available in all, the number of acres of cotton plus those of potatoes cannot be larger than 150, although, as we have seen, it might well be less. Thus, $x + y$ is less than or equal to 150. We can express this restriction as

$$x + y \leq 150$$

because the symbol $\leq$ means "less than or equal to." Remember that the farmer is not permitted to plant more than 60 acres of cotton; we express this restriction as

$$x \leq 60$$

Each acre of cotton uses 30 inches of water, so x acres of cotton will require $30x$ acre-inches. Similarly, y acres of potatoes account for $15y$ acre-inches of water. Thus, the total water requirement is $30x + 15y$ acre-inches, and since there are just 3000 acre-inches available, it must be that

$$30x + 15y \leq 3000$$

Another restriction on x and y is so obvious in the context of the problem that it is easy to overlook it. Since x and y represent numbers of acres, they certainly cannot be negative numbers, but must each be greater than or equal to zero. In symbols, we write $x \geq 0$ and $y \geq 0$ ($\geq$ means "greater than or equal to"). We have to state these facts explicitly, because once the problem is given in purely mathematical terms, there is no way to explain what x and y refer to. If we fail to include this restriction, the mathematics may well come up with an answer involving negative numbers.

The farmer's problem is to make his total profit as large as possible. The profit from 1 acre of cotton is $207, so there is a profit of $207x$ from x acres of cotton. In the same way, the profit from y acres of potatoes is $200y$. Thus, the total profit can be written as

$$\text{Profit} = \$207x + \$200y$$

In purely mathematical terms, then, the problem is to find numbers x and y which obey the restrictions

$$x \geq 0$$

$$y \geq 0$$

$$x \leq 60$$

$$x + y \leq 150$$

$$30x + 15y \leq 3000$$

and make the amount

$$\text{Profit} = 207x + 200y$$

as large as possible.

5-3   **Example**   Express the following problem in mathematical terms: In a certain company, it is expected that a secretary can type 10 pages and file 20 reports in 1 hour and that a clerk can type 4 pages and file 30 reports in 1 hour. In a peak hour, the company may need to have 100 pages of typing done and 300 reports filed. Because of other duties such as making appointments, the company decides it must hire at least 5 secretaries. On the other hand, the company is certain it will not require more than 15 employees of each type. If a secretary is paid \$6 an hour and a clerk \$2 an hour, how many secretaries and how many clerks should the company employ to be able to cope with the volume of work described at minimum cost to the company?

Solution   Let x be the number of secretaries hired and y the number of clerks. Since there must be at least five secretaries, we write $x \geq 5$. There need not be any clerks, but there can hardly be a negative number, so $y \geq 0$. The upper limit of fifteen on the number of employees of each type is expressed by writing $x \leq 15$ and $y \leq 15$. Each secretary can type 10 pages in an hour, so x secretaries can type $10x$ pages. In the same way, y clerks can type $4y$ pages, because each clerk can type 4 pages. Thus, the x secretaries and y clerks can altogether type $10x + 4y$ pages an hour. Since it may be necessary to be able to type 100 pages in a single hour, it is required that

$$10x + 4y \geq 100$$

The x secretaries can file $20x$ reports in an hour, because each secretary can file 20. The y clerks can file $30y$ reports, so the total filing capacity per hour is $20x + 30y$. Because there may be as many as 300 reports to be filed in a single hour, $20x + 30y$ must be at least as large as 300, so

$$20x + 30y \geq 300$$

It costs the company \$$6x$ per hour in salaries for the x secretaries and \$$2y$ an hour for the y clerks, so the total cost per hour to the company in salaries is $6x + 2y$.

Thus, the mathematical problem is to find numbers x and y so that

$$x \geq 5$$

$$y \geq 0$$

$$x \leq 15$$

$$y \leq 15$$

$$10x + 4y \geq 100$$

$$20x + 30y \geq 300$$

and at the same time

$$\text{Salaries} = 6x + 2y$$

is as small as possible.

5-4 **Example** Express the following problem in mathematical terms: A bank grants mortgages and makes business loans. The bank has \$50 million available to invest in this way. Suppose state banking laws require that the bank invest at least three times as much money in mortgages as in business loans. If the bank charges 8% interest on mortgages and 10% on business loans, how much should it invest in each in order to earn as much as possible from its investments?

Solution Let x be the amount of money the bank should invest in mortgages and y the amount in business loans. Certainly, the bank cannot invest negative amounts of money, so $x \geq 0$ and $y \geq 0$. Using \$1 million as the unit of measurement, the fact that the bank has \$50 million to invest means

$$x + y \leq 50$$

The legal requirement that the bank invest at least three times as much in mortgages as in business loans will be easier to express in symbols if we say the same thing in a different way. The law says that even if the amount invested in business loans is multiplied by 3, that amount ($3y$) must still be less than x, the amount invested in mortgages. Therefore, the requirement is

$$3y \leq x$$

For each \$1 million the bank invests in mortgages it will earn 8% of that amount or, in decimal form, 0.08 times the amount invested. Thus, from \$$x$ million invested in mortgages the bank will earn \$$0.08x$ million. Similarly, the earnings from \$$y$ million of business loans will be $0.10y$, so altogether

$$\text{Earnings} = 0.08x + 0.10y$$

The mathematical problem, then, is to find x and y so that

$$x \geq 0$$

$$y \geq 0$$

$$x + y \leq 50$$

$$3y \leq x$$

and at the same time

$$\text{Earnings} = 0.08x + 0.10y$$

are as large as possible.

The expressions $x + y \leq 150$ and $30x + 15y \leq 3000$ from the farmer's problem are both of the form

$$ax + by \leq c$$

where a, b, and c are constants. For example, in the case of $30x + 15y \leq 3000$, we have $a = 30$, $b = 15$, and $c = 3000$. For $x + y \leq 150$, the constants are $a = b = 1$ and $c = 150$. An expression which can be written in the form $ax + by \leq c$ or, like $10x + 4y \geq 100$ in Example 5-5, in the form $ax + by \geq c$, is called a **linear inequality.** Even a condition like $x \geq 0$ is a linear inequality, since we can write

$$x = (1)x + (0)y \geq 0$$

so $x \geq 0$ has the form $ax + by \geq c$ with $a = 1$ and $b = c = 0$. Since the same reasoning applies to the expressions $x \leq 60$ and $y \geq 0$, we see that all the restrictions on x and y in the farmer's problem have been expressed as linear inequalities. A glance at the solution to Example 5-3 should convince you that the conditions on x and y in that example are all linear inequalities also. The restriction

$$3y \leq x$$

in Example 5-4 may look less like a linear inequality than the others, but if we subtract x from both sides, it becomes

$$-x + 3y \leq 0$$

Thus, we still have the form $ax + by \leq c$, where this time, $a = -1$, $b = 3$, and $c = 0$. We conclude that the restrictions in Example 5-4 are linear inequalities.

5-5   **Example**   Are the following expressions linear inequalities: (*a*) $-3x + (1/2)y \leq 2$, (*b*) $x + xy \geq 5$, (*c*) $x \geq y$, (*d*) $x^2 + y^2 \leq 5$, (*e*) $x + 2 \geq 5 - 3y$?

Solution   (*a*) Yes, because $-3x + (1/2)y \leq 2$ is of the form $ax + by \leq c$, with $a = -3$, $b = 1/2$, and $c = 2$.

(*b*) No, since an expression like $ax + by \geq c$ does not involve multiplying x and y together, a linear inequality cannot contain a term like xy.

(*c*) Yes it is. Subtract y from both sides of the inequality to get $x - y \geq 0$. Then we have the form $ax + by \geq c$, with $x = 1$, $y = -1$, and $c = 0$.

(*d*) No, because the terms x^2 and y^2 cannot appear in a linear inequality.

(*e*) This is a linear inequality. Add $3y$ to both sides of the inequality to get $x + 3y + 2 \geq 5$, and then subtract 2 from both sides to obtain $x + 3y \geq 3$, which is now easily recognized as a linear inequality.

When the greater than or equal to sign in $ax + by \geq c$ is instead an equals sign, we have the equation

$$ax + by = c$$

which is, naturally, a **linear equation.** Thus, if we choose a particular level of profit for the farmer, say \$28,592, then the formula for the farmer's profit becomes a linear equation:

$$207x + 200y = 28{,}592.$$

If we start with a linear inequality and change it to a linear equation just by replacing the greater than or equal to sign by an equals sign, we will refer to the equation as "corresponding to" the original linear inequality. For example, the linear equation corresponding to $30x + 15y \leq 3000$ is $30x + 15y = 3000$.

In expressions like

$$\text{Profit} = 207x + 200y \qquad \text{and} \qquad \text{Salaries} = 6x + 2y$$

one side of the equation is of the linear form $ax + by$ and the other side is not constant but depends on the values of x and y chosen. Each of these equations is a **linear function.** Linear programming is so-called because it is concerned with finding the largest or smallest value a *linear* function can achieve when the unknowns are required to satisfy restrictions expressed as *linear* inequalities. We will see in Section 5-2 why expressions like $ax + by \geq c$ and $ax + by = c$ are called "linear."

It is quite possible to learn to solve linear programming problems by mastering purely mechanical manipulations of matrices. That is just what a computer does in solving such problems, as we shall learn in Sections 5-6 and 5-7. However, in order to understand how the manipulations work and why the answers they give are the best possible, it is necessary to have a good way of picturing what is going on. In the next sections, we will discuss the geometric tools which permit us to "see" a linear programming problem.

Summary

The symbol $\leq$ means "less than or equal to" and $\geq$ means "greater than or equal to."

A **linear inequality** is an expression which can be written in the form

$$ax + by \geq c \qquad \text{or} \qquad ax + by \leq c$$

where a, b, and c are constants and x and y are unknowns.

A **linear equation** is an expression which can be written in the form

$$ax + by = c$$

The linear equation corresponding to a linear inequality is obtained by replacing the inequality sign with an equals sign.

An equation in which one side is of the form $ax + by$ and the other side is a variable value that depends on the choice of x and y is called a **linear function.**

Exercises

- *In Exercises 5-1 through 5-3, a farmer has 310 acres of land on which she wishes to grow potatoes and barley. Because of the uncertain price of potatoes, she decides to devote no more than 100 acres to this crop. It requires 15 inches of water per acre to grow potatoes and 6 inches of water per acre to grow barley. The farmer has 2000 acre-inches of water for irrigation. The farmer's profit is estimated to be $200 an acre for potatoes and $30 an acre for barley.*

5-1 Will the conditions permit the farmer to plant the following amounts of each crop: (*a*) 100 acres of potatoes and 100 acres of barley, (*b*) 130 acres of potatoes and no barley, (*c*) no potatoes and 310 acres of barley, (*d*) 15 acres of potatoes and 15 acres of barley?

5-2 How many acre-inches of water are required and how much profit will the farmer make if she plants the following amounts of each crop, each of which requires no more than 310 acres of land: (*a*) 100 acres of potatoes and 83 acres of barley, (*b*) no potatoes and 310 acres of barley, (*c*) 15 acres of potatoes and 295 acres of barley, (*d*) 50 acres of potatoes and 200 acres of barley?

5-3 Express all of the above restrictions on the number of acres of potatoes and the number of acres of barley by means of linear inequalities.

● *Exercises 5-4 through 5-6 concern a manufacturer of furniture who makes only one type of chair and one type of table. The factory consists of a machine shop with 15 employees and an assembly and finishing line with 25 employees. Each employee works 8 hours a day. It requires 2 man-hours of machine shop labor and 1½ man-hours of assembly and finishing labor to produce a chair, and 1¾ man-hours of machine shop effort and 4 man-hours of assembly and finishing time to make a table. The manufacturer can sell no more than 60 chairs and no more than 45 tables a day to wholesalers. The manufacturer's profit is $75 for each chair and $100 for each table.*

5-4 Express all these restrictions on the number of chairs and the number of tables by means of linear inequalities. You will need separate inequalities to express the limitation on the number of man-hours of machine shop labor available and on the number of man-hours of assembly and finishing labor available.

5-5 Is it possible for the manufacturer to produce and sell the following numbers of chairs and tables each day: (*a*) 45 chairs and 17 tables, (*b*) no chairs and 50 tables, (*c*) 60 chairs and no tables, (*d*) 20 chairs and 45 tables?

5-6 How much profit would the manufacturer make if the following numbers of chairs and tables were produced: (*a*) 45 chairs and 15 tables, (*b*) no chairs and 45 tables, (*c*) 60 chairs and no tables, (*d*) 25 chairs and 40 tables?

● *In Exercises 5-7 through 5-15, determine whether each of the expressions is a linear equation or inequality. If it is, what are a, b, and c of the standard form? If it is not, explain why it fails to be linear.*

5-7 $5y + 2x \geq 4$

5-8 $x - y \leq 6$

5-9 $x^2 \leq y$

5-10 $y = 3x + 2$

5-11 $4y \geq 2x - 5$

5-12 $x \geq 1$

5-13 $x + 2y \leq 1 + x^3$

5-14 $x + y = 2$

5-15 $x + 2 \geq 2x - 5 + y$

● *In Exercises 5-16 through 5-18, express the problems in mathematical terms by means of linear inequalities and a linear function, but do not attempt to solve them.*

5-16 A woman has no more than 7 hours a week to devote to exercise. She plans to jog and to ride her bicycle. She will jog at least 1 hour a week. Since she prefers bicycling, she wishes to devote at least twice as much time to that activity as to jogging. If jogging consumes 600 calories an hour and bicycling 350 calories an hour, how many hours should she devote to each activity in order to use up as many calories as possible while satisfying her other requirements?

5-17 A man has promised to bake chocolate cookies and vanilla cookies for a charity bake sale. He agreed to bake at least five batches of cookies altogether and to bake at least one more batch of chocolate cookies than vanilla cookies. The man discovers that the cost of sugar will make his charitable offer pretty expensive. If a batch of chocolate cookies requires $1\frac{1}{4}$ cups of sugar and a batch of vanilla cookies requires 1 cup of sugar, how many batches of each type of cookie should he bake in order to keep his promise and yet use as little sugar as possible?

5-18 A farmer's cattle are fed a mixture of two types of feed; call them type A and type B. One unit of type A feed supplies a steer with 25% of its minimum daily requirement of carbohydrates and 10% of its minimum daily requirement of protein. One unit of type B feed supplies the steer with 10% of its minimum daily requirement of carbohydrates and 50% of its requirement of protein. In order to be sure the steer gets certain vitamins which are not present in type A feed, the farmer must give the steer at least 1 unit of type B feed. If 1 unit of type A feed costs 10¢ and 1 unit of type B feed costs 15¢, how much of each type of feed should the farmer give a steer each day so that it gets at least 100% of its minimum daily requirement of carbohydrates and protein while the feeding costs the farmer as little as possible?

5-2 Visualizing Linear Equations

Recall that a **number line** is a straight line marked off at intervals of a fixed length to represent the positive and negative whole numbers and zero, as shown in Figure 5-1. Any fraction or decimal number can be represented as a point on the line by dividing the intervals in an appropriate way. For example, $1\frac{1}{3}$ is found on the number line by dividing the interval from 1 to 2 into equal thirds, as in Figure 5-2. We can locate -2.4 on the number line by dividing the interval from -3 to -2 into ten equal parts, as in Figure 5-3.

Figure 5-1

Figure 5-2

Figure 5-3

On the other hand, every point on the number line represents some number. If we think of the number line as an infinitely long ruler marked off into as small divisions as we wish, then a point on the number line represents a measurement of distance from the point 0 on the number line. If the point is to the right of 0, the corresponding number is positive; if it is to the left, the corresponding number is the negative of the distance to 0.

5-6 **Example** (*a*) Mark off a number line into intervals 1 centimeter long and find the points on the line corresponding to $X = -1.3$, $Y = 2.1$, and $Z = 0.55$. (*b*) Determine the numbers represented by the points A, B, and C on the number line in Figure 5-4.

Solution (*a*) With the aid of a ruler marked off in millimeters, we can find -1.3 and 2.1 without difficulty, as indicated in Figure 5-5. The point is located midway between millimeter marks corresponding to 0.5 and 0.6.

(*b*) Measuring the line with a ruler marked off in millimeters, we find that A is -1.8, B is 1.5, and C is 4.9.

Figure 5-4

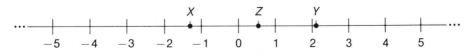

Figure 5-5

A number line gives us a numerical label, or address, for any point on the *line*. But we now require a numerical address for every point on the *plane*. To obtain this, we place two number lines on the plane at right angles to each other, as shown in Figure 5-6. The two number lines are called the **coordinate axes.** By tradition, the horizontal axis is referred to as the **x axis** and the vertical as the **y axis.** The point where the axes meet is called the **origin.** As before, the positive numbers on the horizontal number line lie to the right of 0. It is customary to place the positive numbers on the y axis above the 0, as we have indicated.

A numerical address for each point on the plane consists of an address on the x axis followed by an address on the y axis. Ordered pairs of numbers such as (3, 1), (−1/2, 17/4), or (17/4, −1/2) are used to represent this address. Suppose A is a point on the plane. The two numbers that make up the address of A are determined by passing a vertical line and a horizontal line through A, as in Figure 5-7. The vertical line through A intersects the horizontal x axis at exactly one point—in this case, −2. The horizontal line through A intersects the y axis at the point 3. The ordered pair of numbers corresponding to the point A consists of first, the number where the vertical line through A intersects the x axis, called the **x coordinate** of A, and then the number corresponding to the point where the horizontal line through A crosses the y axis, called the **y coordinate** of A. For the point A in Figure 5-7, the address, or the **coordinates,** for A is the ordered pair (−2, 3).

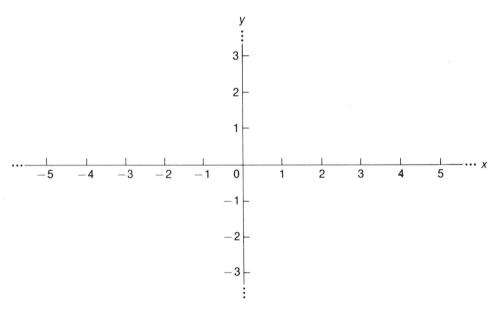

Figure 5-6

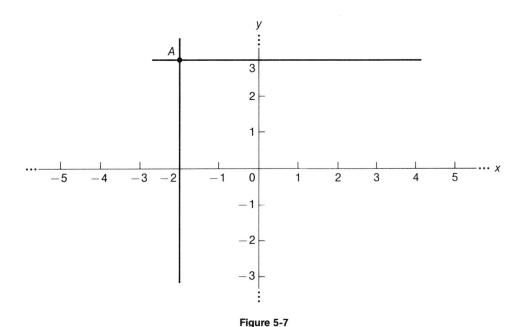

Figure 5-7

5-7 **Example** Determine the coordinates of the points B, C, D, and E in Figure 5-8.

Solution A vertical line through B hits the x axis at the number 3 and a horizontal line through B intersects the y axis at -2, so the coordinates of B are $(3, -2)$. Comparing B to the point A in Figure 5-7 illustrates the importance of the order in which the two coordinates appear. The coordinates of C are $(-1, -1)$ and of D $(-1/2, 3/2)$, as can be seen by constructing the necessary vertical and horizontal lines. A vertical line through E intersects the x axis at the point E itself. Since E corresponds to the number 4 on the x axis number line, the x coordinate is 4. A horizontal line through E is the x axis. The x axis intersects the y axis at the point corresponding to 0, so the coordinates of E are $(4, 0)$.

Just as any number corresponds to some point on a number line, so every ordered pair of numbers corresponds to a point on the plane. To find the position on the plane of an ordered pair, on the point of the x axis that corresponds to the first number of the pair, place a vertical line. Through the point on the y axis that corresponds to the second number, place a horizontal line. The vertical and horizontal lines will meet at the point with the coordinates of the given ordered pair.

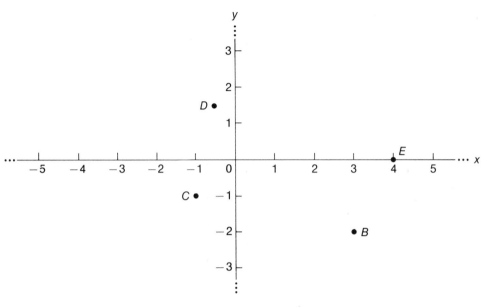

Figure 5-8

5-8   **Example**   Locate the points on the plane corresponding to the ordered pairs (3/2, 3), (−1, 2), (2, −1) and (0, −2).

Solution

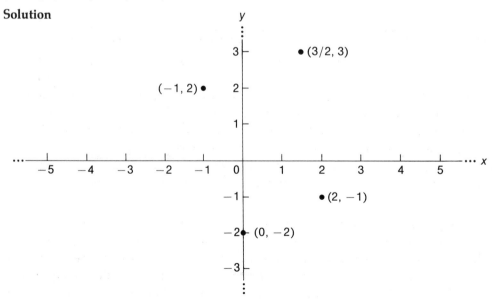

Figure 5-9

Now let us consider a particular linear equation. We choose the equation $x + y = 3$. There are many combinations of values for x and y whose sum is 3, for example, $x = 2$ and $y = 1$, $x = 1$ and $y = 2$, $x = 0$ and $y = 3$, $x = 3$ and $y = 0$, $x = 1/2$ and $y = 5/2$, $x = -1$ and $y = 4$, and so on. We would like to be able to picture *all* the ordered pairs (x, y) for which x and y add up to 3. The set of all such pairs corresponds to a subset of the coordinate plane. This subset is called the **graph** of the linear equation $x + y = 3$. We have identified some pairs with this property, namely, (2, 1), (1, 2), (0, 3), (3, 0), (1/2, 5/2), and (−1, 4). Figure 5-10 pictures the corresponding points in the coordinate plane. Notice that all the points corresponding to pairs (x, y), where $x + y = 3$, appear to lie in a straight line. That is indeed the case; in fact, the straight line through these points consists precisely of those points corresponding to the pairs (x, y) for which $x + y = 3$. Thus, the graph of $x + y = 3$ is the line shown in Figure 5-11.

More generally, given any linear equation $ax + by = c$, the graph of the equation is defined to be the set of points on the coordinate plane corresponding to all ordered pairs (x, y) of numbers for which $ax + by = c$. The graph of a linear equation is always a straight line—and that is why such equations are called "linear."

Accepting the fact that the graph of any linear equation $ax + by = c$ is a straight line, there is an efficient technique for drawing the graph. The procedure depends on the axiom of euclidean geometry that two points determine a line. We will find two points on the graph of the linear equation. Then the unique straight line which passes through both of them is the graph of the equation. For most linear equations, the two points on the graph which are easiest to find are those where the graph intersects the two coordinate axes.

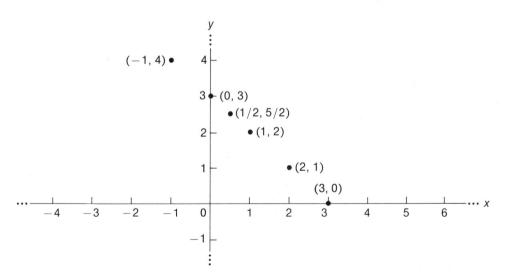

Figure 5-10

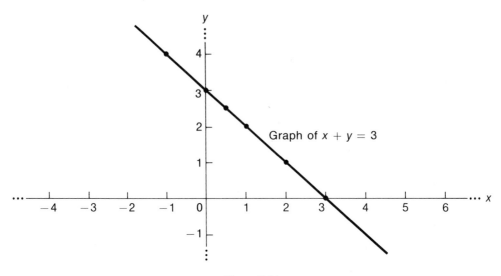

Figure 5-11

A point on the y axis is represented by a pair of numbers of the form $(0, y)$ for some y. If we substitute 0 for x in the linear equation $ax + by = c$, the equation becomes $by = c$, which we can easily solve: $y = c/b$ (provided, of course, that b is not zero). Thus, the pair $(0, c/b)$ represents a point on the graph of $ax + by = c$. Specifically, it is the point where the graph passes through the y axis. Similarly, substituting 0 for y in the equation $ax + by = c$ to find a pair of the form $(x, 0)$ for which the equation is true (that is, a pair corresponding to a point on the x axis), we are left with $ax = c$. Therefore, $x = c/a$ (when a is not zero), which means the point corresponding to the pair $(c/a, 0)$ lies on the graph also.

5-9 **Example** Determine the graph of the equation: (a) $4x - 2y = 5$, (b) $y = -2$.

Solution (a) The linear equation $4x - 2y = 5$ is of the form $ax + by = c$, where a is 4, b is -2, and c is 5. We observe that $(0, c/b) = (0, -5/2)$ and $(c/a, 0) = (5/4, 0)$ represent the points where the graph intersects the coordinate axes, so the straight line they determine is the graph shown in Figure 5-12.

(b) We write the equation $y = -2$ in the form $(0)x + (1)y = -2$. The graph intersects the y axis at $(0, c/b) = (0, -2/1) = (0, -2)$. Since $a = 0$, we cannot locate another point on the graph at $(c/a, 0)$. Instead, letting $x = 1$, the equation becomes $(0)(1) + (1)y = -2$ or $y = -2$, which means that $(1, -2)$ is on the graph. The line containing $(0, -2)$ and $(1, -2)$ is the horizontal line shown in Figure 5-13.

The method for determining the graph of $ax + by = c$ by drawing a line through the points with coordinates $(0, c/b)$ and $(c/a, 0)$ fails not only when either a or b is zero (in which

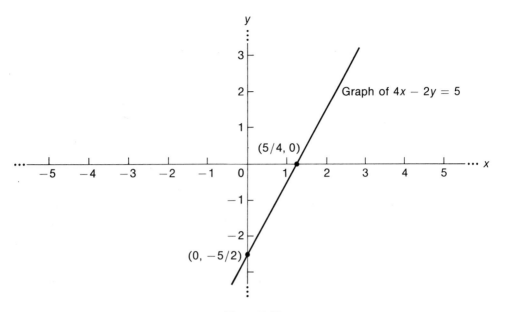

Figure 5-12

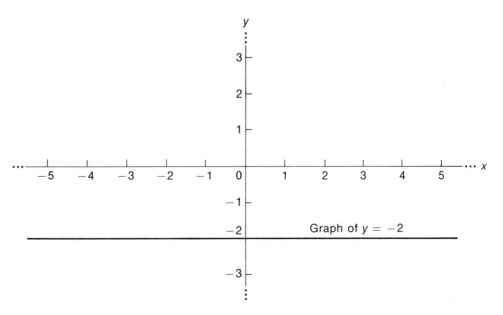

Figure 5-13

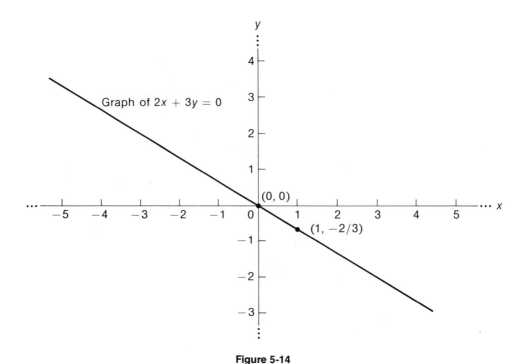

Figure 5-14

case the graph is either a horizontal or vertical line), but also when c is zero. When c is zero, setting either x to zero or y to zero establishes that the graph passes through the origin $(0, 0)$. Thus, for example, the graph of $2x + 3y = 0$ passes through the origin, but we require another point in order to determine which line through the origin it is. One such point can be obtained by letting $x = 1$ in the equation. What we are looking for, then, is a point corresponding to an ordered pair of the form $(1, y)$ which lies on the graph of $2x + 3y = 0$. To find out what y is, we substitute 1 for x and the equation becomes $2 + 3y = 0$ which, when we solve for y, gives $y = -2/3$. We conclude that $(0, 0)$ and $(1, -2/3)$ represent points on the graph of $2x + 3y = 0$, and we can therefore indicate the graph on the plane as in Figure 5-14.

5-10   **Example**   Determine the graph of the equation $4x - y = 0$.

Solution   We know the origin lies on the graph, because $4(0) - 0 = 0$. Setting $x = 1$ turns the equation into $4 - y = 0$, so the point with coordinates $(1, 4)$ is also on the graph. Figure 5-15 shows the solution.

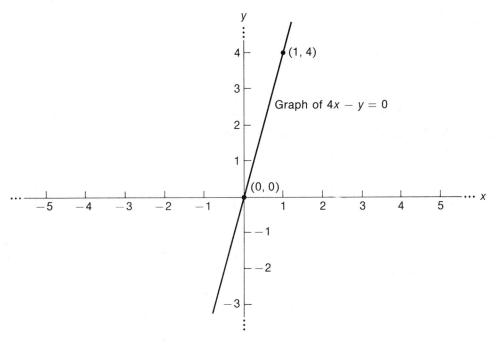

Figure 5-15

Summary

The **number line** is a straight line marked off at intervals of a fixed length. Every number can be represented by a point on a number line. Every point on a number line represents a number.

The **coordinate axes** in a plane consist of a horizontal number line, called the *x* **axis,** and a vertical number line, called the *y* **axis.** The **origin** is the point of intersection of the two coordinate axes.

Every point on the plane can be represented by an ordered pair (x, y) of numbers and every such pair of numbers corresponds to a point on the coordinate plane.

The **graph** of a linear equation $ax + by = c$ is the set of all points in the plane represented by pairs (x, y) for which $ax + by = c$. The graph is always a straight line. Consequently, to find the graph of a linear equation, it is sufficient to locate two points in the plane that lie on the graph. If none of the numbers a, b, and c is zero, then the graph of $ax + by = c$ can be constructed by finding the points corresponding to the pairs $(0, c/b)$ and $(c/a, 0)$.

Exercises

5-19   Represent the following numbers on a number line: (*a*) −2, (*b*) 3/2, (*c*) −13/4, (*d*) 0, (*e*) 7/3.

5-20   Mark off a number line into intervals with lengths equal to 1 centimeter and find the points corresponding to the following numbers: (*a*) 2.2, (*b*) −.3, (*c*) −2.4, (*d*) −.85, (*e*) 1.9.

5-21   Use a ruler divided into millimeters to determine the numbers corresponding to points *A–J* on the number line:

5-22   Determine the coordinates of the points in the figure. (The distance between whole numbers is 1 centimeter.)

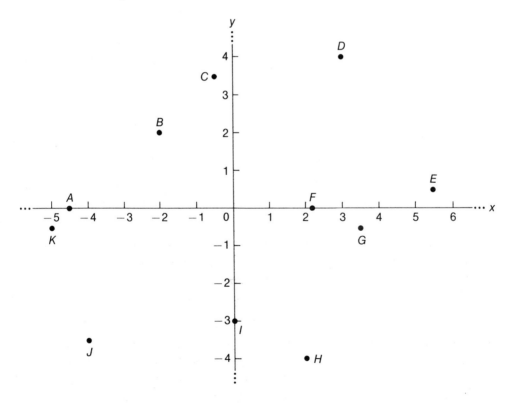

5-23   Locate the points on the plane corresponding to the following ordered pairs: (*a*) (−4, −1), (*b*) (7/2, 0), (*c*) (−1/2, −1), (*d*) (2, 9/4), (*e*) (0, −3/4).

5-24 Draw the graphs of the following linear equations: (a) $x = -1$, (b) $2x + y = -1$, (c) $x - 2y = 0$, (d) $x - y = 1$, (e) $y = 3$, (f) $3x + y = 0$, (g) $-x = -3/2$, (h) $-4x + 4y = 3$.

5-25 Indicate on the plane the set of points corresponding to ordered pairs (x, y) of numbers where y is twice as large as x.

5-26 If coffee is \$2 a pound and tea is \$3 a pound, different quantities of coffee and tea can produce a total cost of \$10. Indicate these quantities by means of a portion of a line in the coordinate plane.

5-3 Visualizing Linear Inequalities

In Section 5-2, we defined the graph of a linear equation $ax + by = c$ to be the set of points in the plane representing pairs (x, y) for which $ax + by = c$. It is consistent, therefore, to define the graph of a linear inequality $ax + by \geq c$ to be the set of points in the plane representing all ordered pairs (x, y) for which $ax + by \geq c$. The graph of a linear equation is a straight line. What sort of geometric figure is the graph of a linear inequality?

As an example, we will use the linear inequality $2x + y \geq 1$. First we draw the graph of the corresponding linear equation, $2x + y = 1$. Recall from Section 5-2 that the line which is the graph of the equation intersects the coordinate axes at $(0, 1)$ and $(1/2, 0)$, as shown in Figure 5-16. We choose some points of the plane around the graph of $2x + y = 1$, as indicated in Figure 5-17. Now, if we substitute $x = 1$ and $y = 0$ (for point A) on the left side of the equation, since $2(1) + 0 = 2 > 1$, the inequality $2x + y \geq 1$ is true for $(x, y) = (1, 0)$. Therefore, A is in the graph of $2x + y \geq 1$. Calculating that $2(1/2) + 3/2 = 5/2 \geq 1$, shows that B is also in the graph. You can check that C and D are also in the graph of $2x + y \geq 1$ in the same way. On the other hand, $2(1) + (-2) = 0$ is not greater than or equal to 1, so the point E corresponding to $(1, -2)$ does not belong to the graph of $2x + y \geq 1$. Neither do the points F and G lie in the graph of $2x + y \geq 1$. We notice that A, B, C, and D (the points that lie on the graph of $2x + y \geq 1$) are all on one side of the line which is the graph of $2x + y = 1$, while the points E, F, and G (the points that are not on the graph of $2x + y \geq 1$) are on the other side.

We claim that just as in the example we have studied, the graph of any linear inequality $ax + by \geq c$ (or, of course, $ax + by \leq c$) consists of the straight line which is the graph of the linear equation $ax + by = c$ together with all the points of the plane that lie to one side of the line. To be more precise about what we mean by the sides of a line in the plane, we use the axiom of euclidean geometry called the "plane separation postulate." This axiom states that given a line in the plane, the set of points not on the line form two disjoint subsets (i.e., they have no point in common) called **half-planes.** Two points not on the line are in the same half-plane if the line segment that may be drawn between them fails

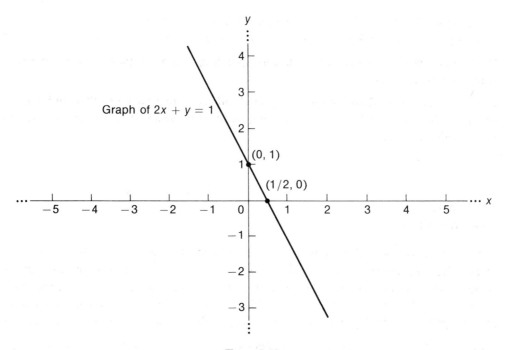

Figure 5-16

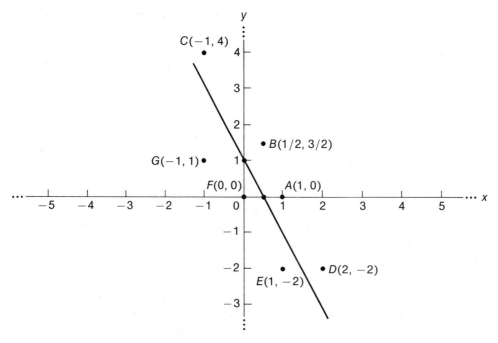

Figure 5-17

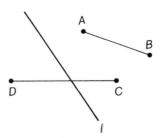

Figure 5-18

to intersect the line. The two points are in different half-planes if the line segment does intersect the line. Thus, in Figure 5-18, the points A and B lie in the same half-plane determined by l, while C and D lie in different half-planes.

The plane separation postulate thus makes precise the meaning of a side of a line: It is a half-plane determined by the line. A line together with one of its half-planes is a **closed half-plane.** The graph of a linear inequality $ax + by \geq c$ is a closed half-plane of the line which is the graph of the corresponding linear equation $ax + by = c$.

The problem of finding the graph of a linear inequality amounts to first drawing the graph of the corresponding linear equation and then establishing which of its two half-planes is in the graph of the inequality. For this purpose, it is sufficient to test just one point that is not on the graph of the linear equation. If the point belongs to the graph, then all the points in the same half-plane will also. If the point does not belong to the graph, then the other half-plane does. If the origin does not lie on the graph of the linear equation (that is, if c is not 0), then the origin is a particularly easy point to check.

5-11   **Example**   (*a*) Draw the graph of $2x - (1/2)y \geq 1$. (*b*) Draw the graph of $3x + y \leq 0$.

Solution   (*a*) The graph of $2x - (1/2)y = 1$ intersects the axes at the points corresponding to $(1/2, 0)$ and $(0, -2)$. Since $2(0) - (1/2)(0) = 0$ is smaller than one, the origin does not lie in the graph of $2x - (1/2)y \geq 1$, so we use the other half-plane for the graph of $2x - (1/2)y \geq 1$ (the shaded area in Figure 5-19).

(*b*) The graph of the linear equation $3x + y = 0$ goes through the origin and the point representing $(-1, 3)$. Since $3(-1) + 0 = -3 \leq 0$, the point corresponding to the pair $(-1, 0)$ lies in the graph of the linear inequality $3x + y \leq 0$ as do all other points in the same half-plane (see Figure 5-20).

Now we can begin to apply our geometric information to the linear programming problems we described earlier. Let us recall first the mathematical form of the farm problem we discussed in Section 5-1. The problem was to find numbers x and y which obey the restrictions

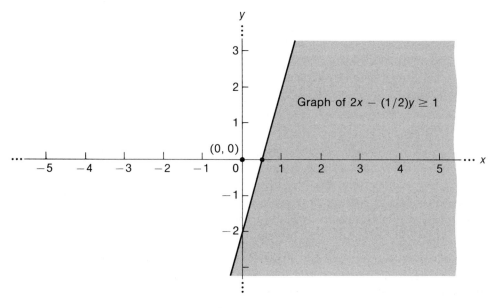

Figure 5-19

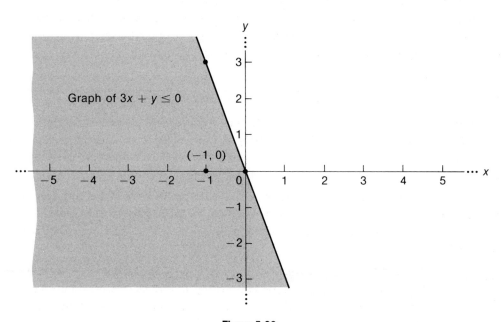

Figure 5-20

$$x \geq 0$$

$$y \geq 0$$

$$x \leq 60$$

$$x + y \leq 150$$

$$30x + 15y \leq 3000$$

and make the value

$$\text{Profit} = 207x + 200y$$

as large as possible. We will construct a picture of the set of numbers x and y which obey all the restrictions described by the five inequalities above.

The graph of the linear inequality $x \geq 0$ is just the closed half-plane to the right of the y axis (the graph of $x = 0$), while $y \geq 0$ is represented by the half-plane above the x axis, as shown in Figure 5-21. We require that x and y satisfy *both* inequalities, that is, x must be greater than or equal to zero and, at the same time, y must be greater than or equal to zero. Thus, a pair (x, y) satisfying both conditions must be represented by a point in the plane which lies both in the graph of $x \geq 0$ and in the graph of $y \geq 0$. Now, the graphs of $x \geq 0$ and $y \geq 0$ are subsets of the plane, and the points in both subsets form the intersection of the two subsets. Thus, the points representing pairs satisfying both conditions are precisely those points which are in the intersection of the sets and so describe the two conditions. In the case of $x \geq 0$ and $y \geq 0$, these points lie in the darkly shaded portion of the plane in Figure 5-22. Or, we can simplify this, as shown in Figure 5-23.

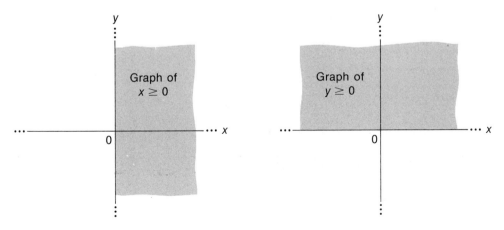

Figure 5-21

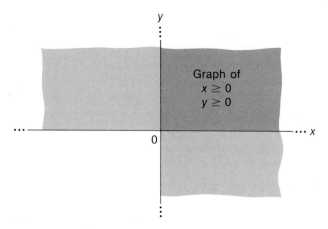

Figure 5-22

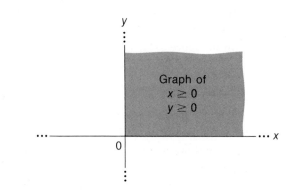

Figure 5-23

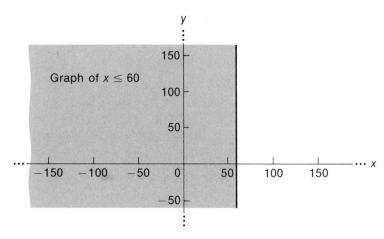

Figure 5-24

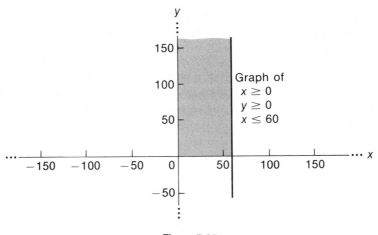

Figure 5-25

But the pairs (x, y) which can be used for the linear programming problem must also satisfy other restrictions, for example, $x \leq 60$. The graph of $x \leq 60$ is shown in Figure 5-24. Since x and y must satisfy the requirement $x \leq 60$ in addition to the restrictions $x \geq 0$ and $y \geq 0$, we take the intersection of the graph we just drew and the region representing $x \geq 0$ and $y \geq 0$ (Figure 5-23) in order to represent the pairs (x, y) which obey all three restrictions (see Figure 5-25).

Continuing in the same way, the graph of $x + y \leq 150$ is determined (see Figure 5-26) by noting that the line representing $x + y = 150$ intersects the axes at $(0, 150)$ and $(150, 0)$ and that the origin lies in the graph. To add the condition $x + y \leq 150$ to the three already imposed on x and y, we intersect the graph of $x + y \leq 150$ (Figure 5-26) with the portion

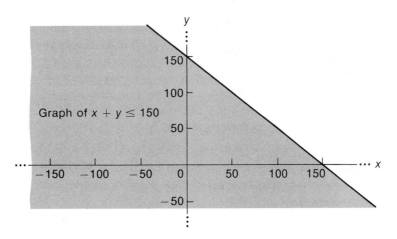

Figure 5-26

of the plane that we constructed before (Figure 5-25). The resulting subset of the plane
(see Figure 5-27) represents all pairs for which all four restrictions hold.

Finally, we graph $30x + 15y \leq 3000$ (Figure 5-28) and intersect it with the previously
constructed region to produce Figure 5-29. The shaded region enclosed by the five sided
figure represents all pairs (x, y) of numbers which satisfy all the linear inequalities of the
linear programming problem. We call the region the **domain** of the linear programming

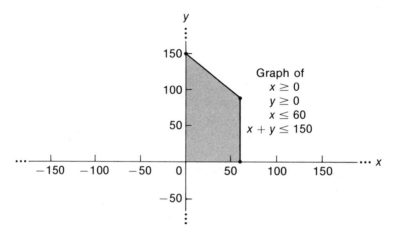

Figure 5-27

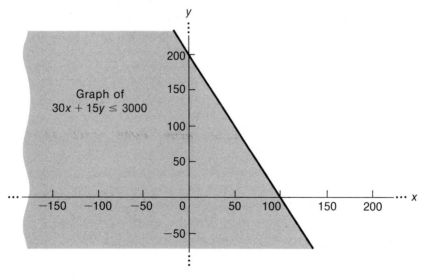

Figure 5-28

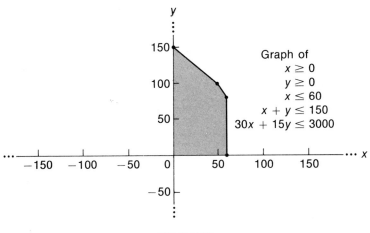

Figure 5-29

problem. The object of linear programming is to find the point in the domain corresponding to the pair (x, y) which will make the linear function larger (or smaller) than will any other pair represented by a point in the domain. In this particular problem, we seek the pair (x, y) in the domain for which the value of

$$\text{Profit} = 207x + 200y$$

is as large as possible.

In general, then, given a linear programming problem involving two unknowns x and y, the **domain** of the problem is the subset of the plane obtained by forming the intersection of all the closed half-planes which are the graphs of the linear inequalities that described the restrictions on the values of x and y.

5-12   **Example**   Draw the domain of the following linear programming problem (the secretaries and clerks problem of Example 5-3): Find numbers x and y so that

$$x \geq 5$$

$$y \geq 0$$

$$x \leq 15$$

$$y \leq 15$$

$$10x + 4y \geq 100$$

$$20x + 30y \geq 300$$

and at the same time

$$\text{Salaries} = 6x + 2y$$

is as small as possible.

Solution   The graphs of $x \geq 5$ and $y \geq 0$ are shown in Figure 5-30. Their intersection is

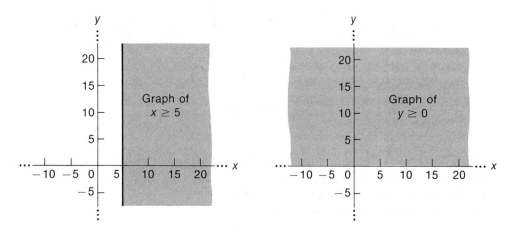

Figure 5-30

shown in Figure 5-31. We next intersect with the graph of $x \leq 15$, which appears in Figure

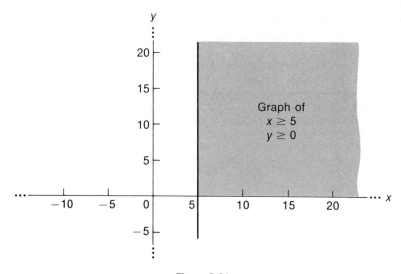

Figure 5-31

5-32, to produce the intersection of the three inequalities in Figure 5-33. Similarly, the

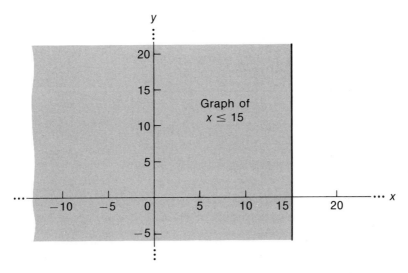

Figure 5-32

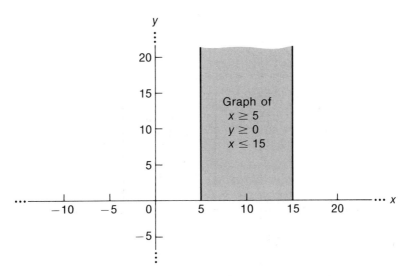

Figure 5-33

graph of $y \le 15$ is given in Figure 5-34, and when we intersect it, we get Figure 5-35. Next,

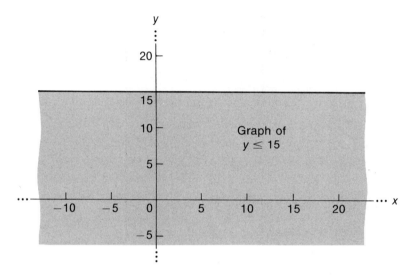

Figure 5-34

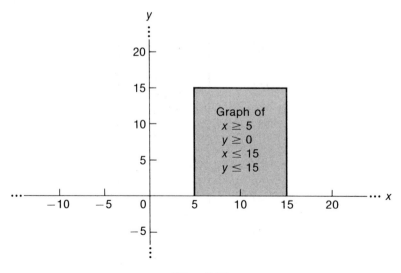

Figure 5-35

we find the graph of $10x + 4y \geq 100$ (Figure 5-36) and intersect it with our previous graph to obtain Figure 5-37. Finally, we graph $20x + 30y \geq 300$ in Figure 5-38 and intersect it with Figure 5-37. We find the domain to be the shaded region in Figure 5-39.

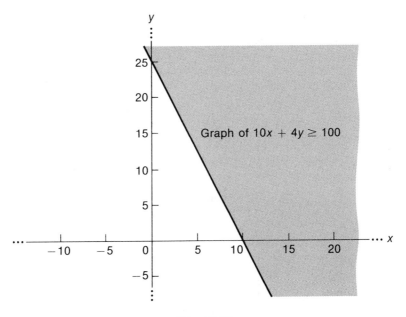

Figure 5-36

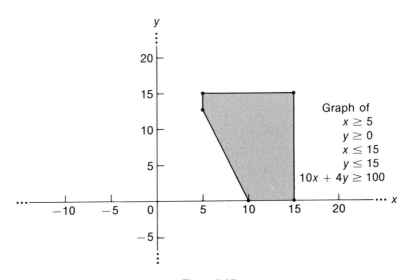

Figure 5-37

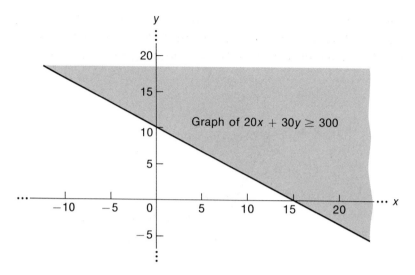

Figure 5-38

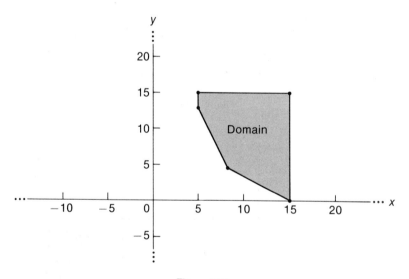

Figure 5-39

Summary

The **graph** of a linear inequality $ax + by \geq c$ is the set of points in the plane representing all pairs (x, y) for which the number $ax + by \geq c$.

Plane Separation Postulate

Given a line l in the plane, the set of points not on l form two disjoint subsets called half-planes. Two points not on l are in the same half-plane if, and only if, the line segment between them does not intersect l.

A **closed half-plane** is a line in the plane together with one of its half-planes.

The graph of a linear inequality is one of the closed half-planes of the graph of the corresponding linear equation.

The **domain** of a linear programming problem involving two unknowns x and y is the set of points in the plane representing pairs (x, y) for which all the inequalities of the problem are true. The domain is therefore the intersection of all the closed half-planes which are the graphs of the linear inequalities of the problem.

Exercises

5-27   Are the following pairs represented by points on the graph of the inequality $-3x + y \leq 2$: (a) $(1, 2)$, (b) $(-1, 2)$, (c) $(-1, -2)$, (d) $(0, 0)$, (e) $(1, 5)$, (f) $(1, 0)$?

●   *In Exercises 5-28 through 5-36, draw the graphs of the given linear inequalities.*

5-28 $x \leq -1$

5-29 $2x - y \geq 0$

5-30 $-3x + 2y \leq 2$

5-31 $2x - y \geq -1$

5-32 $y \geq -3$

5-33 $x - y \leq 1$

5-34 $-2x - y \leq 10$

5-35 $2x + 4y \leq 10$

5-36 $3x \geq y$

5-37   Graph the points on the plane which represent all pairs (x, y) satisfying both the inequality $x \leq 4$ and the inequality $x + y \geq 2$.

5-38   Graph the points in the plane which represent all pairs (x, y) where $x \geq 0$, $y \geq 0$, and $2x - 3y \leq 4$.

5-39   Graph the domain for the following linear programming problem: Find x and y satisfying

$$x \geq 0$$

$$y \geq 2$$

$$x + y \leq 4$$

$$6x + y \leq 6$$

which make the value of Profit $= 3x + 4y$ as large as possible.

5-40   Graph the domain of the following linear programming problem: Find x and y satisfying

$$x \geq 0$$

$$y \geq 0$$

$$x \leq 5$$

$$y \leq 5$$

$$6x + 5y \geq 30$$

$$x + 6y \geq 6$$

for which Cost $= x + 8y$ is as small as possible.

5-4   The Plane Geometry of Linear Programming

A linear function such as Salaries $= 6x + 2y$ in Examples 5-3 and 5-12 is more difficult to visualize on the plane than a linear equation or a linear inequality, but we can do it to an extent which is sufficient for linear programming problems. If we choose a particular salary total, say 60, then there are many combinations of the values of x and y which will make $6x + 2y$ total 60, for example, $x = 10$ and $y = 0$, $x = 8$ and $y = 6$, and so on. If we just think of x and y as representing abstract numbers rather than numbers of secretaries and clerks, then there are an infinite number of pairs (x, y) for which $6x + 2y = 60$, namely, those pairs which are represented by the graph (shown in Figure 5-40) of the linear equation $6x + 2y = 60$.

If we consider various total salary levels, say, 40, 50, 60, 75, 90, we get the corresponding graphs of $6x + 2y = 40$, $6x + 2y = 50$, and so on. We draw them all on the same axes in Figure 5-41. Notice that choosing different total salary levels produces a family of parallel lines. Notice also that as the family of parallel lines goes up and to the right, the total salary represented by the line gets larger. Moving down and to the left, the total becomes smaller.

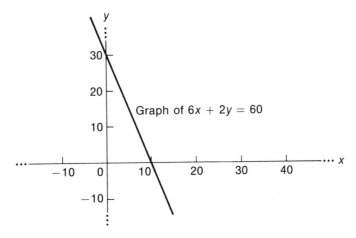

Figure 5-40

The picture we have of the linear function Salary $= 6x + 2y$ is typical of what we would find for any linear function. The linear function is described by an expression of the form Function $= ax + by$. Choosing a value c for the function, we obtain a linear equation $ax + by = c$ whose graph is, of course, a straight line. As we let c vary from one value to another, we produce a family of parallel lines which are the graphs of the linear equations thus defined. The lines are called the **level curves** of the linear function. Let us fix one

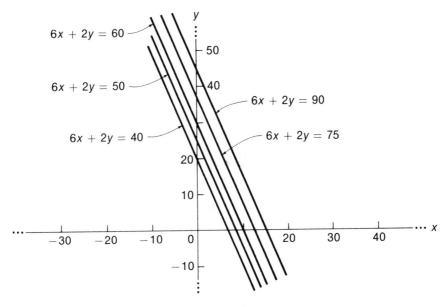

Figure 5-41

such value for the function; denote that value by c^*. All lines parallel to the graph of $ax + by = c^*$ lie in either one half-plane or the other of that line. From Section 5-3, we know that all the points in one half-plane must represent pairs (x, y) with $ax + by \geq c^*$ and all the points in the other half-plane represent pairs satisfying $ax + by \leq c^*$. Thus, all the parallel lines in one half-plane are graphs of $ax + by = c$ where c is greater than c^*, while all the parallel lines in the other half-plane are graphs of $ax + by = c$ where c is less than c^*.

5-13   **Example**   Draw level curves of the linear function Profit $= 207x + 200y$, where the profit has values 25,000, 28,000, 30,000, and 33,000.

Solution

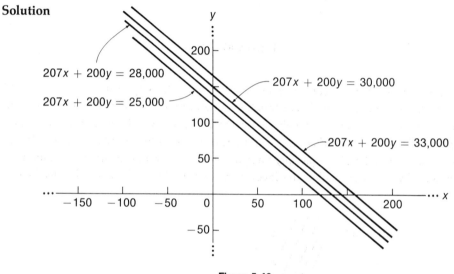

Figure 5-42

Now we are in a position to visualize an entire linear programming problem. For example, the secretaries and clerks problem of Example 5-3 was to find numbers x and y satisfying

$$x \geq 5$$

$$y \geq 0$$

$$x \leq 15$$

$$y \leq 15$$

$$10x + 4y \geq 100$$

$$20x + 30y \geq 300$$

which make

$$\text{Salaries} = 6x + 2y$$

as small as possible. In Figure 5-43, we superimpose level curves of the linear function Salaries $= 6x + 2y$ (Figure 5-41) on a plane containing the domain of the problem (Figure 5-39). The intersection of a level curve of the function and the domain is a set of points representing pairs (x, y) which satisfy all the conditions of the problem and which all produce the same value when they are substituted into the expression $6x + 2y$. If a level curve misses the domain entirely, than no pair (x, y) of numbers will satisfy the conditions of the problem and produce a salary total equal to that represented by the line. Thus, for example, no point in the domain represents a pair (x, y) for which $6x + 2y = 40$.

In general, if a level curve of the linear function does intersect the domain of the linear programming problem, as in the situation pictured in Figure 5-44, then level curves to one side of it will represent higher values of the function and those to the other side will represent lower values. Thus, the points in the domain that lie on the level curve cannot produce either a largest or smallest possible value for the function. However, if all the level curves on one side, that is, in one half-plane, miss the domain entirely, as in Figure 5-45, then all possible pairs (x, y) lie on level curves which correspond to smaller values or larger values of the function exclusively. Therefore, the single point where the level curve from the

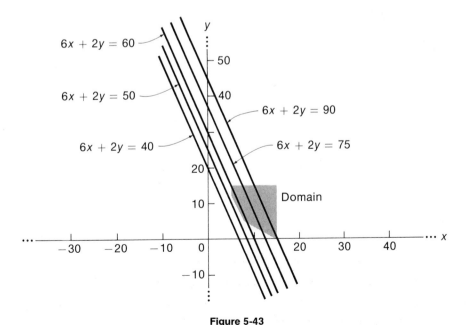

Figure 5-43

Figure 5-44   Figure 5-45

equation $ax + by = c^*$ intersects the domain represents the pair (x, y) where the function is as large or as small as possible.

It is intuitively clear, and a geometric fact, that a level curve which passes through the inside of the domain cannot contain the solution to the linear programming problem, because nearby level curves on both sides will intersect the domain. Thus, the solution must be represented by a point which is on a level curve that intersects the domain only on its boundary.

5-14   **Example**   Decide whether the solution to a linear programming problem whose domain is indicated could lie on the given line.

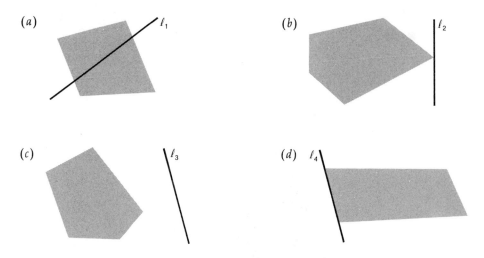

Solution   (a) No, since l_1 intersects the inside of the domain.

(b) Yes, since the domain lies entirely in one closed half-plane of l_2.

(c) No, since the line l_3 does not intersect the domain, no point on l_3 will represent a pair (x, y) that satisfies the requirements of the problem.

(d) Yes, since all lines parallel to and to the left of l_4 miss the domain.

Looking at the pictures above, we see that the pair (x, y) for which the linear function attains its largest or smallest possible value is usually represented by a corner point of the domain, that is, a point where two edges of the domain come together. Such points are called **extreme points** of the domain. The technical definition of an extreme point of a domain is one with the property that there is a line in the plane (not necessarily a level curve) which interescts the domain only in that point. Thus, the line l_1 in Figure 5-46 shows that A is an extreme point of the domain and l_2 establishes that B is also an extreme point. There are five extreme points of this domain in all.

In the case illustrated by Figure 5-47 [this is the same as Example 5-14, part (d)], every point on the line segment AB where l_4 intersects the domain might give the smallest possible value for the linear function. However, since the extreme points A and B are included in the line segment, it is still correct to state the following:

> The largest and smallest values that the linear function may achieve among the values of x and y satisfying the linear inequalities of a linear programming problem are represented by extreme points of the domain of the problem.

Thus, in order to solve a linear programming problem, we need not be concerned with the value the linear function takes on each of the infinite number of pairs represented by the points of the domain of the problem. Instead, we need only calculate the value of the function at each of the finite number of pairs (x, y) represented by extreme points of the domain. The solution to the problem is the pair which produces the largest (or, if we are seeking it, the smallest) value of the function, from among the pairs represented by extreme points. Therefore, the solution to a linear programming problem involving two unknowns x and y consists of two steps:

1. Determine the pairs represented by extreme points of the domain.

2. Calculate the value of the linear function at each such pair.

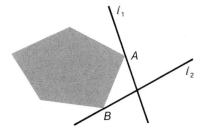

Figure 5-46

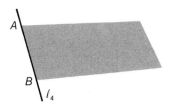

Figure 5-47

Since the second step is just a matter of simple arithmetic, the solution to the linear programming problem really amounts to finding out what pairs are represented by the extreme points. To illustrate the technique for determining these pairs, let us recall the domain from Figure 5-29 of the farming problem we described in Section 5-1. The inequalities were

$$x \geq 0$$

$$y \geq 0$$

$$x \leq 60$$

$$x + y \leq 150$$

$$30x + 15y \leq 3000$$

The domain is shown again here in Figure 5-48. We observe that there are five extreme points in this domain, labeled A, B, C, D, and E. It is clear that A represents the pair $(0, 150)$, B is the origin $(0, 0)$, and that C represents $(60, 0)$, but it is not evident what pairs (x, y) are represented by the extreme points D and E. Notice that E is the point where the lines which are the graphs of the linear equations $x + y = 150$ and $30x + 15y = 3000$ meet. The problem is to find numbers x and y for which both equations are true.

We can replace the two ordinary equations

$$x + y = 150$$

$$30x + 15y = 3000$$

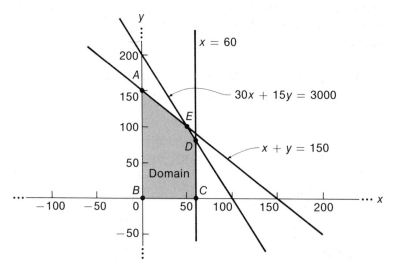

Figure 5-48

with the single matrix equation $AX = B$, where

$$A = \begin{bmatrix} 1 & 1 \\ 30 & 15 \end{bmatrix} \qquad X = \begin{bmatrix} x \\ y \end{bmatrix} \qquad B = \begin{bmatrix} 150 \\ 3000 \end{bmatrix}$$

Thus, we can see that the ordered pair corresponding to the point E can be described as the solution to a matrix equation of the sort we studied in Chapter 3. Following the technique of that chapter, we form the augmented matrix

$$[A|B] = \begin{bmatrix} 1 & 1 & 150 \\ 30 & 15 & 3000 \end{bmatrix}$$

Adding -30 times the first row to the second gives us

$$\begin{bmatrix} 1 & 1 & 150 \\ 0 & -15 & -1500 \end{bmatrix}$$

Dividing the second row through by -15, we get

$$\begin{bmatrix} 1 & 1 & 150 \\ 0 & 1 & 100 \end{bmatrix}$$

If we then subtract the second row from the first, the final result is

$$\begin{bmatrix} 1 & 0 & 50 \\ 0 & 1 & 100 \end{bmatrix}$$

We have found that

$$X = \begin{bmatrix} 50 \\ 100 \end{bmatrix}$$

In other terms, the point E corresponds to the ordered pair (50, 100).

Similarly, the point D corresponds to the ordered pair (x, y) that satisfies the two equations $x = 60$ and $30x + 15y = 3000$. If we rewrite $x = 60$ in the form $ax + by = c$ and list the equations, they are

$$(1)x + (0)y = 60$$

$$30x + 15y = 3000$$

Therefore, (x,y) can be calculated as the solution to the matrix equation $AX = B$, where

$$A = \begin{bmatrix} 1 & 0 \\ 30 & 15 \end{bmatrix} \qquad B = \begin{bmatrix} 60 \\ 3000 \end{bmatrix}$$

A straightforward application of the technique of Chapter 3 tells us that the ordered pair corresponding to D is $(60, 80)$.

More generally, in order to determine the pairs represented by the extreme points of a linear programming problem, we will be required to find the pair represented by the point where the graphs of two linear equations intersect. The linear equations correspond to linear inequalities that describe the problem. The two linear equations can be written as a single matrix equation $AX = B$, where A is a 2×2 matrix and B is 2×1. The matrix equation can be solved by the method of Chapter 3 and the solution

$$X = \begin{bmatrix} x \\ y \end{bmatrix}$$

produces the ordered pair (x, y) corresponding to the extreme point.

Let us continue to study the farm problem. We want to find (x, y) for which

$$\text{Profit} = 207x + 200y$$

is as large as possible. We know that the solution is represented by one of the extreme points of the domain of the problem. We have already determined that the extreme points are the pairs $(0, 150)$, $(0, 0)$, $(60, 0)$, $(60, 80)$, and $(50, 100)$. To solve the problem, it remains only to calculate the profit obtained from each pair. The pair which gives the largest profit is the solution to the linear programming problem. Substituting x and y from each pair into the expression $207x + 200y$, we find

$$207(0) + 200(150) = 30{,}000$$
$$207(0) + 200(0) = 0$$
$$207(60) + 200(0) = 12{,}420$$
$$207(60) + 200(80) = 28{,}420$$
$$207(50) + 200(100) = 30{,}350$$

Thus, we see that the pair $(50, 100)$ is the solution to the problem. That is, if the farmer plants 50 acres of cotton and 100 acres of potatoes, then the profit he obtains, $30,350 is the largest possible under the conditions of the problem.

5-15 **Example** Find the pairs (x, y) represented by the extreme points of the domain of Examples 5-3 and 5-12 (the secretaries and clerks problem) pictured in Figure 5-49.

Solution An examination of Figure 5-49 indicates that there are extreme points A representing the pair $(15, 0)$, B representing $(15, 15)$, and C representing $(5, 15)$. There is also an extreme point D where the graphs of the equations

$$x = 5$$

$$10x + 4y = 100$$

intersect. Thus, we have the matrix equation $AX = B$, where

$$A = \begin{bmatrix} 1 & 0 \\ 10 & 4 \end{bmatrix} \qquad B = \begin{bmatrix} 5 \\ 100 \end{bmatrix}$$

We solve the matrix equation by the methods of Chapter 3 to find that D represents the pair $(5, 25/2)$. The remaining extreme point, E, represents the intersection of

$$10x + 4y = 100$$

$$20x + 30y = 300$$

We solve by matrix techniques and calculate that E represents the pair $(90/11, 50/11)$.

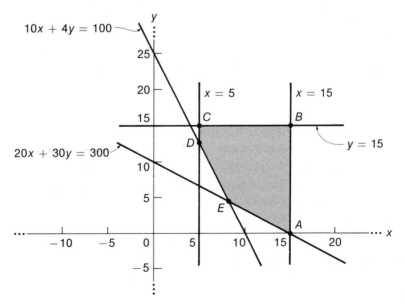

Figure 5-49

5-16   **Example** Solve the secretaries and clerks linear programming problem graphed in Figure 5-49 for x and y values that will make

$$\text{Salaries} = 6x + 2y$$

is as small as possible.

Solution   In Example 5-15, we established that the extreme points of the domain of the problem represented the pairs $(15, 0)$, $(15, 15)$, $(5, 15)$, $(5, 25/2)$, and $(90/11, 50/11)$. Substituting each pair (x, y) into the expression $6x + 2y$, we find

$$6(15) + 2(0) = 90$$

$$6(15) + 2(15) = 120$$

$$6(5) + 2(15) = 60$$

$$6(5) + 2\left(\frac{25}{2}\right) = 55$$

$$6\left(\frac{90}{11}\right) + 2\left(\frac{50}{11}\right) = \frac{640}{11} = 58.18 \quad \text{(approximately)}$$

Therefore, the solution is $(5, 25/2)$. That is, if the company hires five secretaries, twelve full-time clerks, and one half-time clerk, then the required amount of work will be done at the lowest possible cost to the company. Note that if the salary function had been such that the pair $(90/11, 50/11)$ was the solution to the problem, the answer could have been interpreted in terms of full- and part-time employees. If the company had insisted that the solution involve only full-time employees, that is, that the answer be in whole numbers, then the problem would no longer be a linear programming problem but rather what is called an "integer programming problem," and the techniques of solution are quite different

Summary

The **level curves** of a linear function, Function $= ax + by$, are the graphs of the linear equations $ax + by = c$ for various constant values c. The level curves are parallel lines. For a fixed constant value c^*, the level curves corresponding to values c greater than c^* are all in one half-plane of the graph of $ax + by = c^*$, while the level curves corresponding to values c less than c^* are all in the other half-plane.

An **extreme point** of the domain of a linear programming problem in two unknowns x and y is a point in the domain for which there is a line in the plane which intersects the domain only at that point.

The solution to a linear programming problem in the plane, that is, the pair (x, y) satisfying the conditions of the problem and making the linear function as large (or small) as possible, is represented by an extreme point of the domain.

To solve a linear programming problem involving two unknowns x and y:

1. Carefully construct a picture of the domain of the problem.

2. Identify the extreme points of the domain and calculate the pairs (x, y) representing the extreme points.

3. Calculate the value of the linear function for each of the pairs represented by an extreme point.

The extreme points will be the intersections of the graphs of two linear equations corresponding to linear inequalities that describe the problem. The two linear equations are written as a single matrix equation $AX = B$, and the solution

$$X = \begin{bmatrix} x \\ y \end{bmatrix}$$

obtained by the methods of Chapter 3 is the same as the ordered pair (x, y) corresponding to the extreme point.

Exercises

5-41   Graph level curves of the linear function

$$\text{Cost} = 10x + 7y$$

corresponding to the values $c = 30, 35, 40, 50, 55$.

5-42   Graph level curves of the linear function

$$\text{Sales} = 3x + y$$

corresponding to the values $c = 0, 2, 4, 8, 12$.

5-43   Graph the domain of the linear programming problem and the level curves corresponding to $c = 0, 3, 5, 8$, and give the solution to the problem when the problem is to find x and y satisfying the conditions

$$x \geq 0$$

$$y \geq 0$$

$$x \leq 2$$

$$x + y \leq 3$$

such that

$$\text{Output} = 2x + y$$

is as large as possible.

5-44   Graph the domain of the linear programming problem and the level curves corresponding
to the values $c = 5, 7, 9, 14, 20, 25$, for the following linear programming problem and
state the solution to the problem: Find x and y satisfying the conditions

$$x \geq 0$$

$$y \geq 0$$

$$x \leq 5$$

$$y \leq 10$$

$$x + y \geq 7$$

so that

$$\text{Cost} = x + 2y$$

is as small as possible.

5-45   Graph the domain of the linear programming problem with the conditions stated below
and determine the pairs (x, y) represented by the extreme points:

$$x \geq 2$$

$$y \geq 0$$

$$x \leq 10$$

$$y \leq 7$$

$$x + 2y \geq 10$$

5-46   Graph the domain of the linear programming problem with the conditions stated below
and determine the pairs (x, y) represented by the extreme points:

$$x \geq 0$$

$$y \geq 0$$

$$x + y \leq 2$$

$$3x + y \leq 3$$

5-47   Graph the domain of the linear programming problem with the conditions stated below and determine the pairs (x, y) represented by the extreme points:

$$x \geq 0$$

$$y \geq 0$$

$$x \leq 4$$

$$y \leq 4$$

$$x + y \geq 3$$

$$4x + y \geq 4$$

$$x + 4y \geq 4$$

5-48   Graph the domain of the linear programming problem with the conditions stated below and determine the pairs (x, y) represented by the extreme points:

$$x \geq 0$$

$$y \geq 0$$

$$x \leq 10$$

$$y \leq 10$$

$$x + y \geq 5$$

$$10x + y \geq 10$$

$$x + 15y \geq 15$$

5-49   Find the pair (x, y) which satisfies the conditions of Exercise 5-45 and makes the value of

$$\text{Man-hours} = 10x + 25y$$

as small as possible.

5-50   Find the pair (x, y) which satisfies the conditions of Exercise 5-46 and makes the value of

$$\text{Output} = 2x + y$$

as large as possible.

5-51   Find the pair (x, y) which satisfies the conditions of Exercise 5-47 and makes the value of

$$\text{Pollution} = 2x + 3y$$

as small as possible.

5-52   Find the pair (x, y) which satisfies the conditions of Exercise 5-48 and makes the value of

$$\text{Taxes} = 10x + 5y$$

as small as possible.

5-53   A copper company has two smelters. The smaller smelter can refine 5000 tons of copper ore per hour and the larger refines 10,000 tons per hour. Each smelter must operate at least 8 hours a day to justify the investment the company has in the equipment (and, of course, nothing can last more then 24 hours a day). The company must process at least 150,000 tons of ore a day. If the smaller smelter consumes 6 megawatts of energy an hour and the larger smelter consumes 11 megawatts, how many hours should each smelter operate each day so as to consume as little energy as possible?

5-54   A baker plans to bake cakes and cookies. Each cake requires 5/2 cups of flour and 2 cups of sugar, while a batch of cookies uses 1 cup of flour and 1/2 cup of sugar. The baker wishes to use no more than 70 cups of flour and 50 cups of sugar in all. If he can sell each cake for $10 and each batch of cookies for $3, how many cakes and how many batches of cookies should he sell in order to make the greatest income?

5-55   A taxi with a conventional diesel engine is equipped so that it also can burn bottled natural gas. It causes twice as much pollution when it burns diesel fuel as it does when it burns natural gas. The taxi can average 30 miles per hour when it burns diesel fuel, but because of poorer engine performance with natural gas, it averages only 20 miles per hour under this fuel. The taxi must drive as least 200 miles a day to be profitable. If the driver cannot work more than 10 hours a day, how many hours under each type of fuel should she drive in order to cause as little pollution as possible?

5-56   Repeat Exercise 5-55, but now assume that the driver can work 8 hours a day at most.

5-57   A bank grants mortgages and makes business loans. It has $500 million to invest. Suppose government regulations require the bank to invest at least $3 in mortgages for each $1 in business loans. If the bank charges 8% interest on mortgages and 10% interest for business loans, how much money should it devote to each type of investment in order to make the greatest profit?

Essay on the Geometry of Linear Programming

Let us return once again to the example from Section 5-1 of the farmer whose farm consists of 150 acres of land for which 3000 acre-inches of irrigation water are available. With a 60 acre limit on cotton and an expected profit of $207 an acre for cotton and $200 an acre for potatoes, we found in Section 5-4 that the farmer can make the greatest profit by devoting 50 acres to cotton and 100 acres to potatoes (taking into account the limitation imposed by the fact that an acre of cotton requires 30 inches of water and an acre of potatoes 15 inches).

Now suppose that the farmer's profit from potatoes turned out to be considerably less than he expected, so he decides that, in the future, it would be unwise to plant more than 50 acres to this rather speculative crop. Since the farmer is still restricted to 60 acres of cotton, he will require another crop if he is to make good use of his 150 acres. He decides to raise barley, estimating that it will require 6 inches of water per acre for irrigation and that he will receive a profit of $29 an acre from this crop. Still assuming $207 an acre profit from cotton and hoping for $200 an acre profit from potatoes, how many acres should the farmer devote to each of the three crops in order to make as large a total profit as possible?

We express the farmer's new problem in mathematical terms in order to compare it with what we had before. Again, let x be the number of acres devoted to cotton and y the number used for potatoes. There is now a third unknown quantity, the number of acres that should be used to grow barley. We denote the number of acres of barley by the letter z. We have the obvious restrictions:

$$x \geq 0$$
$$y \geq 0$$
$$z \geq 0$$

The government restrictions on cotton acreage and the farmer's self-imposed limit on potatoes are expressed by

$$x \leq 60$$
$$y \leq 50$$

The farm still contains 150 acres of land, so the number of acres devoted to each of the three crops must add up to no more than 150:

$$x + y + z \leq 150$$

Each acre of barley requires 6 inches of water, so z acres will use $6z$ acre-inches. This amount must be added to the $30x + 15y$ acre-inches required to grow x acres of cotton and y acres of potatoes in calculating the total irrigation requirement. Since there are just 3000 acre-inches of water available, it must be that

$$30x + 15y + 6z \leq 3000$$

The profit from z acres of barley is $\$29z$, so the expression for the total profit from all three crops will be

$$\text{Profit} = 207x + 200y + 29z$$

Thus, the mathematical problem is to find numbers x, y, and z satisfying the restrictions

$$x \geq 0$$

$$y \geq 0$$

$$z \geq 0$$

$$x \leq 60$$

$$y \leq 50$$

$$x + y + z \leq 150$$
$$30x + 15y + 6z \leq 3000$$

that make

$$\text{Profit} = 207x + 200y + 29z$$

as large as possible.

We claim that the only important difference between this problem and the ones we discussed previously is that now there are three unknown quantities instead of two. The inequalities involve up to three unknowns, but they are of a familiar form. Notice that $x + y + z \leq 150$ and $30x + 15y + 6z \leq 3000$ are both of the form $ax + by + cz \leq d$,

where a, b, c, and d are constants. Even an inequality like $x \leq 60$ can be viewed as being of this form if we write x as $x + 0y + 0z$, so that a is 1, b and c are 0, and d is 60. Thus, all the inequalities which describe the restrictions on x, y, and z are of the form $ax + by + cz \leq d$ (or $\geq d$). An inequality of this form is still called a "linear inequality." We obtain the corresponding linear equation by replacing the inequality sign to form an equation of the form $ax + by + cz = d$. An expression like Profit $= 207x + 200y + 29z$ is an example of a function of the form Function $= ax + by + cz$, which is again called a "linear function." A linear programming problem in three unknowns, then, requires us to find the values of the unknowns among those satisfying a set of requirements described by linear inequalities which will make a linear function as large, or as small, as possible. But, except for the number of unknowns, that is precisely what a linear programming problem in two unknowns requires of us.

Not only is the mathematical form of a linear programming problem involving three unknowns pretty much the same as one involving two unknowns, but the geometry is similar as well. A list of values for the three unknowns can be expressed as an ordered triple (x, y, z). By setting up three number lines in space at right angles to each other, it is possible to represent each ordered triple by a point in space so that the triple is the address of the point. The "graph" of a linear equation of the form

$$ax + by + cz = d$$

is defined to be the set of points in space which represent triples (x, y, z) for which $ax + by + cz$ adds up to d. Geometrically, the graph is a plane in space.

A plane in space divides the set of points in space not on the plane into two subsets, called "half-spaces," determined by the plane. The graph of a linear inequality in three unknowns,

$$ax + by + cz \leq d$$

that is, the points representing triples (x, y, z) for which $ax + by + cz \leq d$, is a closed half-space of the graph of the corresponding linear equation. In other words, the graph of the linear inequality $ax + by + cz \leq d$ consists of the graph of $ax + by + cz = d$ together with one of its half-spaces.

The restrictions on the triples (x, y, z) which are possible solutions to a linear programming problem in three unknowns are expressed by linear inequalities; the graphs of these are closed half-spaces. Thus, the domain of a problem, that is, the points representing triples which satisfy all the restrictions at once, is the intersection of these half-spaces. It is a solid figure something like the one shown in Figure 5-50.

We can visualize a linear function such as Profit $= 207x + 200y + 29z$

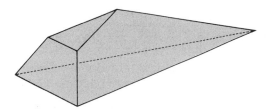

Figure 5-50

in space by choosing various profit totals and then drawing the graphs of the linear equations which result. These graphs form a family of parallel planes called the "level surfaces" of the linear function. If we choose a constant d^* to obtain from the linear function the linear equation $ax + by + cz = d^*$, then the parallel planes in one half-space of the graph represent the graphs of the equations $ax + by + cz = d$, where d is larger than d^*, and the other half-space contains the graphs when d is smaller than d^*.

If a level surface intersects the domain of the linear programming problem in some inside points of the domain, as in Figure 5-51, then the linear function cannot attain either its largest

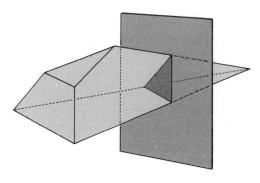

Figure 5-51

or smallest value on the domain at a point of that level surface, because level surfaces to both sides of that one will also intersect the domain. However, if the level surface intersects the domain in the manner indicated in Figure 5-52 so that the domain lies entirely in one closed half-space of the level surface, then all the parallel planes in the other half-space miss the domain entirely. Thus, for instance, the single point where this level surface intersects the domain represents the triple satisfying the conditions of the problem for which the linear function is, say, as small as possible. A point in a domain for which there is a plane in space (not necessarily a level surface) that intersects the domain only at that point is called an "extreme point" of the domain. A domain has only a finite number of extreme points and, as the preceding discussion has tried to suggest, the largest and smallest values that the linear function can attain for triples represented by points of the domain come from triples represented by extreme points.

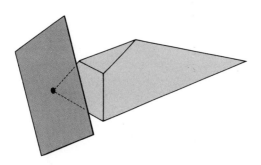

Figure 5-52

In principle, then, the solution to a linear programming problem involving three unknowns can be found by determining the triples (x, y, z) represented by extreme points of the domain and then calculating for which triple the linear function is largest or smallest. In practice, however, such an approach is usually worthless, because of the difficulty of drawing an accurate picture of a three-dimensional domain. Consequently, we require methods of solution for linear programming problems that do not depend on pictures.

In the farm management problem from Kern County, California, described in the introductory essay for this chapter, the farmer could choose from among five different crops. As we shall see later, this problem just amounts to finding the largest possible value of a linear function (involving five unknowns) where the solution must satisfy a list of linear inequalities. The actual geometric realization of such a problem is certainly not possible, but techniques do exist for studying the "geometry" in spite of that fact. These techniques establish what we hope you would by now expect, namely, that no matter how many unknowns are involved, the solution to the linear programming problem amounts to determining which of a finite number of possible best solutions is the correct one— where the possible best solutions are represented by extreme points of some suitably defined domain. There is no restriction on the number of unknowns for the matrix methods we will describe. The methods work because the geom-

etry of a linear programming problem is essentially the same no matter how many unknowns there are.

5-5 Matrices and Linear Programming

We showed in Chapter 3 that a set of linear equations such as

$$c + 2b = 15$$

$$8c + 20b = 132$$

could be written as the single matrix equation $AX = B$, where

$$A = \begin{bmatrix} 1 & 2 \\ 8 & 20 \end{bmatrix} \qquad X = \begin{bmatrix} c \\ b \end{bmatrix} \qquad B = \begin{bmatrix} 15 \\ 132 \end{bmatrix}$$

We now wish to express a set of linear inequalities such as

$$x \leq 60$$

$$x + y \leq 150$$

$$30x + 15y \leq 3000$$

as a single matrix inequality.

We can express a set of linear equations as a single matrix equation, because matrix equality is defined as follows: Two matrices are equal if and only if they are the same size and the numbers in each location are equal. Thus, in the example above,

$$AX = \begin{bmatrix} 1 & 2 \\ 8 & 20 \end{bmatrix} \begin{bmatrix} c \\ b \end{bmatrix} = \begin{bmatrix} c + 2b \\ 8c + 20b \end{bmatrix}$$

so the equation $AX = B$ means that $c + 2b = 15$ and $8c + 20b = 132$.

The useful definition of matrix inequality is, as we might expect, analogous to the definition of matrix equality. We say, for matrices A and B, that A is *less than or equal to* B, written $A \leq B$, if A and B are the same size and, at each location, the number in A is less than or equal to the number in B.

5-17   **Example**   Is $A \leq B$ in each of the following cases?

(a) $A = \begin{bmatrix} -1 & 2 & 1 \\ 3 & 0 & 2 \end{bmatrix}$ $B = \begin{bmatrix} 0 & 3 & 2 \\ 3 & 1 & 2 \end{bmatrix}$

(b) $A = \begin{bmatrix} -1 & -2 \\ 0 & 1 \end{bmatrix}$ $B = \begin{bmatrix} 0 & 0 & 1 \\ 1 & 2 & 0 \end{bmatrix}$

(c) $A = \begin{bmatrix} 1 & 2 \\ -1 & -2 \end{bmatrix}$ $B = \begin{bmatrix} 2 & 3 \\ -2 & -1 \end{bmatrix}$

(d) $A = \begin{bmatrix} 1 & 2 \\ 2 & -1 \end{bmatrix}$ $B = \begin{bmatrix} 1 & 2 \\ 2 & -1 \end{bmatrix}$

(e) $A = \begin{bmatrix} 1 \\ 2 \\ -1 \end{bmatrix}$ $B = [2 \quad 3 \quad 0]$

Solution   (a) Yes, since both matrices are 2×3 and $-1 \leq 0, 2 \leq 3, 1 \leq 2, 3 \leq 3, 0 \leq 1,$ $2 \leq 2$.

(b) No, because A and B are different sizes.

(c) No, because in the second row and first column of A the number is -1, while -2 appears in the same location in the matrix B.

(d) Yes, because if $A = B$, then $A \leq B$.

(e) No, because A is 3×1 and B is 1×3, so they are of different sizes.

Similarly, a matrix A is *greater than or equal to* a matrix B, written $A \geq B$, if A and B are the same size and, at each location, the number in A is greater than or equal to the number in B. If no number in a matrix A is negative, then $A \geq B$, where B is the zero matrix of the same size. We will write $A \geq 0$, where 0 denotes the zero matrix of the appropriate size.

Now let us turn to the matrix formulation of linear programming problems. We use as our example the farming problem from Section 5-1. That problem, in mathematical terms, was to find x and y satisfying the inequalities

$$x \geq 0$$

$$y \geq 0$$

$$x \leq 60$$

$$x + y \leq 150$$

$$30x + 15y \leq 3000$$

which make

$$\text{Profit} = 207x + 200y$$

as large as possible. We first form the matrix of unknowns,

$$X = \begin{bmatrix} x \\ y \end{bmatrix}$$

The first two conditions, $x \geq 0$ and $y \geq 0$, become the matrix inequality $X \geq 0$. Since the solution to a linear programming problem is always required to be a nonnegative number, this type of inequality will always be a part of the problem. We next express the remaining conditions

$$x \leq 60$$

$$x + y \leq 150$$

$$30x + 15y \leq 3000$$

as a matrix inequality of the form $AX \leq B$. We have already defined X and it is natural to let

$$B = \begin{bmatrix} 60 \\ 150 \\ 3000 \end{bmatrix}$$

So, setting

$$A = \begin{bmatrix} 1 & 0 \\ 1 & 1 \\ 30 & 15 \end{bmatrix}$$

we see that

$$AX = \begin{bmatrix} 1 & 0 \\ 1 & 1 \\ 30 & 15 \end{bmatrix} \begin{bmatrix} x \\ y \end{bmatrix} = \begin{bmatrix} x \\ x + y \\ 30x + 15y \end{bmatrix} \leq \begin{bmatrix} 60 \\ 150 \\ 3000 \end{bmatrix} = B$$

The linear function Profit $= 207x + 200y$ also lends itself to matrix formulation. Let

$$C = [207 \quad 200]$$

Then

$$CX = [207 \quad 200]\begin{bmatrix} x \\ y \end{bmatrix} = [207x + 200y]$$

so the function can be written Profit $= CX$. To summarize, in matrix language the farmer's problem is to find a matrix

$$X = \begin{bmatrix} x \\ y \end{bmatrix}$$

such that

$$X \geq 0 \qquad AX \leq B$$

where

$$A = \begin{bmatrix} 1 & 0 \\ 1 & 1 \\ 30 & 15 \end{bmatrix} \qquad B = \begin{bmatrix} 60 \\ 150 \\ 3000 \end{bmatrix}$$

which makes the function

$$\text{Profit} = CX$$

for

$$C = [207 \quad 200]$$

as large as possible. A linear programming problem which consists of finding a matrix X satisfying

$$X \geq 0$$

$$AX \leq B$$

and making

$$\text{Function} = CX$$

as large as possible is called a **maximum problem.**

5-18 **Example** Express as a maximum problem: Find x and y satisfying the conditions

$$x \geq 0$$
$$y \geq 0$$
$$x \leq 2$$
$$x + y \leq 3$$

such that

$$\text{Output} = 2x + y$$

is as large as possible.

Solution Let

$$X = \begin{bmatrix} x \\ y \end{bmatrix} \qquad A = \begin{bmatrix} 1 & 0 \\ 1 & 1 \end{bmatrix} \qquad B = \begin{bmatrix} 2 \\ 3 \end{bmatrix} \qquad C = [2 \quad 1]$$

Then the problem is to find X such that $X \geq 0$, $AX \leq B$, and Output $= CX$ is as large as possible.

The form of the farmer's problem described in the preceding essay (p. 316) involved three unknowns, x, y, and z, representing the amount of acreage to be devoted to cotton, potatoes, and barley, respectively. The mathematical formulation of the problem was to find x, y, and z so that

$$x \geq 0$$
$$y \geq 0$$
$$z \geq 0$$
$$x \leq 60$$
$$y \leq 50$$
$$x + y + z \leq 150$$
$$30x + 15y + 6z \leq 3000$$

and

$$\text{Profit} = 207x + 200y + 29z$$

is as large as possible. To turn this into a matrix problem, we set up the matrix of unknowns as

$$X = \begin{bmatrix} x \\ y \\ z \end{bmatrix}$$

and note that the inequality $X \geq 0$ is still required. To express the remaining conditions,

$$x \leq 60$$

$$y \leq 50$$

$$x + y + z \leq 150$$

$$30x + 15y + 6z \leq 3000$$

as a single matrix inequality, we let

$$A = \begin{bmatrix} 1 & 0 & 0 \\ 0 & 1 & 0 \\ 1 & 1 & 1 \\ 30 & 15 & 6 \end{bmatrix}$$

Then

$$AX = \begin{bmatrix} 1 & 0 & 0 \\ 0 & 1 & 0 \\ 1 & 1 & 1 \\ 30 & 15 & 6 \end{bmatrix} \begin{bmatrix} x \\ y \\ z \end{bmatrix} = \begin{bmatrix} x \\ y \\ x + y + z \\ 30x + 15y + 6z \end{bmatrix}$$

So if we set

$$B = \begin{bmatrix} 60 \\ 50 \\ 150 \\ 3000 \end{bmatrix}$$

then the matrix inequality is again $AX \leq B$. Finally, $\text{Profit} = CX$, where

$$C = [207 \quad 200 \quad 29]$$

The important observation to make on the basis of this last maximum problem is that, although there were more unknowns and more inequalities than in the previous problems, the matrix formulation is still "Find X so that $X \geq 0$, $AX \leq B$, and Function $= CX$ is as large as possible." The matrix formulation of a maximum problem is always of this type.

5-19   **Example**   Express as a maximum problem: Find w, x, y, and z satisfying the conditions

$$w \geq 0$$

$$x \geq 0$$

$$y \geq 0$$

$$z \geq 0$$

$$w \leq 20$$

$$x + y \leq 100$$

$$z \leq 150$$

$$w + 10x + 3y + 4z \leq 250$$

$$8w + 2y + z \leq 200$$

which make

$$\text{Output} = 8w + 10x + 4y + 2z$$

as large as possible.

Solution   Define

$$X = \begin{bmatrix} w \\ x \\ y \\ z \end{bmatrix}$$

Then, $X \geq 0$ and $AX \leq B$ if we let

$$A = \begin{bmatrix} 1 & 0 & 0 & 0 \\ 0 & 1 & 1 & 0 \\ 0 & 0 & 0 & 1 \\ 1 & 10 & 3 & 4 \\ 8 & 0 & 2 & 1 \end{bmatrix} \qquad B = \begin{bmatrix} 20 \\ 100 \\ 150 \\ 250 \\ 200 \end{bmatrix}$$

while Output $=$ CX, where

$$C = [8 \quad 10 \quad 4 \quad 2]$$

So far in this section, all the linear programming problems we have considered have been of the type which seek to make the linear function as large as possible. The other type of linear programming problem that we discussed in previous sections required the linear function to be made as small as possible. Such a problem is called a **minimum problem.** The matrix formulation of the minimum problem is probably not quite what you would expect, but we will see in Section 5-7 that it is the one which will facilitate the solution to the problem.

We will illustrate the minimum problem by means of the secretaries and clerks problem we first introduced as Example 5-3. The mathematical formulation of the problem is to find x and y which obey the restrictions

$$x \geq 5$$

$$y \geq 0$$

$$x \leq 15$$

$$y \leq 15$$

$$10x + 4y \geq 100$$

$$20x + 30y \geq 300$$

and make

$$\text{Salaries} = 6x + 2y$$

as small as possible. The first step is to write the matrix of unknowns which, for a minimum problem, we denote by Z and write as a row rather than a column:

$$Z = [x \quad y]$$

Just as in a maximum problem, we will always have the condition $Z \geq 0$ (since it is required that $x \geq 5$, it certainly must be that $x \geq 0$). The remaining conditions are

$$x \geq 5$$

$$x \leq 15$$

$$y \leq 15$$

$$10x + 4y \geq 100$$

$$20x + 30y \geq 300$$

We notice that not all the inequalities face the same direction. However, to make use of the definition of matrix inequality, it is necessary that they do so. In a minimum problem, we wish the constant on the right-hand side of each inequality to be smaller than the expression involving unknowns on the left-hand side. (In a maximum problem, we require the constant to be larger.) Thus, we must reverse the direction of the inequalities

$$x \le 15$$

$$y \le 15$$

Recalling that an inequality will be reversed if both sides are multiplied by a negative number, we multiply by -1 to get

$$-x \ge -15$$

$$-y \ge -15$$

and so the set of inequalities becomes

$$x \ge 5$$

$$-x \ge -15$$

$$-y \ge -15$$

$$10x + 4y \ge 100$$

$$20x + 30y \ge 300$$

Letting

$$A = \begin{bmatrix} 1 & -1 & 0 & 10 & 20 \\ 0 & 0 & -1 & 4 & 30 \end{bmatrix}$$

then

$$ZA = \begin{bmatrix} x & y \end{bmatrix} \begin{bmatrix} 1 & -1 & 0 & 10 & 20 \\ 0 & 0 & -1 & 4 & 30 \end{bmatrix}$$

$$= \begin{bmatrix} x & -x & -y & 10x + 4y & 20x + 30y \end{bmatrix}$$

So if we set

$$C = \begin{bmatrix} 5 & -15 & -15 & 100 & 300 \end{bmatrix}$$

then the five inequalities become the single matrix inequality $ZA \ge C$. Finally, let

$$B = \begin{bmatrix} 6 \\ 2 \end{bmatrix}$$

Then

$$\text{Salaries} = ZB = \begin{bmatrix} x & y \end{bmatrix} \begin{bmatrix} 6 \\ 2 \end{bmatrix} = \begin{bmatrix} 6x + 2y \end{bmatrix}$$

In general, a minimum problem is written in matrix form by seeking a $1 \times n$ matrix Z satisfying conditions $Z \geq 0$ and $ZA \geq C$, which will make Function $= ZB$ as small as possible. We have interchanged the roles of the matrices called B and C between the maximum and the minimum problems intentionally. This will make it possible in the next sections to describe a single method for solving both types of problems.

5-20   **Example**   Express as a minimum problem: Find x, y, and z so that

$$x \geq 0$$
$$y \geq 0$$
$$z \geq 2$$
$$x + y + z \leq 10$$
$$x + 2y \geq 3$$
$$2x + y + 3z \geq 4$$
$$3x + z \geq 4$$

which make

$$\text{Pollution} = 5x + 10y + 3z$$

as small as possible.

Solution   Let

$$Z = \begin{bmatrix} x & y & z \end{bmatrix}$$

Then it must be that $Z \geq 0$. The inequalities

$$z \geq 2$$
$$-x - y - z \geq -10$$

$$x + 2y \geq 3$$

$$2x + y + 3z \geq 4$$

$$3x + z \geq 4$$

become the matrix inequality $ZA \geq C$, where

$$A = \begin{bmatrix} 0 & -1 & 1 & 2 & 3 \\ 0 & -1 & 2 & 1 & 0 \\ 1 & -1 & 0 & 3 & 1 \end{bmatrix} \qquad C = \begin{bmatrix} 2 & -10 & 3 & 4 & 4 \end{bmatrix}$$

and Pollution $= ZB$, where

$$B = \begin{bmatrix} 5 \\ 10 \\ 3 \end{bmatrix}$$

Summary

A matrix A is *less than or equal to* a matrix B, written $A \leq B$, if A and B are the same size and, at each location, the number in A is less than or equal to the number in B. A matrix A is *greater than or equal to* a matrix B, written $A \geq B$, if A and B are the same size and, at each location, the number in A is greater than or equal to the number in B. In particular, $A \geq 0$ if every number in A is greater than or equal to zero.

A **maximum problem** in linear programming is to find an $n \times 1$ matrix X such that $X \geq 0$ and $AX \leq B$, which makes Function $= CX$ as large as possible. A **minimum problem** in linear programming is to find a $1 \times n$ matrix Z such that $Z \geq 0$ and $ZA \geq C$, which makes Function $= ZB$ as small as possible.

Exercises

- *Determine whether $A \leq B$, $A \geq B$, or neither is true in each of Exercises 5-58 through 5-63.*

5-58 $A = \begin{bmatrix} -1 & 0 \\ 2 & 1 \end{bmatrix} \qquad B = \begin{bmatrix} 0 & 1 \\ 2 & -1 \end{bmatrix}$

5-59 $A = \begin{bmatrix} 5 \\ 4 \\ 0 \end{bmatrix}$ $B = \begin{bmatrix} 1 \\ 2 \\ -1 \end{bmatrix}$

5-60 $A = \begin{bmatrix} 4 & 5 & 8 \\ 5 & 6 & 7 \end{bmatrix}$ $B = \begin{bmatrix} 2 & 1 \\ 1 & -1 \\ 2 & 0 \end{bmatrix}$

5-61 $A = \begin{bmatrix} -1 & -2 \\ -3 & 0 \end{bmatrix}$ $B = \begin{bmatrix} 0 & 0 \\ 0 & 0 \end{bmatrix}$

5-62 $A = \begin{bmatrix} 1 & 2 \\ -1 & -1 \\ 1 & 3 \end{bmatrix}$ $B = \begin{bmatrix} -1 & 2 \\ -1 & -2 \\ 1 & -3 \end{bmatrix}$

5-63 $A = \begin{bmatrix} 0 & 0 & 0 \end{bmatrix}$ $B = \begin{bmatrix} 1 & 2 & 3 \end{bmatrix}$

●   *In Exercises 5-64 through 5-74, write the linear programming problem in matrix form as a maximum or minimum problem, but do not attempt to solve it.*

5-64   Find x and y satisfying

$$x \geq 0$$
$$y \geq 0$$
$$x + y \leq 4$$
$$6x + y \leq 6$$

which make Profit $= 3x + 4y$ as large as possible.

5-65   Find x and y satisfying

$$x \geq 0$$
$$y \geq 0$$
$$x \leq 5$$
$$y \leq 5$$
$$6x + 5y \geq 30$$
$$x + 6y \geq 6$$

which make Cost $= x + 8y$ as small as possible.

5-66 Find x and y satisfying

$$x \geq 1$$
$$y \geq 2$$
$$2x + 3y \leq 10$$
$$x + 2y \leq 7$$
$$3x + y \leq 12$$

which make

$$\text{Output} = 5x + 9y$$

as large as possible.

5-67 Find x and y satisfying

$$x \geq 0$$
$$y \geq 0$$
$$x + y \leq 12$$
$$3x + 2y \geq 8$$
$$5x + 6y \geq 15$$

which make

$$\text{Taxes} = 40x + 35y$$

as small as possible.

5-68 Find x, y, and z satisfying

$$x \geq 3$$
$$y \geq 2$$
$$z \geq 3$$
$$x + y + z \leq 100$$
$$2x + y \leq 40$$
$$2y + z \leq 60$$

which make

$$\text{Profit} = 10x + 12y + 15z$$

as large as possible.

5-69   Find x, y, and z satisfying

$$x \geq 0$$

$$y \geq 0$$

$$z \geq 0$$

$$x + y + 30z \geq 207$$

$$y + 15z \geq 200$$

which make

$$\text{Cost} = 50x + 150y + 3000z$$

as small as possible.

5-70   Find w, x, y, and z satisfying

$$w \geq 0$$

$$x \geq 0$$

$$y \geq 0$$

$$z \geq 0$$

$$w \leq 12$$

$$x \leq y$$

$$x + y + z \leq 15$$

which make

$$\text{Value} = 2w + 3x + y + z$$

as large as possible.

5-71   Suppose a military emergency ration is made up of three types of food concentrates: call them type x, type y, and type z. One unit of type x concentrate weighs 1 ounce and contains 1/4 of a soldier's minimum daily requirement (MDR) of protein, 1/8 MDR of carbohydrates, and 1/4 MDR of calories. One unit of type y concentrate weighs 2 ounces and contains 1/8 MDR of protein, 1/4 MDR of carbohydrates, and 1/4 MDR of calories. One unit of type z concentrate weights 3/2 ounce and contains 3/8 MDR of protein, 1/8 MDR of carbohydrates, and 3/8 MDR of calories. The emergency ration must contain enough food for 3 days, that is, at least 3 MDR of protein, 3 MDR of carbohydrates, and 3 MDR of calories. How many units of each type of concentrate should be used in the ration in order to meet the nutritional requirements but make the total weight as small as possible?

5-72   A university wishes to invest its $100 million endowment so as to make a good return on its money without taking too many risks. It decides to invest in mutual funds and municipal bonds while keeping at least $15 million in savings accounts. The university estimates it can earn 9% a year from mutual funds, $7\frac{1}{2}$% a year from bonds, and 5% from savings accounts. Because of the risk associated with mutual funds (which invest in the stock market), the university decides to invest no more in mutual funds than half the amount it invests in bonds, and to put no more in mutual funds than it puts in savings accounts. How much money should the university invest in mutual funds, bonds, and savings accounts in order to make the largest possible return on its endowment?

5-73   A jeweler makes rings, earrings, pins, and necklaces. He wishes to work no more than 40 hours a week. It takes him 1 hour to make a ring, $1\frac{1}{2}$ hours to make a pair of earrings, 1/2 hour to make a pin, and 2 hours to make a necklace. He estimates that he can sell no more than ten rings, ten pairs of earrings, fifteen pins, and three necklaces in a week and that he must make at least five each of rings, earrings, and pins in order to have enough variety of stock in his store. The jeweler charges $50 for a ring, $80 for a pair of earrings, $30 for a pin, and $200 for a necklace. How many rings, earrings, pins, and necklaces should the jeweler make in order to earn the most money possible?

5-74   Suppose the jeweler in Exercise 5-73 decides that if he can bring in $1500 a week, then he can afford to stay in business (that is, pay his expenses, buy materials, and make enough profit to live on). How many rings, earrings, pins, and necklaces should he make, subject to the restrictions on the number of such in Exercise 5-73, so he can stay in business but work as few hours as possible?

5-6   The Simplex Method for Nondegenerate Maximum Problems

The solution, in Section 5-4, of a linear programming problem involving two unknowns x and y depended on finding pairs (x, y) which represent extreme points of the domain of the problem. For example, one extreme point in the farming problem we solved in that section was the intersection of the graphs of

$$x + y = 150$$

$$30x + 15y = 3000$$

We recall that we found the pair (x, y) corresponding to this extreme point by using the techniques of Chapter 3 to solve the matrix equation $AX = B$, where

$$A = \begin{bmatrix} 1 & 1 \\ 30 & 15 \end{bmatrix} \qquad B = \begin{bmatrix} 150 \\ 3000 \end{bmatrix}$$

The matrix method for finding ordered pairs corresponding to extreme points is of limited value in solving linear programming problems. However, the mathematics that lies behind the matrix method has been used to create a very effective technique, called the **simplex method,** for solving linear programming problems. The simplex method is a technique of matrix manipulation which automatically finds pairs of linear equations with graphs that intersect at extreme points, solves for the pairs (x, y) that satisfy both equations, and calculates the values of the linear functions for the pairs found. In addition, the simplex method identifies the pair that solves the linear programming problem and, in general, does not need to examine all the extreme points in order to do this. But the most important feature of the simplex method is that it will work for linear programming problems in any number of unknowns, not just two. Furthermore, the method is very well suited to computer implementation, so it is practical to solve very large linear programming problems. We should mention, however, that this method will not solve all linear programming problems and thus, we will need to put some restrictions on the problems we consider in order to be certain the method will work. Fortunately, most linear programming problems that occur in practice are of the type we will study in this section and the next.

Before describing the steps to be performed in the general simplex method, we will illustrate its use in the farming problem from Section 5-1. We will see that the method is essentially an elaboration of the technique we described in Chapter 3 for solving a matrix equation. The farmer's problem, in matrix language, is the maximum problem

$$X \geq 0$$

$$AX \leq B$$

$$\text{Profit} = CX$$

where

$$X = \begin{bmatrix} x \\ y \end{bmatrix} \qquad A = \begin{bmatrix} 1 & 0 \\ 1 & 1 \\ 30 & 15 \end{bmatrix} \qquad B = \begin{bmatrix} 60 \\ 150 \\ 3000 \end{bmatrix} \qquad C = \begin{bmatrix} 207 & 200 \end{bmatrix}$$

The first step in solving a matrix equation is to form the augmented matrix. Similarly, the first step in finding the solution to a linear programming problem given in matrix terms is to form the **simplex tableau:**

$$\left[\begin{array}{c|ccc|c} A & & I & & B \\ \hline -C & 0 & \cdots & 0 & 0 \end{array}\right]$$

Here, I is an identity matrix with as many rows as A has and $-C$ is the matrix C with the sign of every number in C changed. Thus, the simplex tableau of the farmer's problem is

$$\left[\begin{array}{cc|ccc|c} 1 & 0 & 1 & 0 & 0 & 60 \\ 1 & 1 & 0 & 1 & 0 & 150 \\ 30 & 15 & 0 & 0 & 1 & 3000 \\ \hline -207 & -200 & 0 & 0 & 0 & 0 \end{array}\right]$$

You may be pleasantly surprised to find that all the matrices fit in the matrix tableau to form a rectangular array, but this must always be the case, as we will see below (p. 337).

The procedure for solving a matrix equation was to transform the augmented matrix to a more useful form by means of certain permissible operations or moves. We recall that the permissible moves were (1) the interchange of two rows, (2) the multiplication or division of every number in a row by the same constant, and (3) the addition of a multiple of one row to another row. We will transform the simplex tableau by means of exactly the same allowable moves to a matrix of the form

$$\left[\begin{array}{cc|ccc|c} 1 & 0 & * & * & * & x \\ 0 & 1 & * & * & * & y \\ * & * & * & * & * & * \\ \hline * & * & * & * & * & p \end{array}\right]$$

where x and y are the values which solve the farmer's linear programming problem and p is the profit obtained from that solution. The asterisks indicate numbers in the matrix whose values are not part of the solution. The form of the simplex tableau that solves a linear programming problem is not always exactly like this, but later we will see that there are rules that will permit you to recognize the solution in any case.

Although the moves used in the simplex method are just those used to solve a matrix equation, the procedure is somewhat different. Instead of necessarily starting with the left-hand column and moving column-by-column to the right, we examine the numbers in the bottom row of the tableau and choose the number with the largest negative value. The column in which that number lies, called the **pivot column,** is the one we wish to change to a desired form. As it happens here, the largest negative number in the bottom row of the tableau, namely -207, does occur in the left-hand column.

Next, we divide each number in the right-hand column of the tableau by the number in the pivot column which lies in the same row, provided the number in the pivot column is positive, that is, neither negative nor 0. Placing these quotients to the right of the tableau, we have

$$
\begin{bmatrix}
① & 0 & 1 & 0 & 0 & 60 \\
1 & 1 & 0 & 1 & 0 & 150 \\
30 & 15 & 0 & 0 & 1 & 3000 \\
\hline
-207 & -200 & 0 & 0 & 0 & 0
\end{bmatrix}
\qquad
\begin{aligned}
&\frac{60}{1} = 60 \\[4pt]
&\frac{150}{1} = 150 \\[4pt]
&\frac{3000}{30} = 100
\end{aligned}
$$

The number in the pivot column for which the quotient is smallest is called the **pivot element.** In this case, it is the 1 in the first row. It is convenient to circle the pivot element as shown above.

Once we have determined the pivot element, the objective is to convert the pivot column, by means of allowable moves, into a column with a 1 in the pivot element position and 0 elsewhere. The steps are the same as those of Chapter 3. If the pivot element is not already a 1 (it cannot be 0), we divide through its row by itself to make it a 1. Then for each element a in the pivot column, we add $-a$ times the row containing the pivot element to the row containing a, and so change a to 0. Thus, adding -1 times the first row to the second row of the tableau above gives

$$
\begin{bmatrix}
① & 0 & 1 & 0 & 0 & 60 \\
0 & 1 & -1 & 1 & 0 & 90 \\
30 & 15 & 0 & 0 & 1 & 3000 \\
\hline
-207 & -200 & 0 & 0 & 0 & 0
\end{bmatrix}
$$

Adding -30 times the first row to the third row improves the situation to

$$
\begin{bmatrix}
① & 0 & 1 & 0 & 0 & 60 \\
0 & 1 & -1 & 1 & 0 & 90 \\
0 & 15 & -30 & 0 & 1 & 1200 \\
\hline
-207 & -200 & 0 & 0 & 0 & 0
\end{bmatrix}
$$

Finally, 207 times the first row, when added to the fourth row, transforms the tableau into a form which accomplishes our objective:

$$
\begin{bmatrix}
① & 0 & 1 & 0 & 0 & 60 \\
0 & 1 & -1 & 1 & 0 & 90 \\
0 & 15 & -30 & 0 & 1 & 1200 \\
\hline
0 & -200 & 207 & 0 & 0 & 12{,}420
\end{bmatrix}
$$

The presence of -200 in the bottom row indicates that we have not yet found the solution to the problem. Therefore, we begin again by looking for the pivot column, then the pivot element, and converting the pivot column to a column with a 1 in the pivot position and 0 elsewhere. The pivot column is the second from the left, because of the -200, and we form the quotients to the right of the tableau to determine the pivot element:

$$
\begin{bmatrix}
① & 0 & 1 & 0 & 0 & 60 \\
0 & 1 & -1 & 1 & 0 & 90 \\
0 & 15 & -30 & 0 & 1 & 1200 \\
\hline
0 & -200 & 207 & 0 & 0 & 12{,}420
\end{bmatrix}
\qquad
\begin{array}{l}
\\[4pt]
\dfrac{90}{1} = 90 \\[10pt]
\dfrac{1200}{15} = 80
\end{array}
$$

We find that the pivot element is 15. To convert the pivot position to a 1, we divide the third row by 15:

$$
\begin{bmatrix}
① & 0 & 1 & 0 & 0 & 60 \\
0 & 1 & -1 & 1 & 0 & 90 \\
0 & ① & -2 & 0 & \dfrac{1}{15} & 80 \\
\hline
0 & -200 & 207 & 0 & 0 & 12{,}420
\end{bmatrix}
$$

Adding -1 times the third row to the second row and adding 200 times the third row to the last row, produces

$$
\begin{bmatrix}
① & 0 & 1 & 0 & 0 & 60 \\
0 & 0 & 1 & 1 & -\dfrac{1}{15} & 10 \\
0 & ① & -2 & 0 & \dfrac{1}{15} & 80 \\
\hline
0 & 0 & -193 & 0 & 13\tfrac{1}{3} & 28{,}420
\end{bmatrix}
$$

The presence of -193 in the last row indicates that we still have not found the solution to the problem. Therefore, we must identify the new pivot element:

$$
\begin{bmatrix}
① & 0 & 1 & 0 & 0 & 60 \\
0 & 0 & ① & 1 & -\dfrac{1}{15} & 10 \\
0 & ① & -2 & 0 & \dfrac{1}{15} & 80 \\
\hline
0 & 0 & -193 & 0 & 13\frac{1}{3} & 28{,}420
\end{bmatrix}
\quad
\begin{array}{l}
\dfrac{60}{1} = 60 \\[1em]
\dfrac{10}{1} = 10
\end{array}
$$

Transforming the third column produces

$$
\begin{bmatrix}
① & 0 & 0 & -1 & \dfrac{1}{15} & 50 \\
0 & 0 & ① & 1 & -\dfrac{1}{15} & 10 \\
0 & ① & 0 & 2 & -\dfrac{1}{15} & 100 \\
\hline
0 & 0 & 0 & 193 & \dfrac{7}{15} & 30{,}350
\end{bmatrix}
$$

The fact that there are no longer any negative numbers in the last row means that we have reached the correct extreme point. The final step in the solution is to interchange rows so that an identity matrix appears in the upper left-hand corner of the tableau. In this case, we interchange the second and third rows:

$$
\begin{bmatrix}
① & 0 & 0 & -1 & \dfrac{1}{15} & 50 \\
0 & ① & 0 & 2 & -\dfrac{1}{15} & 100 \\
0 & 0 & ① & 1 & -\dfrac{1}{15} & 10 \\
\hline
0 & 0 & 0 & 193 & \dfrac{7}{15} & 30{,}350
\end{bmatrix}
$$

The simplex method has found the solution $x = 50$, $y = 100$, and calculated that the corresponding profit is \$30,350.

Let us now consider the general maximum problem

$$X \geq 0$$

$$AX \leq B$$

$$\text{Function} = CX$$

If the problem has n unknowns, then X is an $n \times 1$ matrix and C must be $1 \times n$, since the value of the function is just a number. Certainly, A must have as many columns as X has rows (n of them) in order to form the product AX, so A is an $n \times m$ matrix for some number m. Then, AX is an $m \times 1$ matrix, which implies that B is the same size by the definition of inequality. The simplex tableau is the $(m + 1) \times (n + m + 1)$ matrix

$$\left[\begin{array}{c|c|c} A & I & B \\ \hline -C & 0 \cdots 0 & 0 \end{array} \right]$$

where I is the $m \times m$ identity matrix. Since C has as many columns as A and B has as many rows as A, the various parts of the tableau must indeed fit in the matrix.

In order to be sure that the simplex method will solve our maximum problem, we require that $B \geq 0$. We will see in Example 5-22 that sometimes a little ingenuity is needed to put the problem in a form which satisfies this requirement.

The steps in the simplex method for solving a maximum problem are:

1. Choose the pivot column to be the column which contains the largest negative number of the bottom row.

2. Divide each number in the right-hand column by the number in the pivot column which lies in the same row, provided the number in the pivot column is positive. The pivot element is the number in the pivot column which produces the smallest quotient (if there are more than one which give the smallest quotient, choose any of them).

3. By means of the method of Chapter 3, transform the pivot column into a column with a 1 in the pivot element position and 0 elsewhere.

4. Repeat Steps 1–3 as long as there are negative numbers in the bottom row.

5. When all the numbers in the last row are either positive or 0, rearrange the rows so that the upper left-hand corner of the tableau is an $n \times n$ identity matrix with pivot elements on the diagonal. The first n numbers in the last column are the values of the solution to the problem, in the order of the unknowns in X, and the greatest possible value of the function is the number in the bottom right-hand corner of the tableau.

In general, it may not be possible to carry out Step 5, because the maximum problem is "degenerate." In this section, we will see only nondegenerate maximum problems, so this difficulty will not arise. In Section 5-7, we will explain the slightly modified version of Step 5 required in the degenerate case.

5-21   **Example**   Solve the form of the farmer's problem described on pp. 313–314, which, in matrix terms, is the maximum problem where

$$X = \begin{bmatrix} x \\ y \\ z \end{bmatrix} \qquad A = \begin{bmatrix} 1 & 0 & 0 \\ 0 & 1 & 0 \\ 1 & 1 & 1 \\ 30 & 15 & 6 \end{bmatrix} \qquad B = \begin{bmatrix} 60 \\ 50 \\ 150 \\ 3000 \end{bmatrix} \qquad C = \begin{bmatrix} 207 & 200 & 29 \end{bmatrix}$$

Solution   The simplex tableau is

$$\begin{bmatrix} \begin{array}{ccc|cccc|c} 1 & 0 & 0 & 1 & 0 & 0 & 0 & 60 \\ 0 & 1 & 0 & 0 & 1 & 0 & 0 & 50 \\ 1 & 1 & 1 & 0 & 0 & 1 & 0 & 150 \\ 30 & 15 & 6 & 0 & 0 & 0 & 1 & 3000 \\ \hline -207 & -200 & -29 & 0 & 0 & 0 & 0 & 0 \end{array} \end{bmatrix}$$

The pivot column is the first and the pivot element is detected by calculating quotients:

$$\begin{bmatrix} \begin{array}{ccc|cccc|c} ① & 0 & 0 & 1 & 0 & 0 & 0 & 60 \\ 0 & 1 & 0 & 0 & 1 & 0 & 0 & 50 \\ 1 & 1 & 1 & 0 & 0 & 1 & 0 & 150 \\ 30 & 15 & 6 & 0 & 0 & 0 & 1 & 3000 \\ \hline -207 & -200 & -29 & 0 & 0 & 0 & 0 & 0 \end{array} \end{bmatrix} \qquad \begin{array}{l} \dfrac{60}{1} = 60 \\[2ex] \\[1ex] \dfrac{150}{1} = 150 \\[2ex] \dfrac{3000}{30} = 100 \end{array}$$

We transform the first column:

$$\begin{bmatrix} \begin{array}{ccc|cccc|c} ① & 0 & 0 & 1 & 0 & 0 & 0 & 60 \\ 0 & 1 & 0 & 0 & 1 & 0 & 0 & 50 \\ 0 & 1 & 1 & -1 & 0 & 1 & 0 & 90 \\ 0 & 15 & 6 & -30 & 0 & 0 & 1 & 1200 \\ \hline 0 & -200 & -29 & 207 & 0 & 0 & 0 & 12{,}420 \end{array} \end{bmatrix}$$

Then we find the next pivot element,

$$\begin{bmatrix} ① & 0 & 0 & 1 & 0 & 0 & 0 & 60 \\ 0 & ① & 0 & 0 & 1 & 0 & 0 & 50 \\ 0 & 1 & 1 & -1 & 0 & 1 & 0 & 90 \\ 0 & 15 & 6 & -30 & 0 & 0 & 1 & 1200 \\ \hline 0 & -200 & -29 & 207 & 0 & 0 & 0 & 12{,}420 \end{bmatrix}$$

$$\frac{50}{1} = 50$$

$$\frac{90}{1} = 90$$

$$\frac{1200}{15} = 80$$

and adjust the pivot column:

$$\begin{bmatrix} ① & 0 & 0 & 1 & 0 & 0 & 0 & 60 \\ 0 & ① & 0 & 0 & 1 & 0 & 0 & 50 \\ 0 & 0 & 1 & -1 & -1 & 1 & 0 & 40 \\ 0 & 0 & 6 & -30 & -15 & 0 & 1 & 450 \\ \hline 0 & 0 & -29 & 207 & 200 & 0 & 0 & 22{,}420 \end{bmatrix}$$

Again we identify the pivot element:

$$\begin{bmatrix} ① & 0 & 0 & 1 & 0 & 0 & 0 & 60 \\ 0 & ① & 0 & 0 & 1 & 0 & 0 & 50 \\ 0 & 0 & ① & -1 & -1 & 1 & 0 & 40 \\ 0 & 0 & 6 & -30 & -15 & 0 & 1 & 450 \\ \hline 0 & 0 & -29 & 207 & 200 & 0 & 0 & 22{,}420 \end{bmatrix}$$

$$\frac{40}{1} = 40$$

$$\frac{450}{6} = 75$$

Finally, we obtain

$$\begin{bmatrix} ① & 0 & 0 & 1 & 0 & 0 & 0 & 60 \\ 0 & ① & 0 & 0 & 1 & 0 & 0 & 50 \\ 0 & 0 & ① & -1 & -1 & 1 & 0 & 40 \\ 0 & 0 & 0 & -24 & -9 & -6 & 1 & 210 \\ \hline 0 & 0 & 0 & 178 & 171 & 29 & 0 & 23{,}580 \end{bmatrix}$$

Since we already have the required identity matrix in the upper left-hand corner of the tableau, the solution tells us that under the terms of this problem, the farmer should plant 60 acres of cotton, 50 acres of potatoes, and 40 acres of barley, from which he will make the largest possible profit, $23,580.

5-22 **Example** An investor has $10,000 from which she would like to make as much money as possible. She plans to invest some money in stocks, some in bonds, and put the rest in a savings account. The investor believes she can earn 8% on the money invested in stocks and 7% on bonds. The savings bank pays 5% interest. Since stocks are a rather risky investment, she decides to invest no more in stocks than half the amount invested in bonds, and no more in stocks than she puts in the bank. The investor also will keep at least $2000 in the bank in case she needs cash on short notice. How much money should she invest in stocks, how much in bonds, and how much should she put in the bank?

Solution Let x be the amount of money to be invested in stocks, y the amount in bonds, and z the amount in the bank, and define

$$X = \begin{bmatrix} x \\ y \\ z \end{bmatrix}$$

There is $10,000 available in all, so it must be that

$$x + y + z \le 10,000$$

The conditions on the amount spent on stocks are

$$x \le 0.5y$$

$$x \le z$$

We rewrite these two restrictions as linear inequalities

$$x - 0.5y \le 0$$

$$x - z \le 0$$

Since at least $2000 is to be left in the bank, we have

$$z \ge 2000$$

This condition is not in the correct form for a maximum problem, because the inequality sign points in the wrong direction (remember that we want $AX \leq B$). If we multiply through the inequality by -1 to change the direction of the inequality,

$$-z \leq -2000$$

we will no longer have the condition $B \geq 0$ which we need to ensure that the simplex method will work, so this is not an acceptable way out of our difficulty. The trick is to observe that since x, y, and z add up to no more than $10,000$, then the requirement that z be at least 2000 is the same as the requirement that $x + y$ be no larger than 8000. Thus, the condition $z \geq 2000$ can be replaced by

$$x + y \leq 8000$$

which is acceptable. To summarize, the restrictions of the problem are

$$x + y + z \leq 10,000$$

$$x - 0.5y \leq 0$$

$$x - z \leq 0$$

$$x + y \leq 8000$$

Consequently, we see that

$$A = \begin{bmatrix} 1 & 1 & 1 \\ 1 & -0.5 & 0 \\ 1 & 0 & -1 \\ 1 & 1 & 0 \end{bmatrix} \qquad B = \begin{bmatrix} 10,000 \\ 0 \\ 0 \\ 8000 \end{bmatrix}$$

The investor will earn 8% of the money invested in stocks, so her earnings from $\$x$ worth of stocks will be $0.08x$. Similarly, her earnings from bonds and the savings account will be $0.07y$ and $0.05z$, respectively. Thus,

$$\text{Earnings} = 0.08x + 0.07y + 0.05z = CX$$

so

$$C = \begin{bmatrix} 0.08 & 0.07 & 0.05 \end{bmatrix}$$

The simplex tableau is

$$\begin{bmatrix} 1 & 1 & 1 & 1 & 0 & 0 & 0 & 10{,}000 \\ 1 & -0.5 & 0 & 0 & 1 & 0 & 0 & 0 \\ 1 & 0 & -1 & 0 & 0 & 1 & 0 & 0 \\ 1 & 1 & 0 & 0 & 0 & 0 & 1 & 8000 \\ \hline -0.08 & -0.07 & -0.05 & 0 & 0 & 0 & 0 & 0 \end{bmatrix}$$

Combining the steps of determining the pivot column and finding the pivot element, we give the steps of the solution without further comment:

$$\begin{bmatrix} 1 & 1 & 1 & 1 & 0 & 0 & 0 & 10{,}000 \\ ① & -0.5 & 0 & 0 & 1 & 0 & 0 & 0 \\ 1 & 0 & -1 & 0 & 0 & 1 & 0 & 0 \\ 1 & 1 & 0 & 0 & 0 & 0 & 1 & 8000 \\ \hline -0.08 & -0.07 & -0.05 & 0 & 0 & 0 & 0 & 0 \end{bmatrix} \quad \begin{matrix} 10{,}000 \\ 0 \\ 0 \\ 8000 \end{matrix}$$

$$\begin{bmatrix} 0 & 1.5 & 1 & 1 & -1 & 0 & 0 & 10{,}000 \\ ① & -0.5 & 0 & 0 & 1 & 0 & 0 & 0 \\ 0 & 0.5 & -1 & 0 & -1 & 1 & 0 & 0 \\ 0 & 1.5 & 0 & 0 & -1 & 0 & 1 & 8000 \\ \hline 0 & -0.11 & -0.05 & 0 & 0.08 & 0 & 0 & 0 \end{bmatrix} \quad \begin{matrix} 6666.67 \\ \\ 0 \\ 5333.33 \end{matrix}$$

$$\begin{bmatrix} 0 & 1.5 & 1 & 1 & -1 & 0 & 0 & 10{,}000 \\ ① & -0.5 & 0 & 0 & 1 & 0 & 0 & 0 \\ 0 & ① & -2 & 0 & -2 & 2 & 0 & 0 \\ 0 & 1.5 & 0 & 0 & -1 & 0 & 1 & 8000 \\ \hline 0 & -0.11 & -0.05 & 0 & 0.08 & 0 & 0 & 0 \end{bmatrix}$$

$$\begin{bmatrix} 0 & 0 & ④ & 1 & 2 & -3 & 0 & 10{,}000 \\ ① & 0 & -1 & 0 & 0 & 1 & 0 & 0 \\ 0 & ① & -2 & 0 & -2 & 2 & 0 & 0 \\ 0 & 0 & 3 & 0 & 2 & -3 & 1 & 8000 \\ \hline 0 & 0 & -0.27 & 0 & -0.14 & 0.22 & 0 & 0 \end{bmatrix} \quad \begin{matrix} 2500 \\ 0 \\ 0 \\ 2666.67 \end{matrix}$$

0	0	①	0.25	0.5	0.75	0	2500	5000
①	0	0	0.25	0.5	0.25	0	2500	5000
0	①	0	0.5	−1	0.5	0	5000	
0	0	0	−0.75	0.5	−0.75	1	500	1000
0	0	0	0.0675	−0.005	0.0175	0	675	

0	0	①	1	0	0	−1	2000
①	0	0	1	0	1	−1	2000
0	①	0	−1	0	−1	2	6000
0	0	0	−1.5	①	−1.5	2	1000
0	0	0	0.06	0	0.01	0.005	680

①	0	0	1	0	1	−1	2000
0	①	0	−1	0	−1	2	6000
0	0	①	1	0	0	−1	2000
0	0	0	−1.5	①	−1.5	2	1000
0	0	0	0.06	0	0.01	0.005	680

Thus, if the investor invests $2000 in stocks, $6000 in bonds, and puts $2000 in a savings account, she will earn as much as possible, $680.

Summary

Given a maximum problem in n unknowns,

$$X \geq 0$$

$$AX \leq B$$

$$\text{Function} = CX$$

where $B \geq 0$, form the **simplex tableau**

$$\left[\begin{array}{c|c|c} A & I & B \\ \hline -C & 0 \cdots 0 & 0 \end{array} \right]$$

and perform the following steps in the **simplex method:**

1. Choose the **pivot column,** the column containing the largest negative number among those in the last row.

2. Divide each number in the last column by the number in the pivot column in the same row, provided the number in the pivot column is positive. The **pivot element** is the number in the pivot column which gives the smallest quotient (if more than one number in the pivot column gives the same smallest quotient, use any one of them for the pivot element).

3. Convert the pivot column into a column with a 1 in the pivot element position and 0 elsewhere by first dividing through the row containing the pivot element by that number itself and then, for each nonzero number a in the pivot column, adding $-a$ times the row containing the pivot element to the row containing a.

4. Repeat Steps 1–3 as long as there are any negative numbers in the botton row of the tableau.

5. When all the numbers in the last row are either positive or 0, interchange rows to form an $n \times n$ identity matrix in the upper left-hand corner of the tableau, where the 1s are pivot elements. The first n numbers in order in the last column are the solution to the problem and the last number in that column is the value of the function corresponding to the solution.

Exercises

● *In Exercises 5-75 through 5-78, solve the maximum problem $X \geq 0, AX \leq B$, Function $= CX$.*

5-75 $X = \begin{bmatrix} x \\ y \end{bmatrix}$ $A = \begin{bmatrix} 2 & 3 \\ 3 & 2 \end{bmatrix}$ $B = \begin{bmatrix} 6 \\ 5 \end{bmatrix}$ $C = \begin{bmatrix} 4 & 5 \end{bmatrix}$

5-76 $X = \begin{bmatrix} x \\ y \\ z \end{bmatrix}$ $A = \begin{bmatrix} 1 & 1 & 0 \\ 0 & 1 & 0 \\ 0 & 1 & 2 \end{bmatrix}$ $B = \begin{bmatrix} 2 \\ 1 \\ 3 \end{bmatrix}$ $C = \begin{bmatrix} \frac{1}{2} & 2 & 1 \end{bmatrix}$

5-77 $X = \begin{bmatrix} x \\ y \end{bmatrix}$ $A = \begin{bmatrix} 1 & 1 \\ 1 & 2 \\ 1 & 0 \\ 0 & 1 \end{bmatrix}$ $B = \begin{bmatrix} 3 \\ 4 \\ \frac{5}{2} \\ 3 \\ 2 \end{bmatrix}$ $C = \begin{bmatrix} 5 & 8 \end{bmatrix}$

5-78 $X = \begin{bmatrix} x \\ y \\ z \end{bmatrix}$ $A = \begin{bmatrix} 1 & 0 & 0 \\ 0 & 4 & 1 \\ 1 & -1 & 0 \\ 0 & 0 & 1 \end{bmatrix}$ $B = \begin{bmatrix} 3 \\ 2 \\ 0 \\ 1 \end{bmatrix}$ $C = \begin{bmatrix} 5 & 1 & 3 \end{bmatrix}$

5-79 Find x and y, both greater than or equal to zero, so that

$$3x + y \le 6$$

$$x + 2y \le 4$$

$$y \le 1$$

which will make

$$\text{Value} = 100x + 150y$$

as large as possible.

5-80 Find x, y, and z, all greater than or equal to zero, so that

$$\frac{1}{2}x + y + z \le 10$$

$$x \le 2y$$

$$2x + 3y \le 80$$

$$x + y \le 6$$

which make

$$\text{Distance} = 7x + 8y + 10z$$

as large as possible.

5-81 Find x and y, both greater than or equal to zero, so that

$$x \le 2y$$

$$3y \le 3$$

$$x + y \le 5$$

$$2x + 3y \le 9$$

$$x \ge y$$

which make

$$\text{Output} = 5x + 4y$$

as large as possible.

5-82 A man can devote up to 1500 square feet of his property to a garden in which he plans to grow vegetables and flowers. He wishes to devote at least as much room to vegetables as to flowers. Furthermore, the number of square feet used for vegetables plus three times the number of square feet of garden used for flowers must add up to no more than 2000 square feet. The man estimates that he gets 3 units of pleasure from growing 1 square foot of flowers and 1 unit of pleasure from growing 1 square foot of vegetables. How many square feet should he devote to vegetables and how many to flowers in order to get as much pleasure as possible from his garden?

5-83   A politician plans to walk the length of her state to attract attention to her candidacy for office, get better acquainted with the problems of her state, and get a chance to talk with many voters. She will spend part of her time in fast walking, part in leisurely strolling, and part in stopping to talk to voters. To balance the time spent in achieving the various goals of her walk, she decides to walk as far as she can each hour while spending at least 1/4 hour in conversation and limiting her time of fast walking to no more than the sum of talking time and strolling time combined. If her fast walking speed is 3 miles per hour, her strolling speed is 1 mile per hour, and she stands still while she is talking to voters, what fraction of each hour should she devote to each activity in order to go as far as possible in the hour?

5-84   An employee of an ice cream shop wishes to make the richest (measured in calories) ice cream soda for his friends that he can fit in a 12 ounce glass. The ingredients of a soda are syrup, cream, soda water, and ice cream. In order to look and taste like a soda, the mixture must contain no more than 4 ounces of ice cream, at least as much soda water as the total amount of syrup and cream combined, and no more than 1 ounce more of syrup than of cream. If the syrup is 75 calories an ounce, cream is 50 calories an ounce, ice cream is 40 calories an ounce, and soda water contains no calories, how many ounces of each ingredient should the employee use?

5-7   The Simplex Method, Continued

In Section 5-6, we described the solution to a maximum problem of the following form: Find the numbers in an $n \times 1$ matrix X of unknowns satisfying

$$X \geq 0$$

$$AX \leq B$$

which will make

$$\text{Function} = CX$$

as large as possible. Recall that in Section 5-5 we defined a minimum problem of linear programming as one of the following form: Find the numbers in a $1 \times n$ matrix Z of unknowns satisfying

$$Z \geq 0$$

$$ZA \geq C$$

which will make

$$\text{Function} = ZB$$

as small as possible. We promised in Section 5-5 that the interchange of the roles of the matrices B and C between the maximum and minimum problems would make it possible to solve both types of problems by means of a single method. That method is, of course, the simplex method, and we are now in a position to make good our promise.

We begin the solution of the minimum problem stated above by forming the simplex tableau, just as in Section 5-6:

$$\left[\begin{array}{c|c|c} A & I & B \\ \hline -C & \boxed{0 \cdots 0} & 0 \end{array} \right]$$

The matrices A, B, and C are now those of the minimum problem. We follow the rules of the simplex method until all the numbers in the last row are positive or zero. The solution to the minimum problem consists of the numbers in the circled portion of the last row (the positions below those in which the identity matrix was originally located). These numbers will be in the same order as the unknowns in Z. The value of the function corresponding to the solution to the minimum problem is the number in the lower right-hand corner of the tableau, just as in the maximum problem.

To illustrate the use of the simplex method to solve minimum problems, we will use the secretaries and clerks example that we have discussed several times. In mathematical terms, the problem is to find x and y satisfying

$$x \geq 0$$

$$y \geq 0$$

$$x \geq 5$$

$$x \leq 15$$

$$y \leq 15$$

$$10x + 4y \geq 100$$

$$20x + 30y \geq 300$$

which will make

$$\text{Salaries} = 6x + 2y$$

as small as possible. Before writing this in matrix form as a minimum problem, we can make an observation that will simplify the solution. Since the object of the problem is to make x and y so small that the total salaries will be small, the restrictions that x and y not exceed fifteen employees each is not really a restriction on the size of x and y. (This is in contrast to a condition like $10x + 4y \geq 100$, which really prevents x and y from being very small.) In the previous discussions of this problem, we included the conditions $x \leq 15$ and $y \leq 15$ only so that its domain would have a familiar form when we constructed it in Section 5-4. Consequently, we will replace the original secretaries and clerks problem by the entirely equivalent, but simpler, problem of finding x and y, both greater than or equal to zero, so that

$$x \geq 5$$

$$10x + 4y \geq 100$$

$$20x + 30y \geq 300$$

which make

$$\text{Salaries} = 6x + 2y$$

as small as possible. If we used the matrix formulation of the problem in Section 5-5, the same procedures, but involving a larger simplex tableau and quite a few additional steps, would produce exactly the same answer.

The formulation of the streamlined secretaries and clerks problem in matrix language consists of letting

$$Z = [x \quad y] \qquad A = \begin{bmatrix} 1 & 10 & 20 \\ 0 & 4 & 30 \end{bmatrix} \qquad C = [5 \quad 100 \quad 300] \qquad B = \begin{bmatrix} 6 \\ 2 \end{bmatrix}$$

and then we must find values for Z satisfying

$$Z \geq 0$$

$$ZA \geq C$$

which make

$$\text{Salaries} = ZB$$

as small as possible. Exactly as in the maximum problems of Section 5-6, we form the simplex tableau:

$$\left[\begin{array}{c|c|c} A & I & B \\ \hline -C & 0\cdots0 & 0 \end{array}\right]$$

In this case, we have

$$\left[\begin{array}{ccc|cc|c} 1 & 10 & 20 & 1 & 0 & 6 \\ 0 & 4 & 30 & 0 & 1 & 2 \\ \hline -5 & -100 & -300 & 0 & 0 & 0 \end{array}\right]$$

Next, we perform the steps from Section 5-6 in order to make all the numbers in the bottom row positive or zero. Combining some steps, these are

$$\left[\begin{array}{ccc|cc|c} 1 & 10 & 20 & 1 & 0 & 6 \\ 0 & 4 & \circled{30} & 0 & 1 & 2 \\ \hline -5 & -100 & -300 & 0 & 0 & 0 \end{array}\right] \begin{array}{c} \dfrac{6}{20} \\[2mm] \dfrac{2}{30} \end{array}$$

$$\left[\begin{array}{ccc|cc|c} 1 & \dfrac{110}{15} & 0 & 1 & -\dfrac{2}{3} & \dfrac{14}{3} \\[2mm] 0 & \dfrac{2}{15} & \circled{1} & 0 & \dfrac{1}{30} & \dfrac{1}{15} \\[2mm] \hline -5 & -60 & 0 & 0 & 10 & 20 \end{array}\right] \begin{array}{c} \dfrac{210}{300} \\[2mm] \dfrac{1}{2} \end{array}$$

$$\left[\begin{array}{ccc|cc|c} 1 & 0 & -55 & 1 & -\dfrac{15}{6} & 1 \\[2mm] 0 & \circled{1} & \dfrac{15}{2} & 0 & \dfrac{1}{4} & \dfrac{1}{2} \\[2mm] \hline -5 & 0 & 450 & 0 & 25 & 50 \end{array}\right]$$

$$\left[\begin{array}{ccc|cc|c} \circled{1} & 0 & -55 & 1 & -\dfrac{15}{6} & 1 \\[2mm] 0 & \circled{1} & \dfrac{15}{2} & 0 & \dfrac{1}{4} & \dfrac{1}{2} \\[2mm] \hline 0 & 0 & 175 & 5 & 12\tfrac{1}{2} & 55 \end{array}\right]$$

As it happens, we end up with an identity matrix in the upper left-hand corner of the simplex tableau after all the numbers in the last row have become positive or zero. Were this not so, it would still be unnecessary to carry out the last step in the simplex method for maximum problems, because its only purpose is to put the answers to the maximum problem, which lie in the last column, in the proper order. The answer to the problem is $x = 5$ and $y = 12\frac{1}{2}$, that is, 5 secretaries and $12\frac{1}{2}$ clerks. This is, of course, the answer we found in Section 5-4 by other means. Observe that the value of the salaries, \$55, appears in the lower right-hand corner of the tableau.

5-23   **Example**   A company hires a lazy but effective salesperson on the understanding that he will contact at least 100 potential customers a week and that at least ten of the contacts will be by personal visit. His other means of contact are through telephone calls and letters. The salesperson decides that he will have to make at least fifteen sales a week in order to make enough money from commissions to support himself. His experience is that, in general, he can make one sale out of two personal visits or from ten telephone calls or from twenty letters. He estimates that it takes him 1 hour to make a personal visit, 15 minutes for a telephone call, and 6 minutes to dictate a letter. How many personal visits, telephone calls, and letters should the salesperson plan on so that he can make enough sales while working as few hours as possible?

Solution   Let x be the number of personal visits the salesperson makes, let y be the number of telephone calls, and let z be the number of letters he writes. Set

$$Z = [x \quad y \quad z]$$

The conditions of his employment are

$$x \geq 10$$

$$x + y + z \geq 100$$

If he makes x personal visits, then he expects to accomplish $(1/2)x$ sales in this way. Since one out of ten telephone calls produces a sale, he will make $(1/10)y$ sales from y telephone calls. Similarly, he expects $(1/20)z$ sales on the basis of his letters. He needs at least fifteen sales in all, so

$$\frac{1}{2}x + \frac{1}{10}y + \frac{1}{20}z \geq 15$$

We can represent these restrictions on x, y, and z through the matrix inequality $ZA \geq C$ by letting

$$A = \begin{bmatrix} 1 & 1 & \dfrac{1}{2} \\[2mm] 0 & 1 & \dfrac{1}{10} \\[2mm] 0 & 1 & \dfrac{1}{20} \end{bmatrix} \qquad C = [10 \quad 100 \quad 15]$$

It takes the salesperson 1 hour to make a visit, so it requires x hours to make x visits. A telephone call requires 15 minutes or 1/4 hour, so y calls take up $(1/4)y$ hours. Since 6 minutes is 1/10 hour, he spends $(1/10)z$ hours writing letters. Therefore, the salesperson's total investment of time is

$$\text{Time} = x + \frac{1}{4}y + \frac{1}{10}z = ZB$$

where

$$B = \begin{bmatrix} 1 \\[2mm] \dfrac{1}{4} \\[2mm] \dfrac{1}{10} \end{bmatrix}$$

The simplex tableau is

$$\begin{bmatrix} \begin{array}{ccc|ccc|c} 1 & 1 & \dfrac{1}{2} & 1 & 0 & 0 & 1 \\[2mm] 0 & 1 & \dfrac{1}{10} & 0 & 1 & 0 & \dfrac{1}{4} \\[2mm] 0 & 1 & \dfrac{1}{20} & 0 & 0 & 1 & \dfrac{1}{10} \\[2mm] \hline -10 & -100 & -15 & 0 & 0 & 0 & 0 \end{array} \end{bmatrix}$$

The main steps in changing all the numbers in the last row to positive numbers or zero are

$$\begin{bmatrix} \begin{array}{ccc|ccc|c} 1 & 0 & \dfrac{9}{20} & 1 & 0 & -1 & \dfrac{9}{10} \\[2mm] 0 & 0 & \dfrac{1}{20} & 0 & 1 & -1 & \dfrac{3}{20} \\[2mm] 0 & ① & \dfrac{1}{20} & 0 & 0 & 1 & \dfrac{1}{10} \\[2mm] \hline -10 & 0 & -10 & 0 & 0 & 100 & 10 \end{array} \end{bmatrix}$$

$$\begin{bmatrix} ① & 0 & \dfrac{9}{20} & 1 & 0 & -1 & \dfrac{9}{10} \\[2ex] 0 & 0 & \dfrac{1}{20} & 0 & 1 & -1 & \dfrac{3}{20} \\[2ex] 0 & ① & \dfrac{1}{20} & 0 & 0 & 1 & \dfrac{1}{10} \\[2ex] \hline 0 & 0 & -\dfrac{11}{2} & 10 & 0 & 90 & 19 \end{bmatrix}$$

$$\begin{bmatrix} ① & -9 & 0 & 1 & 0 & -10 & 0 \\[2ex] 0 & -1 & 0 & 0 & 1 & -2 & \dfrac{1}{20} \\[2ex] 0 & 20 & ① & 0 & 0 & 20 & 2 \\[1ex] \hline 0 & 110 & 0 & 10 & 0 & 200 & 30 \end{bmatrix}$$

Therefore, the salesperson should average $x = 10$ personal visits each week, not make any telephone calls since $y = 0$, and write $z = 200$ letters. These activities will require 30 hours each week.

In Example 5-23, the values of x, y, and z which will make the function as small as possible include the value $y = 0$. A solution to a linear programming problem in which at least one of the numbers is 0 is called a **degenerate** solution. As the example indicates, no difficulty arises when a minimum problem has a degenerate solution; the 0 just appears in the appropriate place in the tableau. However, look at the final form of the tableau of Example 5-23 again. Notice that if this had been a maximum problem rather than a minimum problem, we would be unable to carry out the final step in the simplex method described in Section 5-6, because there is no way that an interchange of rows can produce an identity matrix in the upper left-hand corner of the tableau. In the previous section, we avoided this difficulty by the judicious choice of examples and problems. Now we will see that such a maximum problem is degenerate (at least one of the numbers in the solution is zero), but also we will find that it is no more difficult to solve than are the nondegenerate problems of Section 5-6.

As an example, suppose we wish to find x, y, and z, all greater than or equal to zero, satisfying

$$x + y + 2z \le 8$$
$$x + y \le 5$$
$$2y \le 1$$

which will make

$$\text{Sales} = 3x + 2y + 4z$$

as large as possible. We thus have the maximum problem of finding

$$X = \begin{bmatrix} x \\ y \\ z \end{bmatrix}$$

such that

$$X \geq 0$$

$$AX \leq B$$

which makes

$$\text{Sales} = CX$$

as large as possible, where

$$A = \begin{bmatrix} 1 & 1 & 2 \\ 1 & 1 & 0 \\ 0 & 2 & 0 \end{bmatrix} \qquad B = \begin{bmatrix} 8 \\ 5 \\ 1 \end{bmatrix} \qquad C = \begin{bmatrix} 3 & 2 & 4 \end{bmatrix}$$

We form the simplex tableau,

$$\left[\begin{array}{ccc|ccc|c} 1 & 1 & 2 & 1 & 0 & 0 & 8 \\ 1 & 1 & 0 & 0 & 1 & 0 & 5 \\ 0 & 2 & 0 & 0 & 0 & 1 & 1 \\ \hline -3 & -2 & -4 & 0 & 0 & 0 & 0 \end{array} \right]$$

and perform the required operations:

$$\left[\begin{array}{ccc|ccc|c} 1 & 1 & ② & 1 & 0 & 0 & 8 \\ 1 & 1 & 0 & 0 & 1 & 0 & 5 \\ 0 & 2 & 0 & 0 & 0 & 1 & 1 \\ \hline -3 & -2 & -4 & 0 & 0 & 0 & 0 \end{array} \right]$$

$$\left[\begin{array}{ccc|ccc|c} \frac{1}{2} & \frac{1}{2} & ① & \frac{1}{2} & 0 & 0 & 4 \\ 1 & 1 & 0 & 0 & 1 & 0 & 5 \\ 0 & 2 & 0 & 0 & 0 & 1 & 1 \\ \hline -3 & -2 & -4 & 0 & 0 & 0 & 0 \end{array} \right]$$

$$
\left[
\begin{array}{ccc|ccc|c}
\frac{1}{2} & \frac{1}{2} & ① & \frac{1}{2} & 0 & 0 & 4 \\
① & 1 & 0 & 0 & 1 & 0 & 5 \\
0 & 2 & 0 & 0 & 0 & 1 & 1 \\
\hline
-1 & 0 & 0 & 2 & 0 & 0 & 16
\end{array}
\right]
$$

$$
\left[
\begin{array}{ccc|ccc|c}
0 & 0 & ① & \frac{1}{2} & -\frac{1}{2} & 0 & \frac{3}{2} \\
① & 1 & 0 & 0 & 1 & 0 & 5 \\
0 & 2 & 0 & 0 & 0 & 1 & 1 \\
\hline
0 & 1 & 0 & 2 & 1 & 0 & 21
\end{array}
\right]
$$

At this point, we must stop looking for pivot elements, because all the numbers in the bottom row are positive or zero. However, we are unable to obtain an identity matrix, with pivot elements as the 1s, in the upper left-hand corner of the tableau by means of row interchanges. Instead, let us examine the two rows which do contain pivot elements:

$$
\left[
\begin{array}{ccc|cc|c|c}
0 & 0 & ① & \frac{1}{2} & -\frac{1}{2} & 0 & \frac{3}{2}
\end{array}
\right]
$$

and

$$
\left[
\begin{array}{ccc|ccc|c}
① & 1 & 0 & 0 & 1 & 0 & 5
\end{array}
\right]
$$

Since the pivot element of the first row is in the third column, the rule is that the number at the extreme right of the row (3/2 in this case) is the value of the third unknown in the matrix X. In other words, we have

$$
z = \frac{3}{2}
$$

The pivot element of the second row is in the first column, so 5, the number at the right end of the row, is the value of the first unknown in X:

$$
x = 5
$$

Since no pivot element appears in the second column, the second unknown has the value zero in the solution to the maximum problem, that is,

$$
y = 0
$$

The solution to this degenerate linear programming problem, then, is

$$X = \begin{bmatrix} 5 \\ 0 \\ 3 \\ 2 \end{bmatrix}$$

The value of the function Sales corresponding to the solution still appears in the lower right-hand corner of the tableau, so that value is 21 in this example.

If it is not possible to construct an identity matrix in the upper left-hand corner of the tableau in a maximum problem even after all the numbers in the bottom row become zero or positive, then the problem is degenerate. If there is a pivot element in the kth column (k between 1 and n, the number of unknowns in the problem), then the right-hand number in the row containing that pivot element is the value of the kth unknown in the solution to the problem. If no pivot element lies in the kth column, then the value of the kth unknown in the solution is zero.

5-24 **Example** Solve the maximum problem where

$$A = \begin{bmatrix} 1 & 1 & 0 & 1 \\ 1 & 0 & 0 & -1 \\ 1 & 2 & 2 & 0 \\ 0 & 1 & 0 & 0 \\ 0 & 1 & 1 & 0 \end{bmatrix} \qquad B = \begin{bmatrix} 1 \\ 0 \\ 3 \\ 4 \\ 2 \end{bmatrix} \qquad C = \begin{bmatrix} 3 & 2 & 1 & 4 \end{bmatrix}$$

Solution The simplex tableau is

$$\begin{bmatrix} \begin{array}{cccc|ccccc|c} 1 & 1 & 0 & 1 & 1 & 0 & 0 & 0 & 0 & 1 \\ 1 & 0 & 0 & -1 & 0 & 1 & 0 & 0 & 0 & 0 \\ 1 & 2 & 2 & 0 & 0 & 0 & 1 & 0 & 0 & 3 \\ 0 & 1 & 0 & 0 & 0 & 0 & 0 & 1 & 0 & 4 \\ 0 & 1 & 1 & 0 & 0 & 0 & 0 & 0 & 1 & 2 \\ \hline -3 & -2 & -1 & -4 & 0 & 0 & 0 & 0 & 0 & 0 \end{array} \end{bmatrix}$$

The steps of the simplex method produce:

$$\begin{bmatrix} \begin{array}{cccc|ccccc|c} 1 & 1 & 0 & ① & 1 & 0 & 0 & 0 & 0 & 1 \\ 2 & 0 & 0 & 0 & 1 & 1 & 0 & 0 & 0 & 1 \\ 1 & 2 & 2 & 0 & 0 & 0 & 1 & 0 & 0 & 3 \\ 0 & 1 & 0 & 0 & 0 & 0 & 0 & 1 & 0 & 4 \\ 0 & 1 & 1 & 0 & 0 & 0 & 0 & 0 & 1 & 2 \\ \hline 1 & 2 & -1 & 0 & 4 & 0 & 0 & 0 & 0 & 4 \end{array} \end{bmatrix}$$

$$
\begin{bmatrix}
1 & 1 & 0 & \textcircled{1} & 1 & 0 & 0 & 0 & 0 & 1 \\
2 & 0 & 0 & 0 & 1 & 1 & 0 & 0 & 0 & 1 \\
\dfrac{1}{2} & 1 & \textcircled{1} & 0 & 0 & 0 & \dfrac{1}{2} & 0 & 0 & \dfrac{3}{2} \\
0 & 1 & 0 & 0 & 0 & 0 & 0 & 1 & 0 & 4 \\
-\dfrac{1}{2} & 0 & 0 & 0 & 0 & 0 & -\dfrac{1}{2} & 0 & 1 & \dfrac{1}{2} \\
\hline
\dfrac{3}{2} & 3 & 0 & 0 & 4 & 0 & \dfrac{1}{2} & 0 & 0 & \dfrac{11}{2}
\end{bmatrix}
$$

No negative numbers remain in the bottom row and the problem is clearly degenerate. Write the matrix of unknowns:

$$
X = \begin{bmatrix} w \\ x \\ y \\ z \end{bmatrix}
$$

Since no pivot elements lie in the first or second column, we conclude that $w = 0$ and $x = 0$. In the third column, there is a pivot element in the third row, so the right-hand number in that row, $3/2$, is the value of the third unknown: $y = 3/2$. The pivot element in the fourth column is in the first row, so the 1 at the extreme right of that row means $z = 1$. Therefore, the solution is

$$
X = \begin{bmatrix} 0 \\ 0 \\ \dfrac{3}{2} \\ 1 \end{bmatrix}
$$

Summary

To solve a minimum problem of the following form: Find the $1 \times n$ matrix of unknowns Z satisfying

$$Z \geq 0$$

$$ZA \geq C$$

which make

$$\text{Function} = ZB$$

as small as possible, first form the simplex tableau,

$$\left[\begin{array}{c|c|c} A & I & B \\ \hline -C & \boxed{0 \cdots 0} & 0 \end{array}\right]$$

Follow the rules for the simplex method in Section 5-6 until all the numbers in the bottom row are zero or positive. The solution Z to the minimum problem will occupy the circled region of the tableau, in the same order as the unknowns in Z. The corresponding value of the function will be in the lower right-hand corner of the tableau.

If the solution to a linear programming problem includes a value of zero for at least one of the unknowns, then the problem is said to be **degenerate**.

A maximum problem in n unknowns in which it is impossible to form an $n \times n$ identity matrix with pivot elements for 1s in the upper left-hand corner of the tableau after all the numbers in the last row have become positive or zero is a degenerate problem. If a pivot element of such a problem lies in the kth column of the tableau (k between 1 and n), then the number at the far right of the row containing that pivot element is the value of the kth unknown in the solution. If no pivot element lies in the kth column, then the value of the kth unknown in the solution is zero.

Exercises

- *In Exercises 5-85 through 5-89, solve the minimum problem* $Z \geq 0$, $ZA \geq C$, *Function* $= ZB$.

5-85 $Z = \begin{bmatrix} x & y \end{bmatrix}$ $A = \begin{bmatrix} 6 & 1 \\ 1 & 1 \end{bmatrix}$ $B = \begin{bmatrix} 8 \\ 6 \end{bmatrix}$ $C = \begin{bmatrix} 1 & 2 \end{bmatrix}$

5-86 $Z = \begin{bmatrix} x & y \end{bmatrix}$ $A = \begin{bmatrix} 0 & 2 \\ 1 & 3 \end{bmatrix}$ $B = \begin{bmatrix} 1 \\ 3 \end{bmatrix}$ $C = \begin{bmatrix} 2 & 5 \end{bmatrix}$

5-87 $Z = \begin{bmatrix} x & y \end{bmatrix}$ $A = \begin{bmatrix} 1 & 2 & 0 \\ 3 & 1 & 3 \end{bmatrix}$ $B = \begin{bmatrix} 4 \\ 2 \end{bmatrix}$ $C = \begin{bmatrix} 1 & 2 & 3 \end{bmatrix}$

5-88 $Z = \begin{bmatrix} x & y & z \end{bmatrix}$ $A = \begin{bmatrix} 1 & 0 & 1 & 1 \\ 0 & 0 & 2 & 1 \\ 0 & 4 & 0 & 1 \end{bmatrix}$ $B = \begin{bmatrix} 4 \\ 2 \\ 6 \end{bmatrix}$ $C = \begin{bmatrix} 1 & 2 & 3 & 1 \end{bmatrix}$

5-89 $Z = [x \ y \ z]$ $A = \begin{bmatrix} 1 & 4 & 2 \\ 1 & 0 & 3 \\ 1 & 1 & -1 \end{bmatrix}$ $B = \begin{bmatrix} 2 \\ 1 \\ 4 \end{bmatrix}$ $C = [10 \ 8 \ 0]$

● In Exercises 5-90 through 5-94, solve the (possibly degenerate) maximum problem $X \geq 0$, $AX \leq B$, Function $= CX$.

5-90 $X = \begin{bmatrix} x \\ y \end{bmatrix}$ $A = \begin{bmatrix} 1 & 1 \\ 1 & 1 \end{bmatrix}$ $B = \begin{bmatrix} 2 \\ 3 \end{bmatrix}$ $C = [2 \ 1]$

5-91 $X = \begin{bmatrix} x \\ y \end{bmatrix}$ $A = \begin{bmatrix} 2 & 0 \\ 1 & 1 \\ 2 & 0 \end{bmatrix}$ $B = \begin{bmatrix} 3 \\ 2 \\ 1 \end{bmatrix}$ $C = [3 \ 2]$

5-92 $X = \begin{bmatrix} x \\ y \\ z \end{bmatrix}$ $A = \begin{bmatrix} 1 & 1 & 2 \\ 1 & 1 & 0 \\ 0 & 1 & 1 \end{bmatrix}$ $B = \begin{bmatrix} 3 \\ 2 \\ 1 \end{bmatrix}$ $C = [1 \ 3 \ 1]$

5-93 $X = \begin{bmatrix} x \\ y \end{bmatrix}$ $A = \begin{bmatrix} 1 & 1 \\ 2 & 0 \\ -1 & 1 \\ 3 & 2 \end{bmatrix}$ $B = \begin{bmatrix} 1 \\ 3 \\ 0 \\ 7 \end{bmatrix}$ $C = [2 \ 1]$

5-94 $X = \begin{bmatrix} x \\ y \\ z \end{bmatrix}$ $A = \begin{bmatrix} 0 & 0 & 1 \\ 0 & 1 & -1 \\ 2 & \frac{1}{2} & 0 \end{bmatrix}$ $B = \begin{bmatrix} 1 \\ 0 \\ 2 \end{bmatrix}$ $C = [2 \ 1 \ 3]$

5-95   Find x and y, both greater than or equal to zero, satisfying

$$2x \geq y$$

$$2x + 3y \geq 4$$

which make

$$\text{Cost} = 12x + 15y$$

as small as possible.

5-96   Find x, y, and z, all greater than or equal to zero, satisfying

$$x + y + z \leq 10$$

$$x \leq 2y$$

$$2x + 3y \leq 30$$

$$x + y \geq 6$$

which make

$$\text{Distance} = 7x + 8y + 10z$$

as large as possible.

5-97 Find w, x, y, and z, all greater than or equal to zero, satisfying

$$w + x + y + z \geq 15$$

$$w + y \geq 5$$

$$2x + z \geq 10$$

which make

$$\text{Electricity} = 5w + x + 3y + 2z$$

as small as possible.

5-98 Find x, y, and z, all greater than or equal to zero, satisfying

$$x + y + z \leq 1$$

$$y \leq z$$

$$2x + y \leq 2$$

which make

$$\text{Value} = 2x + y + 3z$$

as large as possible.

5-99 A man plans to plant at least 1000 square feet of his yard with vegetables and perennial flowers. He will devote at least as much space to vegetables as to flowers and at least 200 square feet to flowers. The man estimates that growing vegetables requires 2 units of work for each square foot and flowers require 1 unit of work for each square foot. How many square feet should he devote to growing vegetables and how many to growing flowers so as to do the least possible work?

5-100   A jeweler makes rings, earrings, pins, and necklaces. He wishes to work no more than 40 hours a week. It takes him 2 hours to make a ring, 2 hours to make a pair of earrings, 1 hour to make a pin, and 4 hours to make a necklace. He estimates that he can sell no more than ten rings, ten pairs of earrings, fifteen pins, and three necklaces in a week. The jeweler charges $50 for a ring, $80 for a pair of earrings, $25 for a pin, and $200 for a necklace. How many rings, earrings, pins, and necklaces should the jeweler make in order to make the largest possible gross earnings?

5-101   A woman wishes to design a weekly exercise schedule which will involve jogging, bicycling, and swimming. In order to vary the exercise, she plans to spend at least as much time bicycling as she devotes to jogging and swimming combined. Also, she wishes to swim at least 2 hours a week, because she enjoys that activity more than the others. If jogging consumes 600 calories an hour, bicycling uses up 300 calories an hour, and swimming requires 300 calories an hour, and if she wishes to burn up a total of at least 3000 calories a week through exercise, how many hours should she devote to each type of exercise if she wishes to reach this goal in the fewest number of hours?

5-102   In a week in which the woman of Exercise 5-101 has only 7 hours to spend on exercise, she realizes that she may not be able to burn up her usual 3000 calories, but she retains the other restrictions. If she wishes to use up as many calories as possible within 7 hours, how much time should she devote to each type of exercise? [*Hint*: In order to swim at least 2 hours, she cannot spend more than 5 hours on jogging and bicycling combined.]

Application to a Farm Management Problem (Technical Essay)

In the introductory essay to this chapter, we described the problem of a farmer in Kern County, California, who wanted to plant his 150 acres of land so that it would make as much profit as possible. Recall that the major crops in Kern County are cotton, potatoes, alfalfa, sugar beets, and barley. The estimated dollar profit per acre for each crop is given by the following table:

Cotton	$207
Potatoes	$200
Alfalfa	$ 86
Sugar beets	$136
Barley	$ 29

If we let c denote the number of acres of cotton the farmer plants, p the number of acres of potatoes, a the number of acres devoted to alfalfa, s the number of acres of sugar beets, and b the number of acres of barley, then the total profit the farmer expects to earn from his farm is calculated by the formula

$$\text{Profit} = 207c + 200p + 86a$$
$$+ 136s + 29b$$

Thus, the farmer's problem is to choose c, p, a, s, and b, all greater than or equal to zero, to make Profit as large as possible.

There are, of course, restrictions on the possible values for c, p, a, s, and b that reflect the circumstances under which the farmer has to operate his farm. For example, since he has a total of just 150 acres in all, it must be that

$$c + p + a + s + b \leq 150$$

Other restrictions stated earlier were the following: First, the government had limited to 60 acres the amount of cotton the farmer could grow. Thus, we require that

$$c \leq 60$$

In addition, because of his uncertainty with regard to potato prices, the farmer voluntarily limited the number of acres he would devote to potatoes to no more than 50 acres, so

$$p \leq 50$$

The remaining restrictions arose from the limited availability of water. The growing season was divided into three periods, both because each crop requires different amounts of water at different times during its growing season and because the amount of water available to the farmer depends on the time of year. The needs of each crop for water in each period, in inches per acre, and the amount of water available, in acre-inches, is given in Table 5-1. Thus, for example, in the third growing period, alfalfa requires 11.1 inches of water for each acre from the 730 acre-inches available.

Table 5-1

| | Water needed for crop (acre-inches) | | | | | |
Period	Cotton	Potatoes	Alfalfa	Sugar beets	Barley	Water available (acre-inches)
1	4.0	13.3	15.8	13.0	6.3	2200
2	16.6	0	22.2	42.7	0	2110
3	7.8	0	11.1	3.3	0	730

Source: James N. Boles, "Linear Programming and Farm Management Analysis," *Journal of Farm Economics* 3 (Feb. 1955).

If the farmer plants c acres of cotton, p of potatoes, a of alfalfa, s of sugar beets, and b of barley, then his total water needs during the first growing period will be

$$4.0c + 13.3p + 15.8a + 13.0s + 6.3b$$

acre-inches of water. Since there are 2200 acre-inches of water available, then c, p, a, s, and b must be chosen so that

$$4.0c + 13.3p + 15.8a$$
$$+ 13.0s + 6.3b \leq 2200$$

In the second growing period, the total water needs will be

$$16.6c + 0p + 22.2a + 42.7s + 0b$$

There are 2110 acre-inches of water available, so c, p, a, s, and b must be chosen in such a way that the restriction

$$16.6c + 22.2a + 42.7s \leq 2110$$

is satisfied. Similarly, there is the restriction imposed by the water requirements for the various crops during the third growing period. The table states that 730 acre-inches of irrigation water will be available, so the restriction is

$$7.8c + 11.1a + 3.3s \leq 730$$

Certainly, c, p, a, s, and b must be chosen to satisfy all these require-

ments simultaneously, since there is no profit to be had in planting crops that will grow well in the first growing period and then die later for lack of water.

In mathematical terms, then, the problem is to find numbers c, p, a, s, and b, all greater than or equal to zero and satisfying the restrictions

$$c + p + a + s + b \leq 150$$
$$c \leq 60$$
$$p \leq 50$$
$$4.0c + 13.3p + 15.8a$$
$$+ 13.0s + 6.3b \leq 2200$$
$$16.6c + 22.2a + 42.7s \leq 2110$$
$$7.8c + 11.1a + 3.3s \leq 730$$

which will make

$$\text{Profit} =$$
$$207c + 200p + 86a + 136s + 29b$$

as large as possible.

In order to facilitate the solution to this linear programming problem, we express it in matrix form as the maximum problem of finding the 5×1 matrix of unknowns

$$X = \begin{bmatrix} c \\ p \\ a \\ s \\ b \end{bmatrix}$$

satisfying the restriction

$$\begin{bmatrix} 1 & 1 & 1 & 1 & 1 & 1 & 0 & 0 & 0 & 0 & 0 & 150 \\ 1 & 0 & 0 & 0 & 0 & 0 & 1 & 0 & 0 & 0 & 0 & 60 \\ 0 & 1 & 0 & 0 & 0 & 0 & 0 & 1 & 0 & 0 & 0 & 50 \\ 4.0 & 13.3 & 15.8 & 13.0 & 6.3 & 0 & 0 & 0 & 1 & 0 & 0 & 2200 \\ 16.6 & 0 & 22.2 & 42.7 & 0 & 0 & 0 & 0 & 0 & 1 & 0 & 2110 \\ 7.8 & 0 & 11.1 & 3.3 & 0 & 0 & 0 & 0 & 0 & 0 & 1 & 730 \\ \hline -207 & -200 & -86 & -136 & -29 & 0 & 0 & 0 & 0 & 0 & 0 & 0 \end{bmatrix}$$

$$X \geq 0$$

$$AX \leq B$$

with

$$A = \begin{bmatrix} 1 & 1 & 1 & 1 & 1 \\ 1 & 0 & 0 & 0 & 0 \\ 0 & 1 & 0 & 0 & 0 \\ 4.0 & 13.3 & 15.8 & 13.0 & 6.3 \\ 16.6 & 0 & 22.2 & 42.7 & 0 \\ 7.8 & 0 & 11.1 & 3.3 & 0 \end{bmatrix}$$

$$B = \begin{bmatrix} 150 \\ 60 \\ 50 \\ 2200 \\ 2110 \\ 730 \end{bmatrix}$$

which will make

$$\text{Profit} = CX$$

as large as possible, where

$$C = [207 \quad 200 \quad 86 \quad 136 \quad 29]$$

We form the simplex tableau shown above. The first column is the pivot column, and it is easy to determine that the pivot element is the 1 located at the second row and the first column. The steps of the calculation are the same as those described in Section 5-6. The form the tableau takes when all the numbers in the last row are either positive or zero is shown below (all answers are rounded off to two decimal places of accuracy).

$$\begin{bmatrix} 0 & 0 & 0 & 0 & ① & 1 & -0.28 & -1 & 0 & -0.02 & -0.05 & 4.91 \\ ① & 0 & 0 & 0 & 0 & 0 & 1 & 0 & 0 & 0 & 0 & 60 \\ 0 & ① & 0 & 0 & 0 & 0 & 0 & 1 & 0 & 0 & 0 & 50 \\ 0 & 0 & 0 & 0 & 0 & -6.3 & 9.06 & -7 & 1 & -0.07 & -0.66 & 755.32 \\ 0 & 0 & 0 & ① & 0 & 0 & -0.03 & 0 & 0 & 0.07 & -0.06 & 16.34 \\ 0 & 0 & ① & 0 & 0 & 0 & -0.69 & 0 & 0 & -0.01 & 0.11 & 18.75 \\ \hline 0 & 0 & 0 & 0 & 0 & 29 & 135.33 & 171 & 0 & 2.13 & 0.15 & 26{,}397.13 \end{bmatrix}$$

$$
\left[
\begin{array}{ccccc|cccccc|c}
① & 0 & 0 & 0 & 0 & 0 & 1 & 0 & 0 & 0 & 0 & 60 \\
0 & ① & 0 & 0 & 0 & 0 & 0 & 1 & 0 & 0 & 0 & 50 \\
0 & 0 & ① & 0 & 0 & 0 & -0.69 & 0 & 0 & -0.01 & 0.11 & 18.75 \\
0 & 0 & 0 & ① & 0 & 0 & -0.03 & 0 & 0 & 0.07 & -0.06 & 16.34 \\
0 & 0 & 0 & 0 & ① & 1 & -0.28 & -1 & 0 & -0.02 & -0.05 & 4.91 \\
0 & 0 & 0 & 0 & 0 & -6.3 & 9.06 & -7 & 1 & -0.07 & -0.66 & 755.32 \\
\hline
0 & 0 & 0 & 0 & 0 & 29 & 135.33 & 171 & 0 & 2.13 & 0.15 & 26{,}397.13
\end{array}
\right]
$$

We can see that the problem is nondegenerate, because we may interchange rows to obtain an identity matrix with pivot elements as 1s at the upper left-hand corner of the tableau, as shown below. Reading from the last column of the tableau, we conclude that the farmer would obtain the largest possible profit if he devotes 60 acres to cotton, 50 acres to potatoes, 18.75 acres to alfalfa, 16.34 acres to sugar beets, and 4.91 acres to

barley. In that case, his total profit will be $26,397.13.

In practice, it would probably not be worthwhile to plant in fractions of acres so, instead, the farmer would round off to whole acres and plant 60 acres of cotton, 50 acres of potatoes, 19 acres of alfalfa, 16 acres of sugar beets, and 5 acres of barley, with an expected total profit of $26,375.

Concluding Essay

Linear programming can be used in a variety of situations. One of the earliest applications was the solution of the Berlin airlift problem. When, in July 1948, the Soviet Union blocked all road and rail traffic to West Berlin through the Soviet zone of Germany, some action had to be taken if West Berlin was to survive. The United States Air Force wanted to fly in enough supplies to ensure this survival. The Air Force had to deliver maximum tonnage with the available supply of trained crews, runways, aircraft, and money. This airlift was successful and

continued until the blockade was lifted in May 1949.

Since 1948, linear programming has been applied to a multitude of problems in all phases of industry and agriculture. Besides the more or less obvious production and inventory problems, the method has proved useful in airline scheduling, designing and utilizing communication networks, solving transportation problems, blending gasolines, selecting investment portfolios, choosing locations for new factories, assigning toll collectors to tollbooths, scheduling traffic lights, operating a system of dams on the Missouri river, structural designing, and determining the most efficient routes for traveling salespeople.

Most of these applications would be tedious and time-consuming if the computations had to be done by hand. Fortunately, the parallel development of data processing and computing machinery along with the underlying mathematical theory has made possible the complete utilization of linear programming.

References

The application on which this chapter was based comes from James N. Boles' article "Linear Programming and Farm Management Analysis," which appeared in the *Journal of Farm Economics* (Feb. 1955).

The article "Linear Programming," by W. A. Spivey, which appeared in *Science* (Jan. 5, 1962) has a fine bibliography.

Linear Programming Methods and Applications, by Saul I. Gass (New York: McGraw-Hill, 1958), has a particularly interesting chapter on applications.

For a more up-to-date presentation of the mathematics of linear programming, consult Michel Simmonard's *Linear Programming,* translated by William S. Jewell (Englewood Cliffs, N.J.: Prentice-Hall, 1966).

Chapter 6
Competition
and Cooperation:
Game Theory Is
the Test

**Introductory
Essay**

To study hidden prejudice against blacks, George W. Baxter, Jr. (see References) set up an experiment in social psychology in which 90 white first-year college women played a game known as the Prisoner's Dilemma. Some of the women had a white partner and some a black partner. The experimenter was trying to answer several questions: (1) Does the race of the other player affect the extent of a white subject's cooperation, and if so do northern subjects cooperate more with a black than do southern subjects? (2) Does the information supplied to the subject concerning the other person's cooperative or competitive nature affect the extent of the subject's cooperation? (3) Is there a relationship between the subject's attitudes (specifically those toward racial segregation and about people in general) and her level of cooperation in the game? (4) Will the subject's evaluation of a black player's personality differ from the ratings given a white player? (5) Will the ratings taken before the game differ from the same ratings taken after the game?

The women used as subjects for the experiment were all entering college students whose responses on tests during orientation week showed them to be definitely in favor of school integration and, in their own view at least, free of racial prejudice. They were then divided into six groups of fifteen with five women from the North and ten from the South in each group. Three of the groups were introduced to a black woman as the other player in the game, while the other three groups were introduced to a white woman. One group in each half of the experiment was given a personality profile of their partner indicating a cooperative nature, another group in each half was given a profile indicating their partner had a competitive personality, and the remaining students were given no information at all.

When each woman arrived to take part in the experiment she was shown to a room with a control box which had four lights and two switches. On the box the payoffs shown in the table* were indicated, with the subject's payoffs

*This diagram from "Prejudiced Liberals? Race and Information Effects in a Two-Person Game," by George W. Baxter, Jr., is reprinted from *The Journal of Conflict Resolution* Vol. 17, No. 1 (March 1973), p. 138, by permission of the publisher, Sage Publications, Inc.

		Partner	
		Blue	Red
Subject	Blue	+5¢, +5¢	−4¢, +6¢
	Red	+6¢, −4¢	−3¢, −3¢

to the left of the commas and her partner's to the right. For example, if she pushed her blue switch and her partner also chose blue, they would each receive 5¢. However, if she pushed blue and her partner pushed the red switch, the subject would lose 4¢ while her partner would gain 6¢. A few trial rounds were played; then the subject was given a tally sheet and $1 and was told she could keep her earnings from the game. Altogether, 30 rounds of the game were played by each subject. All the responses the partner supposedly made were in fact those of the experimenter using a 90% cooperative strategy. That is, Baxter made noncooperative (red) responses on trials 6, 14, and 22 out of 30. After the game, the subject was asked to fill out the postexperimental questionnaire, while the experimenter totaled the earnings. Every effort was made to see that the subjects did not know one another by picking them from different dormitories and using freshmen early in the school year when their range of acquaintances would not be very large.

After analyzing his results (based on a game theory model) Baxter concluded that the number of cooperative responses was lower when the partner was black and that the origin (North or South) of the subject made no difference. The effect of the pretest information about the supposed personality of the other player was, as expected, that more cooperative choices were made when playing with a supposedly cooperative personality than when playing with a supposedly competitive personality. One of the most significant results that came out of the experiment was that the subjects cooperated less with a black partner than with a white partner even when they had been told that the player had a cooperative personality profile.

In the postgame questionnaire rating the other player for such qualities as greedy–generous, competitive–cooperative, unlikable–likable, dull–intelligent, the supposedly cooperative partners were rated higher than the competitive players. But in every group, no matter how much information had been given, the white partner was rated higher than the black.

Baxter's article concludes,

The results of this study are especially striking if it is remembered that almost all the subjects scored on the antisegregation end of the School Integration—

Segregation scale. . . . Yet, their behavior in the game was not free of more discriminatory and exploitative choices toward Negro other players than toward white other players. Apparently even "liberal-minded" individuals are not immune to the insidious nature and influence of the race prejudice in our culture.

From the description of this experiment it is not evident what, if any, kind of mathematics is involved. We might guess the experimenter would use statistics to interpret his results and we would be correct. However, the essential feature of the experiment is the use of a game, the Prisoner's Dilemma, which can be studied by means of a branch of mathematics known as game theory.

In this chapter we will explain what a mathematician means by a game and analyze an important class of games. At the end of the chapter we will then be able to discuss the game upon which the experiment was based in order to understand why such a game can provide information on people's attitudes.

6-1   Games

Imagine a contest in which each of the two contestants sits out of sight of the other. There are two buttons, one silver and one gold, in front of each of them. Both contestants can see a display that looks like this:

		Contestant *B*	
		Silver	Gold
Contestant *A*	Silver	−$10	$10
	Gold	0	$5

Each contestant is to push a button. The display shows the players the possible results. If a combination of buttons is pushed that corresponds to a positive number in the display, then contestant B will pay contestant A that amount. For example, if both push their gold buttons, then B must pay $5 to A. A negative entry in the display indicates a payment from contestant A to contestant B. The 0 which results if A pushes gold and B pushes silver means that no payment is made. Thus, the numbers in the display describe the outcomes from contestant A's point of view.

At a signal, the contestants push the button of their choice and when *both* buttons have been pushed, the outcome registers on the display and the payoff is made. In this way, neither player knows what the other's choice is until their own choice has been recorded.

We can analyze this game immediately, because common sense is the only tool we need to decide what will happen. Looking at the display from contestant B's perspective, we realize that there is no reason for this contestant to push the gold button. No matter which button contestant A pushes, a choice of gold by B will cause B to lose money. Thus, we may assume that contestant B will push the silver button. Contestant A can also see that the opposing player has no reason to push the gold button. Therefore, contestant A concludes that the choice of the silver button will cost $10 since the outcome will then be silver–silver. So contestant A pushes the gold button, giving the combination A–gold and B–silver represented by 0 on the display, and consequently no money changes hands.

You will probably agree that, as a spectator sport, watching this game would be somewhat less fun than watching grass grow. However, our motivation in describing this game was not to furnish entertainment but rather to present an example of extreme simplicity from which we can readily isolate those features common to the somewhat more interesting games we will analyze in this chapter. Let us, then, use this example to focus on these features.

The first thing to notice is that there were just two players in the game, contestant A and contestant B. Such a game is called a **two person game.** Game theory encompasses games with more than two contestants, called "n person games," but the subject is considerably more complicated than that of two person games. We will restrict ourselves to the two person case throughout this chapter.

Next, we observe that each contestant has a number of clear-cut choices; in the example they are: Push the silver button, push the gold button. In general, we will put no restriction on the number of such choices, nor will we require that the number or types of choices available to each of the two players be the same. We demand only that all the players' choices, called **strategies** in game theory, be distinct and unambiguously described in the game.

The consequences of every possible combination of strategy choices by the two players were specified in the example by means of the display. The general form of such a display would be

	Contestant B		
	Strategy 1	Strategy 2 . . .	Strategy m
Contestant A Strategy 1			
Strategy 2			
.			
.			
Strategy k			

We recognize this form as a matrix, specifically a $k \times m$ matrix, where k is the number of strategies available to contestant A and m is the number of strategies for contestant B. Since the outcome of play is thought of in terms of gains and losses for the players (though not necessarily as amounts of money), the outcomes are described as **payoffs** and thus the matrix is called the **payoff matrix** of the game. It is an important feature of the sort of game theory we are describing that both players have complete information on the consequences of their own and their opponent's strategy choices as specified by the payoff matrix. In terms of our example, it is crucial that both contestants be able to see the display.

In the general subject of two person games, the payoff matrix is not really a matrix in the same sense as we have been using the word until now—a rectangular display of numbers. The payoff matrix in the game described in the introductory essay looked like this:

		Partner	
		Blue	Red
Subject	Blue	+5¢, +5¢	−4¢, +6¢
	Red	+6¢, −4¢	−3¢, −3¢

If the subject and the parnter both push blue, they both gain 5¢, while a combination of subject–red and partner–blue produces a gain of 6¢ for the subject and a loss of 4¢ for the partner. In the general theory of two person games, the entries in the payoff matrix are ordered pairs of numbers (positive, negative, or zero), where the first number in the pair indicates, for that particular combination of strategies, the gain or loss to the contestant whose strategies correspond to the rows of the matrix and the second number in the pair describes the payoff to the other contestant. Thus, for the little game that began this section, the payoff matrix should look like this:

		Contestant B	
		Silver	Gold
Contestant A	Silver	−10, +10	+10, −10
	Gold	0, 0	+5, −5

However, since the rules of this game require that the losing contestant pay the winner, half the information in this matrix is redundant—contestant A's gain equals contestant B's loss and vice versa. Thus, the payoff matrix we used gave the outcomes just from the point of view of contestant A because we could deduce B's outcomes once we knew the rule "loser pays the winner."

Two person games in which the rules specify that for each combination of strategies the payoff to the winner must equal the loss to the loser, are called **zero—sum** games. Such games occupy a privileged position within the theory, because they can be solved in a sense we will make precise later. The characteristic feature of two person zero—sum games is that once the payoffs to a player are known, the payoffs to the other player are also known; they are just the opposite. Consequently, the payoff matrix need only specify the outcomes for a single player, just like the display in the example. The payoff matrix for a two person zero—sum game, therefore is an ordinary matrix (the entries in it are numbers in each position, not ordered pairs). A two person zero—sum game is also called, more briefly, a **matrix game,** because in such a game the payoff matrix is an ordinary matrix that contains the entire mathematical content of the game.

The two contestants in the game described at the start of this section sat out of sight of each other, and neither knew the other's strategy choice until their own choice had been made. The purpose of these rules was to ensure that there was no communication between the contestants. Game theory permits bargaining, the formation of coalitions in n person games, and other forms of preplay communication, but all this is beyond the scope of our book. There is also a type of communication between players that takes place when a game is repeated many times because each player can observe the other's previous choices of strategy. We will take up this matter later on in the chapter. In the meantime, we will require that there be just a single play of the game with a total absence of communication between the players.

The final feature of games that we wish to isolate by means of our simple example is the motivation and behavior of the contestants. We assume that both contestants are entirely rational and completely selfish. Contestant B rejected the strategy "Push the gold button," because the payoff matrix showed no opportunity for gain in that column. Contestant A gave no hint of being tempted to push the silver button in hopes that the opponent's choice would be gold and so A would gain \$10. Rather, contestant A assumed that contestant B would ignore the gold button as unprofitable so that a silver button strategy by A would lead to a \$10 loss. Thus, not only are both contestants rational and selfish, but they expect their opponent to be the same. Game theorists realize that, in real-life competition, motives and behavior are not that simple. But it seems mathematics is not well-adapted to the analysis of irrational actions, so game theory has a hypothesis that the players are "sensible." Motives other than pure selfishness are not excluded from the general theory, but an elementary treatment had best ignore them.

Let us now describe some competitive situations by means of two person games. The features we have been discussing put severe limitations on what we can deal with in this manner, but considerable variety is possible nevertheless.

6-1   **Example**   A football quarterback must decide whether to call a running play or a pass play. The linebackers on the opposing team must decide as the play begins whether a running play or a pass attempt is taking place. A gain of yardage for one team is a loss for the other, and vice versa. Suppose that if the quarterback calls a running play the average gain is only 2 yards, provided the opposing linebackers expect it. But the average gain is 8 yards if the linebackers think the quarterback is about to attempt a pass. If the quarterback calls a pass play but the linebackers expect a running play, then we suppose that, on the average, the outcome is a 15 yard gain for the quarterback's team. However, if the linebackers expect a pass, the average result will be that the quarterback is thrown for a 6 yard loss. Give the payoff matrix for this game.

Solution   Let the quarterback's choices be represented by the rows of the payoff matrix and the decision of the linebackers on the opposing team by the columns. Then the form of the matrix is

		Linebackers	
		Run	Pass
Quarterback	Run		
	Pass		

Since one team's gain of yardage is the other's loss, this is a zero–sum game, so the entries in the matrix are numbers rather than ordered pairs. The payoffs are always given from the point of view of the contestant whose strategies correspond to rows, so in this example we give the figures in terms of the quarterback's team. From the data in the example, the payoff matrix is

		Linebackers	
		Run	Pass
Quarterback	Run	2	8
	Pass	15	−6

6-2   **Example**   Refrigerator manufacturer A would like to set the price on its best-selling model so as to increase its profits in competition with refrigerator manufacturer B, whose product is more widely advertised. It is known throughout the industry that manufacturer B will set the retail price of its comparable refrigerator at either $350 or $400. Company A's options are to choose one of those prices, $350 or $400, or to select a discount price of $300 which will give company A a rather small profit on each unit, but will produce a large total profit if its sales increase dramatically as a result. Specifically, if company A decides on a $300 price for its product and its competitor chooses to sell its refrigerator for $400, there will be a shift in sales volume between the two companies that results in a $100,000 increase in profits for company A and a $60,000 drop in profits for company B. On the other hand, if company B charges only $350, the price differential will not have much effect on sales, so A's profits will drop $100,000 because of lower per-unit profit, while B's profits remain constant. If both companies charge $350, their profits will be unaffected. If company A charges $350 and company B $400, A's profits will go up by $15,000 due to somewhat increased sales, while B's profit, based on slightly reduced sales, will go up by $50,000, because the high price produces large per-unit profit. If A charges $400 and B $350, there will be a substantial shift of buyers from A to B resulting in a $50,000 loss to A and $75,000 gain to B. Finally, if both companies charge $400, company A will suffer a loss of sales that drops profits $20,000, while company B comes out $100,000 ahead. Find the payoff matrix of this game.

Solution   The form of the payoff matrix will be

		Refrigerator B	
		$350	$400
	$300		
Refrigerator A	$350		
	$400		

The result of a $300 price for A and $400 for B was a $100,000 gain for company A and a $60,000 loss for company B. Taking $1000 as the unit of measurement in the matrix, this outcome is represented by the ordered pair $(+100, -60)$. The outcome when A charges $300 and B $350 was a $100,000 loss for A and no change for B; this is written as the pair $(-100, 0)$. Continuing in this manner and placing the pairs in the appropriate locations, the payoff matrix is

		Refrigerator B	
		$350	$400
	$300	$(-100,0)$	$(+100, -60)$
Refrigerator A	$350	$(0, 0)$	$(+15, +50)$
	$400	$(-50, +75)$	$(-20, +100)$

6-3 **Example** In planning a backpacking trip into the wilderness, a hiker has to decide whether to take along a heavy jacket. If the weather is warm, she estimates the unpleasantness of carrying the extra weight of the jacket will cause her to lose 5 points on her personal scale of happiness resulting from the trip. If the weather is cool, the greater comfort from wearing the jacket part of the time will just balance the bother of carrying it the rest of the time, with no effect on total happiness. However, if the weather is cold, the satisfaction of having made the correct decision together with the greater comfort will add 7 points to her happiness. Alternatively, if it is warm, leaving the jacket behind will please the hiker 5 points worth. If it is cool, there will be no happiness effect if she does not have the jacket. But in cold weather the shivering hiker will give up 10 points of happiness if she has no jacket. Find the payoff matrix of this game.

Solution The hiker is certainly one contestant in this game. Her opponent is the weather. We are only interested in the hiker's gains and losses in happiness, so the convenient approach is to treat this, and all other such games against nature as zero–sum games. The backpacker's gain is the weather's loss, while her loss in happiness is interpreted as a victory for the weather. The hiker's options are to take the jacket along or not, which we denote by Yes and No. The weather has the choice of being warm, cool, or cold. Therefore, the payoff matrix is

		Weather		
		Warm	Cool	Cold
Hiker	Yes	-5	0	7
	No	5	0	-10

Summary

In a **two person game,** each of the two competitors has a number of distinct choices, called **strategies.** The game is played just once. Each player is totally selfish and will choose a strategy or not entirely on the basis of rational criteria. The consequences of each possible combination of strategy choices, one by each player, is specified by a rectangular array of ordered pairs, called the **payoff matrix.** The first number in the pair gives the result of that combination for the competitor whose strategies correspond to the rows of the matrix, while the second number states the result for the opposing player whose strategies are represented by the columns. If, for each combination of strategies (one for each player), the gain to one player is equal to the loss for the other, the game is called a **zero–sum game.** The entries in the payoff matrix for a zero–sum game are single numbers giving the result just for the player whose strategies correspond to the rows of the matrix. A two person zero–sum game is also called a **matrix game.**

Exercises

● *Exercises 6-1 through 6-9 describe competitive situations. Find the payoff matrix for each of these games.*

6-1   A family must decide whether to go to the seashore or the mountains for their vacation. The children discuss the matter among themselves and come to an agreement. The parents also agree with one another. If the parents and the children agree (both groups want the seashore or both choose the mountains), then everyone is happy, so score $+1$ happiness points both for the children and the parents. But if the parents and children disagree, family disharmony will be the result, so both groups will be unhappy to the extent of -1 happiness points for both the children and their parents.

6-2   A gambler is considering which of three bets to make in roulette. These choices are to bet $1 that the wheel selects an even number, $5 that the wheel selects a number in the block of numbers from 1 through 12, and $1 that the wheel selects the number 17. If he bets $1 on the even numbers and if this is a winning bet (an even number is selected), the gambler receives $1 from the casino at which he is gambling. If the even bet is a losing bet, then the gambler must pay $1 to the casino. If the gambler chooses to bet $5 on the block of numbers 1 through 12, then he wins $10 from the casino if one of the numbers from this block is selected. But if he makes this bet and some other number is selected, then the gambler pays $5 to the casino. Finally, if the gambler bets $1 on 17 and if that number is selected, then the casino pays the gambler $36, while if a number other than 17 is selected, the gambler loses and he pays the casino $1.

6-3   An investor has $10,000 to invest in the stock market. If she buys conservative blue chip stocks, then at the end of the year the expected value of her investment will be $8000 if the market prices decline, $10,500 provided the market holds generally steady, and $11,500 if the stock market averages show a substantial increase. However, if she chooses to invest in speculative growth stocks, then the expected value of the investment is $3000 at the end of the year if the stock market registers a decline in prices overall, $9500 in a year of steady market values, and $16,000 if average stock prices increase to a considerable extent.

6-4   In a military exercise, the soldiers are divided into two groups, the green army and the blue army. The green army must send reinforcements to a threatened city. There are three routes available and the green army commander will choose one: over a mountain pass, through a river valley, or over a plain. The opposing blue army has just enough soldiers available to set up an ambush at one location. If the blue army fails to intercept the green army reinforcements, the green army is awarded points and an equal number of points are taken from the blue army. Points are awarded to the green army depending on which route is used (based on the speed with which the reinforcements can reach the city). Since the mountain pass route is fastest, the green army receives 20 points for choosing that route—provided the blue army ambush was set up on another route. Under the same circumstances, the valley is worth 15 points to the green army, and the plain 5 points.

If, however, the blue army ambush is set up on the route the green army commander chooses, the blue army receives points and the same number is taken from their opponents. Points are awarded to the blue army based on the likely effectiveness of the ambush, which depends on the terrain: 30 points for the mountain pass, 20 points for the river valley, and 10 points for the plain.

6-5   A developer buys a large parcel of land in a quiet suburban area. She must decide what to build on it. There is a possibility that a major highway will be routed near this land. If she builds a housing development and the highway is not located nearby, she will make a 40% profit on her investment, but if the highway is nearby, the houses will be less desirable and she will make only 10% on her money. She can build a shopping center, and if there is no highway nearby, more housing will be built for more shoppers, thus giving the developer a 60% profit. But if the highway is close by, the area will become industrial and the shopping center will fail causing a 40% loss to the developer. If she builds an office complex on the land and if there is no highway, then the workers will find it difficult to get to the complex, office space will be hard to rent, and the developer will suffer a 30% loss. But with the highway to speed people to work, the office complex will be a great success and she will make a 75% profit. Her final possibility, an industrial development, will fail if there is no highway nearby for the transportation of materials and the developer will lose 50% of her investment. But if the highway comes close to her land and she has built for industry, she will make a 100% profit.

6-6   A family is planning to have a picnic, but it looks as though it may rain. If they go on the picnic and if it does not rain, the pleasure resulting from the picnic will be worth 5 points, but if they go and if it rains on their picnic, their displeasure will lose them 5 points. If they stay home and if it does rain, the family is warded 2 points for the satisfaction of having made the correct decision. If they stay home when in fact if fails to rain, it costs the family 2 points worth of disappointment.

6-7   Two companies which manufacture similar products are planning major advertising campaigns during the coming year. Each can start their compaign either at the beginning of the year, at midyear, or at the end of the year. The companies will measure the effectiveness of their campaigns in terms of impact points. If both campaigns take place at the same time, the public will be unable to distinguish the products and neither company will gain any impact points. If the campaigns are 6 months apart, the company whose advertising occurred first will gain 3 impact points and the other company 2. If one company runs its campaign at the beginning of the year and the other at the end, both campaigns will have maximum effect, worth 5 points to each company.

6-8   Two airlines share a commuter route between two large cities. Since the number of travelers is essentially constant, a gain in business (measured in average percentage of seats filled) for one is a corresponding loss for the other. Both airlines are considering two innovations: redesigning the interiors of their planes to make them more attractive and serving complimentary snacks. They can adopt both innovations, either, or neither. Each innovation is worth a 5% increase to the airline that introduces it, provided the other airline does not. Thus, for example, if one airline chooses both innovations and the other airline chooses just one, the first gains 5%. However, if one airline redesigns interiors and the other airline serves snacks, the result is no change in passenger business.

6-9   Two television networks must decide on programming for the same three time slots. Each network has a sports show (S), an entertainment show (E), and a cultural show (C) to schedule. The sports and entertainment programs are equally popular and so do not affect the overall ratings of the networks. However, the scheduling of the cultural show at the same time as the entertainment show on the other network will cause the network with the cultural show to lose 3 points in the ratings to the other network, and running the cultural program against the sports show costs the network 5 points to the corresponding gain of its competitor. Suppose, for example, that network A runs the programs in the order SEC, while network B chooses the order ECS. The networks break even on the first time slot (S against E), then network A gains 3 points from network B in the next time slot by running E against C, but network A loses 5 points during the last time slot by running C against S. So the total effect of the schedules SEC for A and ECS for B is that network A loses 2 points to network B in the ratings. [*Hint:* Each network has a choice of six different strategies.]

6-2   Matrix Games

Let us concentrate now on the subject of two person zero–sum games (or matrix games). All the information for such a game is contained in an ordinary $k \times m$ matrix. For example, the rules of the game with the matrix

$$\begin{bmatrix} 1 & 3 & -1 & 0 & -1 \\ 5 & -2 & -1 & 2 & -2 \\ 4 & -1 & -2 & 0 & 0 \\ 3 & 2 & 1 & 2 & 3 \end{bmatrix}$$

are that player A has a choice of four strategies, one for each of the 4 rows, and player B has one strategy corresponding to each of the 5 columns, or five in all. If player A selects strategy (row) 2, while player B chooses strategy (column) 4, then B pays $2 (or, more generally, 2 units, depending on the context of the problem) to player A. If A uses strategy 1 and B strategy 3, then A pays B $1. If A uses strategy 3 and B strategy 5, then there is no payment; and so on as the matrix specifies.

We begin our analysis of this game by studying player A's alternatives. If player A chooses strategy (row) 1, then a number of different payoffs are possible, depending on what player B decides to do. For example, if player B were to choose strategy (column) 2, then player A would receive \$3 from player B; a choice of strategy 5 by player B would result in the loss by player A of \$1. Since, by the conditions of the game described in Section 6-1, player A has had no communication with the opposing player, simple prudence requires that player A assume the worst, that is, if player A chooses strategy 1, then player B will choose a strategy which is as undesirable as possible from player A's point of view. The worst that can happen to player A as a result of choosing row 1 takes place when player B chooses column 3 or column 5. In that event the payoff is -1; player A must pay \$1 to player B. The worst possible payoff in row 1, namely -1, is called player A's **security level** for that strategy, because the player knows that nothing worse can happen. We keep a record of the security level for this strategy by putting a line under the payoffs that equal the security level:

$$[1 \quad 3 \quad \underline{-1} \quad 0 \quad \underline{-1}]$$

Stated more simply, we underline the lowest number in the row where, for negative numbers, lowest is in the sense of most negative; for example, -5 is lower than -2. Thus, by lowest we always mean "furthest to the left on the number line."

Considering each strategy (row) in turn, player A identifies the security level of each, i.e., underlines the lowest numbers in each row, with this result:

$$\begin{bmatrix} 1 & 3 & \underline{-1} & 0 & \underline{-1} \\ 5 & \underline{-2} & -1 & 2 & \underline{-2} \\ 4 & -1 & \underline{-2} & 0 & 0 \\ 3 & 2 & \underline{1} & 2 & 3 \end{bmatrix}$$

If player A ignores all the information in the game matrix except the security levels, it is clear which strategy is best. If player A chooses row 4, that will result in a gain no matter what player B chooses to do, so a prudent contestant should select strategy 4, because every other strategy carries with it the possiblity of loss. More formally, we imagine that player A, in keeping with our requirement of Section 6-1 that the contestants make rational choices motivated only by selfishness, will choose the strategy with the highest security level. Highest for negative numbers means least negative number (-2 is higher than -5) and, in general, "furthest to the right on the number line." The strategy for player A with the highest security level is called the **optimum pure strategy** for player A.

It might be argued that a more adventurous player A might prefer row 2 to row 4, because the most player A can gain from row 4 is \$3, while if player A chooses row 2 and player B selects column 1, then player A will gain \$5. However, two considerations arise from the conditions under which the game is played that eliminate this possibility: Player B has all the informatiin player A has and so, in particular, can see that the selection of column 1 would produce a losing payoff no matter what player A does. Furthermore, we assume that player B is a rational person attempting to do as well as possible in the game. Thus, player A must assume that the opposing player will avoid column 1 and so there is really no reason for player A to be tempted by any strategy in preference to the optimum pure strategy, row 4.

The discovery of an optimum pure strategy for player B involves the same sort of reasoning. Player B examines strategy (column) 1 and, taking the point of view that there is no information available on what player A intends, looks at the matrix to discover what the worst possible outcome would be if column 1 is selected. The worst payoff possible for player B in case of the selection of column 1 is a \$5 loss that results if player A were to choose row 2. Thus, \$5 is the security level for player B associated with strategy 1 and, in general, the **security level** for player B of a column is the highest (in the number line sense) number in the column. Player B records the security level of strategy 1 by putting a line over that payoff:

$$\begin{bmatrix} 1 \\ \overline{5} \\ 4 \\ 3 \end{bmatrix}$$

Player B continues this analysis of the game matrix column-by-column, marking the security level by means of a line over the number, with this result:

$$\begin{bmatrix} 1 & \overline{3} & -1 & 0 & -1 \\ \overline{5} & -2 & -1 & \overline{2} & -2 \\ 4 & -1 & -2 & 0 & 0 \\ 3 & 2 & \overline{1} & \overline{2} & \overline{3} \end{bmatrix}$$

The **optimum pure strategy** for player B is, by analogy with the definition for player A above, that strategy (column) with the lowest security level, because low numbers are to B's advantage. In this example, then, column 3 is the optimum pure strategy for player B, because no matter what strategy player A adopts, player B's loss will be no more than \$1, while the selection of any other strategy by player B might result in a greater loss.

Thus far, we have ignored the fact that each player can, as a consequence of having complete information about the game, identify not only their own optimum pure strategy, but their opponent's optimum pure strategy as well. Will this matter to the players? Player A can, for instance, see that player B's optimum strategy is to select column 3. But certainly that information only reinforces player A's choice of row 4, since any other row choice will lead to a loss if player B does choose column 3. In the same way, player B can observe that row 4 is player A's optimum choice and thus that the selection of any column other than 3 will produce a loss for player B that is two or three times greater. So player B has no temptation to stray from the optimum pure strategy either. Thus, we see that in this example the knowledge of the opponent's optimum pure strategy has the effect of strengthening each player's resolve to employ the optimum strategy.

A game with the feature that the optimum pure strategy for each player remains the most desirable choice assuming that one's opponent will play the opposing optimum pure strategy is called a **strictly determined game.** From a numerical point of view, a matrix game is strictly determined if there is a number in the matrix which is simultaneously the lowest in its row and the highest in its column. The row containing that number is the optimum pure strategy for player A, while the column in which it lies is the optimum pure strategy for player B. The identification of this number and the corresponding optimum pure strategies is a **solution** to a strictly determined game.

It is most efficient to mark the lowest numbers in each row by underlining and the highest numbers in each column with a line above, all on the same matrix. Then a solution to a strictly determined game is identified as a number with lines both above and below it. In the example above, we have

$$
\begin{bmatrix}
1 & \overline{3} & \underline{-1} & 0 & \overline{-1} \\
\overline{5} & \underline{-2} & -1 & \overline{2} & -2 \\
4 & -1 & \underline{-2} & 0 & 0 \\
3 & 2 & \underline{\overline{1}} & \overline{2} & \overline{3}
\end{bmatrix}
$$

It can happen that more than one number in the matrix will have lines both above and below, because in case of ties all highests and lowests are recorded. This means there is more than one solution to the game and, since it can be shown that in such a case the payoff is the same for all, it does not matter which one is selected.

6-4   **Example**   Solve the matrix game:

$$
\begin{bmatrix}
10 & -5 & -5 \\
0 & 10 & -5 \\
0 & 5 & 0 \\
-10 & -10 & 0 \\
5 & 10 & -5 \\
0 & 10 & 0
\end{bmatrix}
$$

Solution   Marking the lowest numbers in each row and the highest in each column give us

$$
\begin{bmatrix}
\overline{10} & \underline{\overline{-5}} & \underline{-5} \\
0 & \overline{10} & -5 \\
\underline{0} & 5 & \overline{0} \\
-10 & -10 & \overline{0} \\
\underline{5} & \overline{10} & \underline{-5} \\
\underline{0} & \overline{10} & \underline{0}
\end{bmatrix}
$$

Since the number 0 in the third and sixth rows of the third column has lines above and below, for an optimum pure strategy, player A may choose either row 3 or row 6, while the optimum pure strategy for player B is column 3.

6-5 **Example** A toy manufacturer must plan early in the calendar year for the dolls it will sell the following Christmas. The manufacturer has three possible marketing plans. The first plan is that the manufacturer will sell the dolls it sold the previous year at relatively low prices. Call this plan I. Another possibility is that the manufacturer will sell the dolls with only small changes in design but spend a lot of money on an aggressive advertising campaign calculated to make the changes seem more important. The advertising campaign will be financed by somewhat higher prices. We call this plan II. A final alternative is that the manufacturer will radically redesign the dolls and, backed by a major advertising program, sell them for considerably more money (plan III). If the national economy is strong next Christmas, the buying public will buy newer appearing products without concern for price, so in that case plan I will cause the price of the manufacturer's stock to drop 10 points ($10 a share), plan II will raise the stock 5 points, and plan III will raise it 20 points. If the economy is weak, price will be a major consideration to the public, so the cheaper dolls of plan I will sell well, bringing the stock price up 15 points, while plan II will produce a 5 point rise in stock value, and plan III will produce a 30 point loss. Finally, if the economy is steady but neither noticeably strong nor weak, plan I will lead to an unchanged stock price, while plans II and III will both bring the stock value up 5 points. Which of the plans should the toy manufacturer adopt?

Solution We will represent the toy manufacturer's plans by the 3 rows of the matrix. The opposing player is the national economy which has strategies strong, weak, and steady, represented by the columns of the matrix. The payoff values are in terms of changes in stock price as described in the example. Thus, the payoff matrix is

		Economy		
		Strong	Weak	Steady
	Plan I	−10	15	0
Manufacturer	Plan II	5	5	5
	Plan III	20	−30	5

If we mark the lowest numbers in the row and the highest in the column, for each row and column, the matrix of the game looks like this

$$\begin{bmatrix} -10 & \overline{15} & 0 \\ \underline{\underline{5}} & 5 & \overline{\underline{5}} \\ \overline{20} & -30 & \overline{\underline{5}} \end{bmatrix}$$

which indicates that the optimum pure strategy for the manufacturer is plan II.

The reader should not be left with the impression that because the payoff matrix for a matrix game is in terms of player A's viewpoint, the solution invariably indicates that an optimum pure strategy by both participants will lead to a gain for player A or, at worst, no payment. The solution to

$$\begin{bmatrix} 5 & 0 & -5 \\ 8 & 20 & -10 \end{bmatrix}$$

is for player A to follow strategy 1, while player B uses strategy 3. Thus, the outcome will be a gain of 5 for player B.

If we analyze the innocuous looking matrix game

$$\begin{bmatrix} \overline{3} & -2 \\ \underline{-1} & \overline{4} \end{bmatrix}$$

we see that row 2 would seem a better choice than row 1 for player A, because its security level, -1, is higher than the security level of row 1. For B, the security level of column 1 (3) is smaller than the security level of column 2, so it would seem that player B should choose column 1. If player A does choose row 2 and player B column 1, the payoff is a 1 unit gain for player B. But now if player A believes that player B will choose the column with the lower security level, player A is strongly tempted to abandon the optimum choice of row 2 and select row 1 instead, because then instead of a 1 unit loss, player A achieves a 3 unit gain. On the other hand, player B will be well aware of player A's temptation and so, in defense, would consider choosing column 2 instead of column 1. If player B chooses column 2 and player A did give in to the temptation to choose row 1, then player B is ahead 2 units. But, then again, perhaps player A will not go with row 1 after all—in which case player B will lose 4 units.

The discussion above indicates that the matrix game

$$\begin{bmatrix} 3 & -2 \\ -1 & 4 \end{bmatrix}$$

represents a highly unstable situation. In contrast to a strictly determined game, neither player has much confidence that the opposing player will choose a strategy just based on a consideration of security levels. Such a game is called a **nonstrictly determined** matrix game and is characterized numerically by the absence of any number in the matrix which is both lowest in its row and highest in its column. In contrast to a strictly determined game, a game of this sort can have no solution in the sense of mutually agreeable optimum pure strategies for both players. Thus, we are confronted with a choice of two approaches: Restrict the theory of matrix games to those that are strictly determined or try to discover a reasonable notion of solution which makes sense for nonstrictly determined games and to seek methods for finding these solutions. Of course, game theorists chose the latter alternative, and our next task will be to discuss what they have determined that a solution should mean.

Summary

In a matrix game, let player A be the competitor whose strategies are represented by the rows of the matrix and let player B be the opponent whose strategies are represented by the columns. The **security level** for player A of each row is the lowest number in the row, where lowest means furthest to the left on the number line. An **optimum pure strategy** for player A is a row with the highest security level, where highest is in the sense of furthest to the right on the number line. The **security level** for player B of each column is the highest number in the column and an **optimum pure strategy** for player B is a column with the lowest security level. A **strictly determined game** is a matrix game in which there appears a number that is both the lowest in its row and the highest in its column. The row containing that number is the optimum pure strategy for player A and the column containing the number is the optimum pure strategy for player B. There is no advantage for either player in a strictly determined game to choose a strategy other than the optimum pure strategy if the opposing player chooses the optimum pure strategy. Thus, the identification of the number which is lowest in its row and highest in its column together with the corresponding optimum pure strategies is a **solution** to a strictly determined game. Many matrix games fail to have solutions in this sense, because they are not strictly determined.

Exercises

- *Exercises 6-10 through 6-17 give the matrices of strictly determined games. Solve the games.*

6-10
$$\begin{bmatrix} -1 & 1 & 0 \\ 0 & 2 & -1 \\ 1 & 1 & 3 \end{bmatrix}$$

6-11 $\begin{bmatrix} 7 & 5 & 4 & 6 \\ 1 & 2 & 0 & 2 \end{bmatrix}$

6-12 $\begin{bmatrix} -2 & 4 & -5 \\ -1 & -3 & -4 \\ -1 & 1 & -1 \\ -2 & -1 & -3 \end{bmatrix}$

6-13 $\begin{bmatrix} 0.12 & 0.76 & 0 & -0.31 & 0 \\ -0.01 & 0 & 0.05 & -0.40 & -0.01 \\ -0.22 & -0.15 & -0.12 & -0.30 & -0.01 \end{bmatrix}$

6-14 $\begin{bmatrix} -1 & 1 \\ 3 & 2 \\ -1 & -1 \\ 0 & 2 \end{bmatrix}$

6-15 $\begin{bmatrix} 0 & \dfrac{1}{2} & 0 & 0 \\ -\dfrac{1}{2} & 0 & -1 & -1 \\ -\dfrac{1}{4} & 1 & 0 & 1 \\ 0 & 1 & 0 & 0 \end{bmatrix}$

6-16 $\begin{bmatrix} -1 & -2 & -2 & -2 \\ -2 & -1 & -3 & -4 \\ -1 & -\dfrac{1}{2} & -2 & -2 \end{bmatrix}$

6-17 $\begin{bmatrix} -\dfrac{1}{4} & \dfrac{2}{3} & -\dfrac{1}{5} \\ -\dfrac{1}{4} & -\dfrac{2}{3} & \dfrac{2}{3} \\ -\dfrac{1}{5} & \dfrac{2}{5} & \dfrac{3}{4} \end{bmatrix}$

6-18 Imagine a game in which player A can choose one of the numbers -1, 0, or 1 and player B can also choose one of the numbers -1, 0, or 1. The payoff to A is calculated by squaring the number chosen by player A and then subtracting the number chosen by player B. What are the optimum strategies for player A and for player B?

6-19   An investor is trying to decide whether to buy a conservative blue chip stock or a speculative growth stock. The investor estimates that if the Dow-Jones industrial average (which measures the stock market as a whole) goes up during the next year, then the value of the blue chip stock will increase by 10%, while the value of the growth stock will increase 30%. If the Dow-Jones average remains constant during the year, then the value of the blue chip stock will remain constant, but the value of the growth stock will decline by 5%. If the Dow-Jones average decreases during the year, then the value of the blue chip stock will go down by 10% and the value of the growth stock will decrease by 20%. Should the investor buy the blue chip stock or the growth stock?

6-20   A movie script shows the bank robbers riding out of town at nightfall. They will go either to their hideout in the hills or across the border into Mexico. The sheriff and his men ride after them but, since it soon becomes dark, they cannot see the robbers' trail. If the sheriff rides to the hideout and the robbers have gone there, he will capture them and thus increase by .10 the probability that he will be reelected as sheriff. However, if he rides to the hideout and the robbers have gone to Mexico, his probability of reelection drops by .10. If the sheriff heads for the border while the robbers go to their hideout, there is a good chance he will realize his mistake in time to come back and capture at least some of the robbers, thus raising his reelection probability .05. Finally, if both the robbers and the sheriff head for the Mexican border, there is a good chance they will miss each other in the dark so, in balance, this combination of outcomes will not influence the sheriff's reelection chances. If the script writer wants the sheriff to make the best possible decision under the circumstances, should the script indicate that the sheriff heads for the hideout or the Mexican border?

6-21   A farmer must decide whether to plant potatoes, sugar beets, or barley. If there is little rain during the growing season, the potato crop will be small, earning the farmer only $25 an acre; the sugar beet crop will fail, causing the farmer to lose $50 an acre; and a somewhat reduced barley crop will bring in $25 an acre. In a year of moderate rain, the potatoes will do well, earning $200 an acre; the sugar beets fairly well, bringing in $100 an acre; and the barley will be worth $50 an acre. In a year of heavy rain, most of the potato crop will rot in the ground, so the little that is left will earn only $20 an acre; the sugar beets will flourish and earn $150 an acre; and the barley crop, somewhat damaged by rain, will earn $25 an acre. If the farmer has decided to plant only one crop, which of the three should it be?

6-22   A family is trying to decide whether to take its vacation at the seashore, in the mountains, or at home. The family estimates the pleasure it will receive from the vacation on a scale from -3 (displeasure) to a $+3$ (pleasure), depending on whether the weather is warm, cool, cold, or rainy during the vacation. If the family goes to the seashore, their estimate of the pleasure they will receive from the vacation is $+2$ if the weather is warm, 0 if it is cool, -2 if it is cold, and -3 if it is rainy. If they go to the mountains, warm weather will produce pleasure worth $+2$, cool weather $+3$, cold weather $+1$, and rainy weather -1.

If the family stays home, the family estimates $+2$ points of pleasure if the weather is warm and $+1$ pleasure points for each of the other three types of weather (cool, cold, or rainy) because they can adjust their activities to the weather. Which vacation should this family take?

6-23 An automobile company plans to open a new manufacturing plant and must decide whether to build its standard-sized car, its compact car, or its subcompact at this plant. The United States government is considering an annual tax on automobiles by size. The tax will be very high in the case of standard-sized cars and quite substantial even for compact cars, but does not apply to subcompacts. If the government does not institute the tax, the plant would normally run at 70% capacity if it manufactures standard-sized cars and 80% capacity if either compacts or subcompacts are built there. However, if the tax law passes, it will have a severe effect on the public's automobile buying habits so that the plant will run at only 30% of capacity if it manufactures the standard model, 50% of capacity with compacts, but 90% of capacity if it makes subcompacts. Which model should the manufacturer build at this plant?

6-24 A defense attorney must advise a client who was caught in a store after hours under suspicious circumstances and has therefore been accused of attempted burglary. However, because of careless handling of the evidence by the arresting officers, there is a chance that the judge will dismiss the case on technical grounds. For that reason, the district attorney has agreed to accept a guilty plea to the lesser charge of breaking and entering. If the client pleads innocent and if the judge is harsh, the judge will let the trial for attempted burglary take place and, if that happens, the defense attorney believes her client will be found guilty and sentenced to 5 years in prison. However, if the judge is lenient, the case will be dismissed and her client will go free. On the other hand, if her client pleads guilty to the breaking and entering charge, a harsh judge will give a 1 year prison sentence and a lenient judge a 3 month sentence. How should the lawyer advise her client to plead?

6-3 Mixed Strategies

We have seen that the matrix game

$$\begin{bmatrix} 3 & -2 \\ -1 & 4 \end{bmatrix}$$

is not strictly determined, so we cannot solve the game by the method of Section 6-2. The trick to analyzing a game of this sort consists of temporarily changing our rules of operation. We are really interested in discussing strategies for the two players in a single play of the game, but for the moment, let us suppose the game is to be played ten times. If each player chooses a strategy once and for all and repeats it each of the ten times the game is played, then the payoff will certainly be the same each time and the total payoff from the ten plays will be ten times that number. For instance, if player *A* decides always to use strategy (row) 2 and player *B* always chooses strategy (column) 1, then player *A* must pay player *B* $1 in each of the ten plays for a total payoff of $10.

A consistent strategy for several plays of the game would probably not be wise. Player *A* would notice that player *B* keeps choosing column 1 and would, after a few plays, be tempted to change from row 2, which costs $1 when player *B* chooses column 1, to row 1, which will produce a gain of $3 if player *B* persists with column 1. However, player *B* might suspect that player *A* is tempted to switch strategies from row 2 to row 1 and thus, on the same play, also switch strategies from column 1 to column 2. If that happens,, player *A* will find that instead of going from a $1 loss to a $3 gain, the outcome is the more substantial loss of $2. It is complications of this sort which encourage us to adopt the one play only rule we have been using for the games we wish to analyze.

Our consideration of multiple plays of a game and the problems associated with a consistent strategy leads us to the question: Should the players vary their choice of strategy from play to play in order to keep their opponent uncertain of their intentions? The point of the rules for a single-play game that we have been using is that neither player has any information on what the opponent will do. Therefore, each player must choose a strategy on the assumption that the opponent will make the best possible choice, from the opponent's point of view. If the players in repeated plays of the game change their choice of strategy from play to play in a random manner, then it will still be true that neither knows what the other intends. Thus, each will choose their overall strategy for the ten plays without taking into account the opponent's prior choices during the course of the game.

Using the game above as an example, suppose player *A* decides to select the strategy (row) for each play on the basis of the result of a coin toss (which, of course, we assume player *B* cannot observe). Player *A* will choose row 1 if heads (H) comes up on the coin and row 2 if tails (T) comes up. Player *A* tosses the coin with the following results: T, T, T, H, H, T, T, T, T, H. Unknown to player *A*, player *B* adopts the overall strategy of rolling a die at each play, out of sight of player *A*, and selecting column 1 if a one or two come up and column 2 otherwise. Ten rolls of the die produce the following results: two, four, two, one, five, six, two, one, five, four. Turning to the payoff matrix of the game, here is what will happen:

Play	1	2	3	4	5	6	7	8	9	10
Player *A*	2	2	2	1	1	2	2	2	2	1
Player *B*	1	2	1	1	2	2	1	1	2	2
Payoff	−1	4	−1	3	−2	4	−1	−1	4	−2

In this case, the total payoff will be a $7 gain to player *A*. Of course, if both players continued their coin toss and die roll procedures for another ten plays of the game, the results of the coin tossing and die rolling activities might be quite different, so the total payoff for the next ten plays could be different as well. In contrast, the constant strategy "player A always chooses row 2 and player B always chooses column 1" results in a payoff of $10 to player B for each ten plays

Our digression into the subject of games with repeated plays was for the purpose of introducing an idea which is central to the subject of nonstrictly determined matrix games: For games with repeated plays, an overall strategy might involve the use of more than just one particular row choice or column choice. In the example above, player A's overall strategy was to give each row a probability 1/2 of being selected at each play, with the choice made at random by means of the coin toss. Thus, the strategies "row 1" and "row 2" both have a role in the overall strategy. Similarly, player B gave column 1 a probability of 1/3 and column 2 a probability of 2/3 of being selected by means of the die rolling rule.

But notice that these overall strategies could be used as well for a game with a single play That is, confronted with the matrix game

$$\begin{bmatrix} 3 & -2 \\ -1 & 4 \end{bmatrix}$$

player A could still adopt the strategy "flip a coin and choose row 1 if heads comes up and row 2 for tails" for the single play of the game, while player B could use "roll a die and choose column 1 if one or two appear and column 2 otherwise." Equivalently, in probability language, player A's strategy is "probability 1/2 for row 1 and probability 1/2 for row 2" (with actual choice at random), while player B's is "probability 1/3 for column 1 and probability 2/3 for column 2."

A strategy like "probability 1/2 for row 1 and probability 1/2 for row 2" or, more generally, "probability p_1 for row 1 and probability p_2 for row 2" is called a **mixed strategy.** The sort of strategy we considered earlier such as "choose row 1" is called a **pure strategy,** although that strategy could as well be viewed as the mixed strategy "probability 1 for row 1 and probability 0 for row 2." The mixed strategy concept is not restricted to 2×2 matrix games. A player with a choice of k different pure strategies can choose probabilities p_1 for selecting row 1, p_2 for row 2, and so on to p_k for the last row. The only restriction is that, since the player must choose some row, the numbers p_1, p_2, and so on to p_k must add up to one.

You might wonder how, in practice, a player could actually implement a strategy like "probability .17 of choosing row 1 and probability .83 of choosing row 2." One way to do it would be to use a spinner on a card (see Figure 6-1). A circle is drawn with center where the spinner rotates and the circumference is divided into two regions, one occupying 17% of the circumference and the other the remaining 83%. The player, out of sight of the opposing player, spins the spinner and chooses row 1 if the spinner stops in the smaller region of the circumference and row 2 otherwise. A player with more than two pure strategies to choose from would simply divide the circumference into more regions.

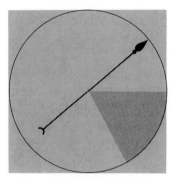

Figure 6-1

Now that we have the concept of a mixed strategy, we are in a position to return to the problem we introduced at the end of Section 6-2: What do we mean by a solution to a nonstrictly determined game? The answer is that we should seek the best possible *mixed* strategies for each player with regard to security levels. A pair of optimum strategies, one for each player, from which there is no temptation for either player to deviate is still considered a solution to the game, just as in the strictly determined case. However, if we are to discuss security levels, the meaning of payoff needs clarification.

In the game we have been using,

$$\begin{bmatrix} 3 & -2 \\ -1 & 4 \end{bmatrix}$$

if player A adopts the pure strategy "choose row 2" and player B chooses the pure strategy "choose column 1," we know that the payoff will be a \$1 gain for player B and that -1 is the security level for the "choose row 2" strategy when we consider only pure strategies. But what if player A's strategy reads "probabilty 1/2 of row 1 and probability 1/2 of row 2" and player B's is "probability 1/3 of column 1 and probability 2/3 of column 2"? What do we mean by "payoff," and thus how do we calculate the security level? If these strategies were followed for several plays of the game, the amount of money changing hands after each play would vary, as we saw above.

Since the mixed strategy concept depends on probability, you may suspect that the correct approach to a definition of payoff is by means of the theory of probability. You would be right, as we illustrate by means of the example discussed above. There are four possible outcomes for a play of that game:

row 1   and   column 1

row 1   and   column 2

row 2   and   column 1

row 2   and   column 2

We can calculate the probability that each of the four possibilities will actually occur as follows: Player A flips a coin (or spins a spinner) to select a row and player B rolls a die in order to choose a column. Neither outcome depends on the other; in fact, neither player knows what the other has done until the play is recorded. In the language of Chapter 2, the selection of the row and the selection of the column are independent events. In that case, as we recall, the probability is the product of the two probabilities. Thus, for example, since the probability is 1/2 that player A will choose row 1 and the probability is 1/3 that player B will choose column 1, then

$$p(\text{row 1 and column 1}) = p(\text{row 1})p(\text{column 1}) = \left(\frac{1}{2}\right)\left(\frac{1}{3}\right) = \frac{1}{6}$$

Similarly,

$$p(\text{row 2 and column 1}) = \left(\frac{1}{2}\right)\left(\frac{2}{3}\right) = \frac{1}{3}$$

and so on, to produce

$$p(\text{row 1 and column 1}) = \frac{1}{6}$$

$$p(\text{row 1 and column 2}) = \frac{1}{3}$$

$$p(\text{row 2 and column 1}) = \frac{1}{6}$$

$$p(\text{row 2 and column 2}) = \frac{1}{3}$$

Once we know the probabilities of the four possible outcomes, we can employ a concept from Chapter 2 which is a suitable notion of payoff in a probability sense. If we have a list of types of outcomes, if we have a value (monetary or otherwise) corresponding to each, and if we know the probability that the outcome will be of each possible type, then we calculate the expected value by multiplying the value of each type of outcome times the probability that the type of outcome will take place and adding up the results. The values of a matrix game are the numbers in the payoff matrix, so the expected value, E, of the mixed strategies we have been studying for the matrix game

$$\begin{bmatrix} 3 & -2 \\ -1 & 4 \end{bmatrix}$$

is

$$E = (3)\left(\frac{1}{6}\right) + (-2)\left(\frac{1}{3}\right) + (-1)\left(\frac{1}{6}\right) + (4)\left(\frac{1}{3}\right) = 1$$

For the pure strategies "player A chooses row 2 and player B chooses column 1" viewed as mixed strategies, the probabilities are $p(\text{row } 1) = 0$, $p(\text{row } 2) = 1$, $p(\text{column } 1) = 1$, and $p(\text{column } 2) = 0$. The payoff for these strategies in the expected value sense is

$$E = (3)(0) + (-2)(0) + (-1)(1) + (4)(0) = -1$$

which is just the number in the second row and first column of the payoff matrix. As this example indicates, when we consider pure strategies, the probability theory definition of payoff gives the corresponding entry in the payoff matrix. In the pure strategy case, then, the two concepts of payoff are the same.

We may interpret the expected value notion of payoff in a manner which appeals to our intuition. Suppose each player in a matrix game adopts a certain mixed strategy and retains it through many repetitions of the game. We imagine, then, that each player constructs a spinner in the way we described above and uses the same spinner over and over in order to decide at each play which pure strategy to employ. Now suppose that a record is kept of the total payoff for all plays of the game and, after each play, the total payoff thus far is divided by the number of times the game has been played. The quotient will be the average payoff per play for all plays up to that point. It can be shown that after many plays of the game, the average payoff per game will be very close to the expected value payoff we computed directly from the probabilities without playing the game at all. We emphasize that this statement is correct only if both players are consistent about using the same spinner (mixed strategy) for every play. In practice, even a mixed strategy can be discovered if it is used consistently in enough repetitions of the game, so it is unlikely that such a sequence of plays would really be carried out.

Now that we have introduced the concept of payoff for mixed strategies through the expected value definition in a particular game, we must next give a general definition. A modest amount of abstract notation will make this task easier. A matrix game is specified by a $k \times m$ matrix G which gives, for each row i and column j, the payoff g_{ij} that occurs if player A chooses row i and player B chooses column j. Thus, g_{ij} is the value of the outcome "row i and column j." Player A has a choice of k different pure strategies (rows of the matrix). A **mixed strategy** for player A is a list of numbers $p_1, p_2, \ldots, p_k$, where p_i is the probability that player A will select the ith row when the game (which, we emphasize, is played just once) takes place. Thus, a mixed strategy is just an ordered k-tuple of numbers, all of them greater than or equal to zero. Since one pure strategy must be selected, the probabilities must add up to one, that is,

$$p_1 + p_2 + \cdots + p_k = 1$$

We represent a particular k-tuple for player A by the symbol P. Similarly, player B has a choice of m distinct pure strategies (columns). An ordered m-tuple $q_1, q_2, \ldots, q_m$ of numbers greater than or equal to zero and adding up to one is a **mixed strategy** for player B. We will use the symbol Q for B's mixed strategy. Once player A has selected a mixed strategy P and player B a mixed strategy Q, the **payoff** for these strategies is calculated by means of the formula for expected value. We will use the symbol $E(P, Q)$ for this quantity. The probability that player A will choose row i using the mixed strategy P is p_i, while the probability that the mixed strategy Q will direct player B to choose column j is q_j. Since we are assuming that the events "row choice" and "column choice" are independent by the rules of the game, the probability of the outcome "row i and column j" is $p_i q_j$ according to the principles of probability. The value of the outcome "row i and column j" is the number g_{ij} from the game matrix, as we observed above. Expected value is defined by multiplying the value of each outcome by its probability and then adding up all the products. Consequently,

$$E(P, Q) = g_{11}(p_1 q_1) + g_{12}(p_1 q_2) + \cdots + g_{1m}(p_1 q_m) + g_{21}(p_2 q_1) + \cdots + g_{km}(p_k q_m)$$

It is possible, and neater, to state the formula for $E(P, Q)$ in matrix terms. Think of the strategy P, an ordered k-tuple, as a $1 \times k$ matrix

$$P = [p_1 \quad p_2 \cdots p_k]$$

Thus, the number in the ith *column* of the matrix P is the probability that player A will choose the ith *row* of the game matrix. Write the m-tuple Q as the $m \times 1$ matrix

$$Q = \begin{bmatrix} q_1 \\ q_2 \\ \cdot \\ \cdot \\ \cdot \\ q_m \end{bmatrix}$$

so that the number in the jth *row* of Q is the probability that player B will choose the jth *column* of the game matrix. The game matrix G is a $k \times m$ matrix. The product PG is thus a $1 \times m$ matrix, and performing the matrix multiplication $(PG)Q$ produces a 1×1 matrix, a number, which is in fact $E(P, Q)$. To summarize, the payoff $E(P, Q)$ can be calculated by means of the matrix multiplication formula

$$E(P, Q) = PGQ$$

To illustrate this formula, we return again to the game

$$\begin{bmatrix} 3 & -2 \\ -1 & 4 \end{bmatrix}$$

player A's mixed strategy "probability 1/2 of row 1 and probability 1/2 of row 2," and player B's strategy "probability 1/3 of column 1 and probability 2/3 of column 2." In this case, we write

$$P = \begin{bmatrix} \frac{1}{2} & \frac{1}{2} \end{bmatrix} \qquad Q = \begin{bmatrix} \frac{1}{3} \\ \frac{2}{3} \end{bmatrix}$$

so

$$E(P, Q) = PGQ = \begin{bmatrix} \frac{1}{2} & \frac{1}{2} \end{bmatrix} \begin{bmatrix} 3 & -2 \\ -1 & 4 \end{bmatrix} \begin{bmatrix} \frac{1}{3} \\ \frac{2}{3} \end{bmatrix}$$

$$= \begin{bmatrix} 1 & 1 \end{bmatrix} \begin{bmatrix} \frac{1}{3} \\ \frac{2}{3} \end{bmatrix} = 1$$

as before.

6-6 **Example**   In the matrix game

$$\begin{bmatrix} .15 & -.31 & 0 & -.55 \\ .76 & .02 & -.62 & -.10 \end{bmatrix}$$

player A's strategy consists of drawing a card from a standard deck and choosing row 1 if a spade appears and row 2 otherwise. Player B flips a nickel and a dime, chooses column 1 if both are heads, column 2 if both are tails, column 3 if the nickel is heads and the dime tails, and column 4 in the remaining case. What is the payoff for these strategies?

Solution   Since there are 13 spades in the 52 card standard deck, $p(\text{row } 1) = 13/52 = .25$ and therefore $p(\text{row } 2) = .75$. Each of the four possible outcomes from flipping the nickel and the dime is equally likely, so $p(\text{column } j) = .25$ for each j. In matrix form,

$$P = [.25 \quad .75] \qquad Q = \begin{bmatrix} .25 \\ .25 \\ .25 \\ .25 \end{bmatrix}$$

and

$$E(P, Q) = PGQ = [.25 \quad .75] \begin{bmatrix} .15 & -.31 & 0 & -.55 \\ .76 & .02 & -.62 & -.10 \end{bmatrix} \begin{bmatrix} .25 \\ .25 \\ .25 \\ .25 \end{bmatrix}$$

$$= -.033125$$

Suppose that player A has chosen a particular mixed strategy $P^{(0)}$. There are many mixed strategies available to player B and the payoff $E(P^{(0)}, Q)$ will depend on which strategy Q player B selects. Since the game matrix G is written from player A's point of view, with positive numbers indicating a profit for player A and negative numbers a loss, player A would prefer player B to choose a strategy Q that makes $E(P^{(0)}, Q)$ a large positive number. In other words, if player B were trying to decide between strategies $Q^{(1)}$ and $Q^{(2)}$ and it were true that

$$E(P^{(0)}, Q^{(1)}) \le E(P^{(0)}, Q^{(2)})$$

then player A would certainly prefer player B to select strategy $Q^{(2)}$. But player A has no information on what player B intends to do, so, as in Section 6-2, we conclude that player A had better assume the worst. Among all the mixed strategies Q available to player B, there is a strategy, call it $Q^{(0)}$, which from player A's viewpoint is the worst response player B could make to the choice of mixed strategy $P^{(0)}$. That is,

$$E(P^{(0)}, Q) \ge E(P^{(0)}, Q^{(0)})$$

for all mixed strategies Q available to player B. In the absence of any information, player A must assume that a choice of mixed strategy $P^{(0)}$ will be met by the response $Q^{(0)}$ and therefore the payoff will be $E(P^{(0)}, Q^{(0)})$, which we abbreviate as $v^{(0)}$. We call $v^{(0)}$ the **security level** for player A's mixed strategy $P^{(0)}$.

6-7 **Example** Suppose that in the game

$$\begin{bmatrix} -1 & 0 & 1 \\ 0 & -1 & 0 \\ -1 & 0 & 1 \end{bmatrix}$$

player A chooses the mixed strategy

$$p(\text{row } 1) = p(\text{row } 2) = p(\text{row } 3) = \frac{1}{3}$$

Which of the following strategies for player B is least desirable from player A's point of view? (a) $p(\text{column } 1) = p(\text{column } 2) = 1/2$, $p(\text{column } 3) = 0$; (b) $p(\text{column } 1) = p(\text{column } 2) = p(\text{column } 3) = 1/3$; (c) $p(\text{column } 1) = 1/2$, $p(\text{column } 2) = 2/5$, $p(\text{column } 3) = 1/10$; (d) a pure strategy.

Solution Since $E(P, Q) = PGQ$, it is convenient to calculate

$$PG = \begin{bmatrix} \frac{1}{3} & \frac{1}{3} & \frac{1}{3} \end{bmatrix} \begin{bmatrix} -1 & 0 & 1 \\ 0 & -1 & 0 \\ -1 & 0 & 1 \end{bmatrix} = \begin{bmatrix} -\frac{2}{3} & -\frac{1}{3} & \frac{2}{3} \end{bmatrix}$$

Then, for part (a),

$$E(P, Q) = PGQ = \begin{bmatrix} -\frac{2}{3} & -\frac{1}{3} & \frac{2}{3} \end{bmatrix} \begin{bmatrix} \frac{1}{2} \\ \frac{1}{2} \\ 0 \end{bmatrix} = -\frac{1}{2}$$

For part (b),

$$E(P, Q) = \begin{bmatrix} -\frac{2}{3} & -\frac{1}{3} & \frac{2}{3} \end{bmatrix} \begin{bmatrix} \frac{1}{3} \\ \frac{1}{3} \\ \frac{1}{3} \end{bmatrix} = -\frac{1}{9}$$

In case (c),

$$E(P, Q) = \begin{bmatrix} -\dfrac{2}{3} & -\dfrac{1}{3} & \dfrac{2}{3} \end{bmatrix} \begin{bmatrix} \dfrac{1}{2} \\ \dfrac{2}{5} \\ \dfrac{1}{10} \end{bmatrix} = -\dfrac{6}{15}$$

The pure strategy for player B, "choose column 1," is written in matrix form as

$$Q = \begin{bmatrix} 1 \\ 0 \\ 0 \end{bmatrix}$$

and similarly for the other pure strategies, so we calculate for part (d),

$$E(P, Q) = \begin{bmatrix} -\dfrac{2}{3} & -\dfrac{1}{3} & \dfrac{2}{3} \end{bmatrix} \begin{bmatrix} 1 \\ 0 \\ 0 \end{bmatrix} = -\dfrac{2}{3}$$

$$E(P, Q) = \begin{bmatrix} -\dfrac{2}{3} & -\dfrac{1}{3} & \dfrac{2}{3} \end{bmatrix} \begin{bmatrix} 0 \\ 1 \\ 0 \end{bmatrix} = -\dfrac{1}{3}$$

$$E(P, Q) = \begin{bmatrix} -\dfrac{2}{3} & -\dfrac{1}{3} & \dfrac{2}{3} \end{bmatrix} \begin{bmatrix} 0 \\ 0 \\ 1 \end{bmatrix} = \dfrac{2}{3}$$

Thus, the pure strategy for player B, "choose column 1," is the worst choice among those given from player A's point of view, because in that case $E(P, Q) = -2/3$—the lowest payoff among the Qs considered in the problem.

If player A chooses strategy $P^{(0)}$, then the worst that can happen is that player B will choose strategy $Q^{(0)}$ and the payoff will be the security level $v^{(0)}$. Player A might, however, choose a different mixed strategy $P^{(1)}$. In that case, there is a different worst possible outcome, because there is some strategy $Q^{(1)}$ for which

$$E(P^{(1)}, Q) \geq E(P^{(1)}, Q^{(1)})$$

no matter what mixed strategy Q player B chooses. Now, player A can compare $E(P^{(1)}, Q^{(1)}) = v^{(1)}$, the security level of mixed strategy $P^{(1)}$, with the security level $v^{(0)}$ of $P^{(0)}$. Suppose that

$v^{(1)}$ is larger than $v^{(0)}$. This means the worst that can happen to player A if strategy $P^{(1)}$ is chosen is not as bad as the worst that can occur if $P^{(0)}$ is chosen. Since player A has no information on what player B will do, and therefore must suppose that the worst outcome will take place, we see that player A will choose strategy $P^{(1)}$ in preference to $P^{(0)}$.

6-8   **Example**   In the game

$$\begin{bmatrix} 3 & -2 \\ -1 & 4 \end{bmatrix}$$

suppose that each player were to consider only the pure strategies and the mixed strategy obtained by flipping a coin to determine row (or column) choice. What are the possible payoffs?

Solution   If each player selects a pure strategy, the payoffs are given by the numbers in the game matrix. If player A uses the strategy

$$P = \begin{bmatrix} \dfrac{1}{2} & \dfrac{1}{2} \end{bmatrix}$$

that arises from flipping the coin, then

$$PG = \begin{bmatrix} \dfrac{1}{2} & \dfrac{1}{2} \end{bmatrix} \begin{bmatrix} 3 & -2 \\ -1 & 4 \end{bmatrix} = \begin{bmatrix} 1 & 1 \end{bmatrix}$$

The strategies player B considers are

$$\begin{bmatrix} 1 \\ 0 \end{bmatrix} \qquad \begin{bmatrix} 0 \\ 1 \end{bmatrix} \qquad \begin{bmatrix} \dfrac{1}{2} \\ \dfrac{1}{2} \end{bmatrix}$$

and the payoff is $E(P, Q) = 1$ in each case. If player B uses the coin flipping strategy to determine which column to select, then for each of the pure strategies available to player A, we calculate

$$E(P, Q) = \begin{bmatrix} 1 & 0 \end{bmatrix} \begin{bmatrix} 3 & -2 \\ -1 & 4 \end{bmatrix} \begin{bmatrix} \dfrac{1}{2} \\ \dfrac{1}{2} \end{bmatrix} = \dfrac{1}{2}$$

$$E(P, Q) = \begin{bmatrix} 0 & 1 \end{bmatrix} \begin{bmatrix} 3 & -2 \\ -1 & 4 \end{bmatrix} \begin{bmatrix} \frac{1}{2} \\ \frac{1}{2} \end{bmatrix} = \frac{3}{2}$$

To summarize, if

$$P = \begin{bmatrix} 1 & 0 \end{bmatrix}$$

then the payoffs are

$$3 \qquad -2 \qquad \frac{1}{2}$$

depending on whether B chose one of the pure strategies or the mixed coin flipping strategy. If

$$P = \begin{bmatrix} 0 & 1 \end{bmatrix}$$

the corresponding payoffs are

$$-1 \qquad 4 \qquad \frac{3}{2}$$

while if

$$P = \begin{bmatrix} \frac{1}{2} & \frac{1}{2} \end{bmatrix}$$

the payoffs are

$$1 \qquad 1 \qquad 1$$

You might suspect that in Example 6-8 the mixed strategy $P = \begin{bmatrix} 1/2 & 1/2 \end{bmatrix}$ has a higher security level, and is thus more desirable for player A than is either pure strategy. In order to compute security levels, of course we must consider *all* possible mixed strategies for player B, not just these three. If this is done, it turns out that these special cases tell the whole story, namely, the security level for $P = \begin{bmatrix} 1 & 0 \end{bmatrix}$ is -2, for $P = \begin{bmatrix} 0 & 1 \end{bmatrix}$ it is -1, but if $P = \begin{bmatrix} 1/2 & 1/2 \end{bmatrix}$, then $E(P, Q) = 1$ no matter what mixed strategy player B chooses. Thus, the mixed strategy $\begin{bmatrix} 1/2 & 1/2 \end{bmatrix}$ is indeed more desirable for player A than is either pure strategy.

As player A considers each possible mixed strategy P, there is a security level v associated with it. The value of v changes from strategy to strategy and there is a highest such v, which we write as v^*. That is, there is a strategy P^* for player A with the property that the worst that can happen to player A using strategy P^* is, considering all choices of mixed strategy available to player B, better than the worst that can happen to player A if player A uses any other strategy. The mixed strategy P^* is called an **optimum strategy** for player A.

In the same way, if player B chooses a particular strategy $Q_{(0)}$, then there is a worst possible response from player A; a strategy $P_{(0)}$ for which

$$E(P, Q_{(0)}) \leq E(P_{(0)}, Q_{(0)})$$

no matter what mixed strategy P player A might use. Let

$$E(P_{(0)}, Q_{(0)}) = v_{(0)}$$

the **security level** for player B associated with the mixed strategy $Q_{(0)}$, and think of all the values that v can take as player B evaluates different mixed strategies. Among all those security levels there is one, v_*, which is least bad from player B's point of view, that is, lowest. The corresponding mixed strategy, written Q_*, the one for which

$$E(P, Q_*) \leq v_*$$

no matter what strategy P player A uses, is called the **optimum strategy** for player B.

What happens if both players use their optimum strategy? It can be shown that

$$E(P^*, Q_*) = v^*$$

and also

$$E(P^*, Q_*) = v_*$$

so v^* and v_* are in fact the same number. That payoff, called the **value of the game,** occurs when both players use their optimum strategies. Thus, neither player is tempted to depart from their optimum strategy, because, if the opposing player employs the op- timum strategy, then a less desirable payoff than $v^* = v_*$, from the point of view of the player who strays from the optimum, will take place. The fact that

$$E(P^*, Q_*) = v^* = v_*$$

therefore implies that every matrix game has a **solution** in the sense of stable optimum mixed strategies for both players. Although this fact is fundamental to the subject of matrix games, its demonstration is beyond the scope of this book. However, we will be able to describe how to find the solution to a matrix game in Section 6-4, now that we know that such a solution exists.

Before turning to the matter of calculating solutions to nonstrictly determined games, let us examine how the reasoning above with respect to optimum mixed strategies relates to our earlier discussion of the solution to a strictly determined game. We consider only pure strategies for the moment. As we observed earlier, a pure strategy can also be thought of as a mixed strategy P with one p_i equal to one and the rest equal to zero (thus, the pure strategy would be "choose row i"). Similarly, for the pure strategy Q "choose column j," we would have $q_j = 1$ and all other qs are zero. All the numbers in the sum that defines $E(P, Q)$ in this case will be zero except for the one that involves p_i and q_j, which are both one, so $E(P, Q) = g_{ij}$, the number in the ith row and jth column of the game matrix G.

If player A chooses a pure strategy (row) $P^{(0)}$ and compares $E(P^{(0)}, Q)$ for all responses (columns) available to player B (recall that we have restricted ourselves to pure strategies in this discussion), then player A is in fact comparing all the numbers in a particular row of the matrix G and so $v^{(0)}$ is the lowest number in the row. Then v^* is found by comparing the lowest numbers from each row and choosing the highest of these. The optimum pure strategy P^* for player A is the selection of the row in which v^* lies. Similarly, player B finds the highest number in each column and for v_* uses the lowest of these. The choice of the column containing v_* is the optimum pure strategy Q_* for player B.

Now a strictly determined game is, by definition, one in which there is a number that is both lowest in its row and highest in its column. That number will satisfy the conditions both for v^* and for v_*. Thus, the solution to a strictly determined game is the pair of optimum pure strategies (P^*, Q_*) given by the row and column of the number which is simultaneously lowest in its row and highest in its column. This is the same solution as the one given in Section 6-2.

You might argue that, in finding optimum strategies for a strictly determined game, we considered only pure strategies even though, in general, optimum strategies are chosen from among all possible mixed strategies. However, it is known that in a strictly determined game, there is a pure strategy for each player which is preferable to any mixed strategy, so in seeking the optimum in that case, it is sufficient to examine only pure strategies.

Summary

In a matrix game in which player A has a choice of k rows and player B has a choice of m columns, let G be the $k \times m$ matrix of the game. A **mixed strategy** for player A is a $1 \times k$ matrix P with entries that are greater than or equal to zero and that add up to one. The number in the ith column of the matrix represents the probability that player A will choose row i. A **mixed strategy** for player B is an $m \times 1$ matrix Q with numbers that are all greater than or equal to zero and that add up to one. The number in the jth row represents the probability that player B will choose column j. The **payoff** if player A uses mixed strategy P and player B uses mixed strategy Q, written $E(P, Q)$, is defined by

$$E(P, Q) = PGQ$$

A strategy in the sense of Section 6-2 is a pure strategy and is represented by a matrix in which there is a single 1 and the rest of the numbers are 0s.

Let $P^{(0)}$ be a mixed strategy for player A and suppose $Q^{(0)}$ is a mixed strategy for player B with the property that, for all Q,

$$E(P^{(0)}, Q) \geq E(P^{(0)}, Q^{(0)})$$

Then $E(P^{(0)}, Q^{(0)}) = v^{(0)}$ is the **security level** for player A's strategy $P^{(0)}$. A mixed strategy P^* for player A with security level v^* as high as possible is called an **optimum strategy** for player A. Similarly, if $Q_{(0)}$ is a mixed strategy for player B and there is a mixed strategy $P_{(0)}$ for player A so that

$$E(P, Q_{(0)}) \leq E(P_{(0)}, Q_{(0)}) = v_{(0)}$$

no matter which mixed strategy P player A adopts, then $v_{(0)}$ is the **security level** of mixed strategy $Q_{(0)}$ for player B. A mixed strategy Q_* for player B is an **optimum strategy** if its security level v_* is the lowest possible among all mixed strategies available to player B. It can be shown that

$$E(P^*, Q_*) = v^* = v_*$$

and this number is called the **value of the game**. A **solution** to a nonstrictly determined matrix game is a pair (P^*, Q_*) of optimum (mixed) strategies for player A and player B, respectively. If the game is strictly determined, this solution is still the one found by locating the number in the matrix which is both lowest in its row and highest in its column.

Exercises

● *In Exercises 6-25 through 6-32, calculate the payoff for the given mixed strategies.*

6-25 $G = \begin{bmatrix} 1 & -2 \\ -1 & 2 \end{bmatrix}$ $P = \begin{bmatrix} \frac{1}{2} & \frac{1}{2} \end{bmatrix}$ $Q = \begin{bmatrix} \frac{1}{3} \\ \frac{2}{3} \end{bmatrix}$

6-26 $\quad G = \begin{bmatrix} \dfrac{1}{2} & 1 & 0 & -\dfrac{1}{2} \\ \dfrac{1}{4} & \dfrac{1}{4} & -\dfrac{1}{4} & \dfrac{1}{4} \\ 1 & -1 & 2 & -2 \end{bmatrix}$ $\quad P = \begin{bmatrix} \dfrac{1}{2} & \dfrac{1}{4} & \dfrac{1}{4} \end{bmatrix}$ $\quad Q = \begin{bmatrix} 0 \\ \dfrac{1}{8} \\ \dfrac{1}{8} \\ \dfrac{3}{4} \end{bmatrix}$

6-27 $\quad G = \begin{bmatrix} -1 & 2 & 0 \\ 1 & 3 & -1 \end{bmatrix}$ $\quad P = [.2 \quad .8]$ $\quad Q = \begin{bmatrix} .7 \\ .1 \\ .2 \end{bmatrix}$

6-28 $\quad G = \begin{bmatrix} 1 & 0 & -1 \\ 1 & 2 & -1 \\ -1 & -2 & 0 \end{bmatrix}$

Player A will choose a card from a standard deck and select row 1 if the card is red, row 2 if a black face card is chosen, or row 3 otherwise. Player B will use the pure strategy "choose column 3."

6-29 $\quad G = \begin{bmatrix} .1 & .2 & 0 \\ .4 & .3 & .1 \\ 0 & 0 & -.4 \end{bmatrix}$ $\quad P = [.7 \quad .1 \quad .2]$ $\quad Q = \begin{bmatrix} .5 \\ .4 \\ .1 \end{bmatrix}$

6-30 $\quad G = \begin{bmatrix} -1 & 2 & 1 \\ 0 & 0 & -1 \end{bmatrix}$

Player A rolls two dice and chooses row 1 if a sum of 7 or 11 appears or row 2 otherwise. Player B flips a coin and chooses column 1 if heads appears or column 3 if tails appears.

6-31 $\quad G = \begin{bmatrix} 1 & 0 & -1 \\ \dfrac{1}{2} & \dfrac{1}{2} & -1 \\ -\dfrac{1}{2} & 2 & -\dfrac{1}{2} \\ 1 & -1 & 2 \end{bmatrix}$ $\quad P = \begin{bmatrix} \dfrac{1}{5} & \dfrac{2}{5} & \dfrac{1}{10} & \dfrac{3}{10} \end{bmatrix}$ $\quad Q = \begin{bmatrix} \dfrac{1}{5} \\ 4 \\ 5 \\ 0 \end{bmatrix}$

6-32 $\quad G = \begin{bmatrix} 2 & 1 & 1 & -4 \\ -3 & -2 & -1 & 3 \end{bmatrix}$

Player A rolls a die and chooses row 1 if a one appears; otherwise row 2 is selected. Player B rolls a die and divides the number on the die by 4. If the remainder is 1, 2, or 3, then player B chooses column 1, 2, or 3 accordingly. If the number is divisible by 4, then the last column is selected.

6-33   In the game

$$\begin{bmatrix} -2 & 0 \\ 0 & 1 \end{bmatrix}$$

player A plans to use the strategy $P = [1/3 \quad 2/3]$. Which of the following strategies for player B is least desirable from player A's point of view?

$$Q = \begin{bmatrix} 1 \\ 0 \end{bmatrix} \qquad Q = \begin{bmatrix} 0 \\ 1 \end{bmatrix} \qquad Q = \begin{bmatrix} \dfrac{1}{3} \\ \dfrac{2}{3} \end{bmatrix} \qquad Q = \begin{bmatrix} \dfrac{1}{2} \\ \dfrac{1}{2} \end{bmatrix}$$

6-34   In the game

$$\begin{bmatrix} -1 & 2 \\ 0 & 0 \\ 1 & -1 \end{bmatrix}$$

player A chooses the pure strategy "row 2." Which of the following strategies for player B is least desirable from player A's point of view?

$$Q = \begin{bmatrix} \dfrac{1}{2} \\ \dfrac{1}{2} \end{bmatrix} \qquad Q = \begin{bmatrix} \dfrac{1}{10} \\ \dfrac{9}{10} \end{bmatrix} \qquad Q \text{ is a pure strategy}$$

6-35   In the game

$$\begin{bmatrix} -1 & 2 \\ 0 & 0 \\ 1 & -1 \end{bmatrix}$$

player B chooses the strategy

$$Q = \begin{bmatrix} \dfrac{3}{4} \\ \dfrac{1}{4} \end{bmatrix}$$

Which of the following strategies for player A is least desirable from player B's point of view?

$$P = \begin{bmatrix} \frac{1}{3} & \frac{1}{3} & \frac{1}{3} \end{bmatrix} \qquad P = \begin{bmatrix} \frac{1}{2} & 0 & \frac{1}{2} \end{bmatrix} \qquad P \text{ a pure strategy}$$

6-36 In the game

$$\begin{bmatrix} 2 & 1 & 1 & -4 \\ -3 & -2 & -1 & 3 \end{bmatrix}$$

player A will use the strategy of tossing three coins and choosing row 2 if all are heads and row 1 otherwise. Which of the following strategies for player B is least desirable from player A's point of view: (a) all columns are equally likely, (b) toss a coin and choose column 1 if heads and column 4 if tails, (c) choose column 3?

6-37 In the game

$$\begin{bmatrix} 0 & 1 & 1 \\ -1 & 0 & 1 \\ -1 & -1 & 0 \end{bmatrix}$$

player B decides on the strategy that consists of rolling two dice and choosing column 1 if the sum of the numbers on the dice is even, choosing column 3 if the sum equals 3, and choosing column 2 otherwise. Which of the following strategies for player A would be least desirable from player B's point of view: (a) toss a coin and choose row 1 if heads or row 2 if tails, (b) all rows equally likely, (c) choose row 2?

6-38 In the game

$$\begin{bmatrix} 3 & -2 \\ -1 & 4 \end{bmatrix}$$

give an example of a strategy Q for player B which demonstrates that the security level of player A's strategy

$$P = \begin{bmatrix} \frac{1}{4} & \frac{3}{4} \end{bmatrix}$$

is lower than one.

6-39 Each of two players holds a playing card and may place the card either face up or face down under a cloth unseen by the other player. If, when the cloths are removed, both cards are face up or both face down, then player A wins $1. If one card is face up and the other is face down, then player B wins $1. Suppose each player considers only pure strategies or the strategy consisting of flipping a coin and placing the card face up for heads and face down for tails. What are the possible payoffs?

6-4 Games and Linear Programming

In order to complete our analysis of matrix games, we need a technique for solving such games. Given the matrix G of the game, we must be able to calculate the optimum mixed strategies P^* and Q_* for players A and B, respectively.

We illustrate the method by means of the game represented by the matrix

$$G = \begin{bmatrix} 6 & 1 \\ 2 & 7 \end{bmatrix}$$

Notice that all the numbers in the matrix are positive (greater than zero). Thus, the payoff is a gain for player A for any choice of pure strategies. A matrix in which every number is positive is called, naturally, a **positive matrix.** We will make essential use of this property of the matrix G above in our discussion, so, initially, the discussion will apply only to games with positive matrices. Actually, once we can solve any game with a positive matrix, the extension of the method to all matrix games will require one easy additional step, as we shall soon see.

Let

$$P = [p_1 \quad p_2]$$

be any mixed strategy for player A in the game above and

$$Q = \begin{bmatrix} q_1 \\ q_2 \end{bmatrix}$$

any mixed strategy for player B. Recall that $E(P, Q)$, the payoff for these strategies, was defined in Section 6-3 to be the sum of all products of the form $g_{ij}(p_i q_j)$, where g_{ij} is the number in row i and column j of the matrix G. Since $p_1 + p_2 = 1$, it must be that at least one of the numbers p_1 and p_2 is positive. For the same reason, at least one of q_1 and q_2 is greater than zero. The matrix G is positive, so all the g_{ij} are positive and therefore at least one of the terms $g_{ij}(p_i q_j)$ is positive (the others might all be zero). We conclude that, if the matrix G is positive, then the payoff $E(P, Q)$ is also positive for any strategies P and Q. We wish to exploit this property of matrix games with positive matrices.

Choose any mixed strategy $P = [p_1 \quad p_2]$ for player A and let v be the security level for that strategy. This means that

$$E(P, Q) \geq v$$

no matter what strategy Q player B chooses. Recall the matrix definition of the payoff:

$$E(P, Q) = PGQ$$

So for this particular strategy, we see that

$$PGQ \geq v$$

for any strategy Q available to player B. Since we know G in this example, we can calculate

$$PG = [p_1 \quad p_2] \begin{bmatrix} 6 & 1 \\ 2 & 7 \end{bmatrix} = \begin{bmatrix} 6p_1 + 2p_2 \\ p_1 + 7p_2 \end{bmatrix}$$

Next we consider two particular payoffs for player A with this strategy P:

$$PGQ = \begin{bmatrix} 6p_1 + 2p_2 \\ p_1 + 7p_2 \end{bmatrix} \begin{bmatrix} 1 \\ 0 \end{bmatrix} = 6p_1 + 2p_2 \quad \text{if } B \text{ chooses} \quad Q = \begin{bmatrix} 1 \\ 0 \end{bmatrix}$$

and

$$PGQ = \begin{bmatrix} 6p_1 + 2p_2 \\ p_1 + 7p_2 \end{bmatrix} \begin{bmatrix} 0 \\ 1 \end{bmatrix} = p_1 + 7p_2 \quad \text{if } B \text{ chooses} \quad Q = \begin{bmatrix} 0 \\ 1 \end{bmatrix}$$

Now, since the relationship $PGQ \geq v$ is true for all strategies Q, then

$$6p_1 + 2p_2 \geq v$$
$$p_1 + 7p_2 \geq v$$

Thus, we have discovered some new restrictions on the choice of p_1 and p_2. Not only do we require $p_1 \geq 0$, $p_2 \geq 0$, and $p_1 + p_2 = 1$, because we are dealing with probabilities, but also we require the restrictions given by the linear inequalities above.

However, in developing a method for calculating optimum strategies, we cannot make any direct use of the inequalities

$$6p_1 + 2p_2 \geq v$$
$$p_1 + 7p_2 \geq v$$

because there is no direct way to determine the security level v. In principle at least, we would have to calculate $E(P, Q)$ for all possible Q and identify the smallest value. To get around this difficulty, we replace the linear inequalities by the equivalent ones obtained by dividing the inequalities by v. Since G is positive, we have observed that $E(P, Q)$ is always positive so, in particular, v is positive. Consequently, dividing through the inequality by v makes sense (which it would not if v were zero) and keeps the inequality sign pointed in the same direction. (Recall that dividing both sides of an inequality by a negative number reverses the direction of the inequality sign.) The first inequality becomes

$$\frac{6p_1 + 2p_2}{v} \geq \frac{v}{v} = 1$$

We rewrite the left-hand side in a convenient form:

$$\frac{6p_1 + 2p_2}{v} = \frac{6p_1}{v} + \frac{2p_2}{v} = 6\frac{p_1}{v} + 2\frac{p_2}{v}$$

So we have

$$6\frac{p_1}{v} + 2\frac{p_2}{v} \geq 1$$

In the same way, the second linear inequality becomes

$$\frac{p_1}{v} + 7\frac{p_2}{v} \geq 1$$

Next, we simplify the notation, replacing p_1/v by the symbol z_1 and p_2/v by z_2. Observe that since $p_1 \geq 0, p_2 \geq 0$, and $v > 0$, then $z_1 \geq 0$ and $z_2 \geq 0$ as well. The linear inequalities now become

$$6z_1 + 2z_2 \geq 1$$

$$z_1 + 7z_2 \geq 1$$

Notice what happens when we add z_1 to z_2:

$$z_1 + z_2 = \frac{p_1}{v} + \frac{p_2}{v} = \frac{p_1 + p_2}{v} = \frac{1}{v}$$

because $p_1 + p_2 = 1$. Thus, $z_1 + z_2$ is the reciprocal of the security level of the mixed strategy $P = [p_1 \quad p_2]$, a fact we express by the equation

$$\text{Reciprocal security} = z_1 + z_2$$

Now we are in a position to give a useful description of the problem of finding an optimum strategy for player A in the game with matrix

$$\begin{bmatrix} 6 & 1 \\ 2 & 7 \end{bmatrix}$$

By definition, an optimum mixed strategy $P^* = [p_1 \quad p_2]$ is one for which the security level v is the highest possible. Since we know that v is positive, this means that we want v to be as large as possible. The larger v is, the smaller positive number its reciprocal $1/v$ is. Thus, in the z notation we introduced above, the problem of finding an optimum strategy P^* for player A in our sample game can be stated as: Find $z_1 \geq 0$ and $z_2 \geq 0$ so that

$$6z_1 + 2z_2 \geq 1$$

$$z_1 + 7z_2 \geq 1$$

and

$$\text{Reciprocal security} = z_1 + z_2$$

is as small as possible.

Once the required z_1 and z_2 have been found, the numbers in $P^* = [p_1 \quad p_2]$ are easily calculated as follows: Since

$$z_1 + z_2 = \frac{1}{v}$$

then taking the reciprocal of both sides, we have

$$v = \frac{1}{z_1 + z_2}$$

By definition,

$$z_1 = \frac{p_1}{v} \qquad z_2 = \frac{p_2}{v}$$

which, when we multiply by v become the equations

$$p_1 = vz_1 \qquad p_2 = vz_2$$

Substituting the expression for v in terms of z_1 and z_2 above, produces

$$p_1 = \frac{z_1}{z_1 + z_2} \qquad p_2 = \frac{z_2}{z_1 + z_2}$$

Thus, once we know z_1 and z_2, we can calculate v, p_1, and p_2 by means of simple formulas.

We hope that you have recognized this problem as being of a familiar type. Perhaps aided by the title of this section, you should recall that it is a linear programming problem— specifically a minimum problem. In matrix form, a minimum problem amounts to finding a $1 \times k$ matrix Z such that $Z \geq 0$ and $ZA \geq C$ that make

$$\text{Function} = ZB$$

as small as possible. The problem is solved by forming the simplex tableau

$$\left[\begin{array}{c|c|c} A & I & B \\ \hline -C & (0 \cdots 0) & 0 \end{array} \right]$$

and applying the simplex method. At the conclusion of the simplex method, the solution $Z = [z_1 \cdots z_k]$ appears in the circled area of the tableau.

For the minimum problem above, the inequalities

$$6z_1 + 2z_2 \geq 1$$

$$z_1 + 7z_2 \geq 1$$

can be expressed by matrix multiplication as

$$[z_1 \quad z_2] \begin{bmatrix} 6 & 1 \\ 2 & 7 \end{bmatrix} \geq [1 \quad 1]$$

which has the form $ZA \geq C$, where

$$A = G = \begin{bmatrix} 6 & 1 \\ 2 & 7 \end{bmatrix} \qquad C = [1 \quad 1]$$

Again using matrix multiplication,

$$\text{Reciprocal security} = z_1 + z_2 = [z_1 \quad z_2] \begin{bmatrix} 1 \\ 1 \end{bmatrix}$$

so

$$\text{Reciprocal security} = ZB$$

where

$$B = \begin{bmatrix} 1 \\ 1 \end{bmatrix}$$

Therefore, the tableau is

$$\left[\begin{array}{cc|cc|c} 6 & 1 & 1 & 0 & 1 \\ 2 & 7 & 0 & 1 & 1 \\ \hline -1 & -1 & 0 & 0 & 0 \end{array}\right]$$

Before applying the simplex method to this tableau in order to determine P^*, let us turn to the other part of the solution to a matrix game, the calculation of an optimum strategy Q_* for player B. As you may well suspect, the story is much the same as before, so we will go through it rather quickly. A mixed strategy for player B is of the form

$$Q = \begin{bmatrix} q_1 \\ q_2 \end{bmatrix}$$

where $q_1 \geq 0, q_2 \geq 0$, and $q_1 + q_2 = 1$. If v is the security level for player B associated with the strategy Q, then

$$E(P, Q) = PGQ \leq v$$

no matter which mixed strategy P player A chooses. Since G is positive, we already know that $E(P, Q)$ is positive for all P and Q so, in particular, v is greater than zero. Calculating

$$GQ = \begin{bmatrix} 6 & 1 \\ 2 & 7 \end{bmatrix} \begin{bmatrix} q_1 \\ q_2 \end{bmatrix} = \begin{bmatrix} 6q_1 + q_2 \\ 2q_1 + 7q_2 \end{bmatrix}$$

and taking the two pure strategies available to player A, gives us the linear inequalities

$$E(P, Q) = \begin{bmatrix} 1 & 0 \end{bmatrix} \begin{bmatrix} 6q_1 + q_2 \\ 2q_1 + 7q_2 \end{bmatrix} = 6q_1 + q_2 \leq v$$

$$E(P, Q) = \begin{bmatrix} 0 & 1 \end{bmatrix} \begin{bmatrix} 6q_1 + q_2 \\ 2q_1 + 7q_2 \end{bmatrix} = 2q_1 + 7q_2 \leq v$$

Again using the fact that v is positive, we divide both sides of the inequalities and the direction of the inequality sign does not change. So we obtain the linear inequalities

$$6\frac{q_1}{v} + \frac{q_2}{v} \leq 1$$

$$2\frac{q_1}{v} + 7\frac{q_2}{v} \leq 1$$

Let

$$x_1 = \frac{q_1}{v} \qquad x_2 = \frac{q_2}{v}$$

and notice that

$$x_1 + x_2 = \frac{q_1}{v} + \frac{q_2}{v} = \frac{q_1 + q_2}{v} = \frac{1}{v}$$

because $q_1 + q_2 = 1$. An optimum strategy for player B is one with the lowest possible security level. Since v is positive, this means a strategy in which v is as small as possible or, equivalently, one in which $x_1 + x_2 = 1/v$ is as large as possible. Expressing the situation in terms of x_1 and x_2 produces the problem: Find $x_1 \geq 0$, $x_2 \geq 0$ satisfying

$$6x_1 + x_2 \leq 1$$
$$2x_1 + 7x_2 \leq 1$$

so that

$$\text{Reciprocal security} = x_1 + x_2$$

is as large as possible.

Once the solution

$$X = \begin{bmatrix} x_1 \\ x_2 \end{bmatrix}$$

to the maximum linear programming problem is known, the calculation of Q_* is easy, because

$$q_1 = vx_1 \qquad q_2 = vx_2$$

where

$$v = \frac{1}{q_1 + q_2}$$

When we express the maximum problem above in matrix form, it becomes: Find a $k \times 1$ matrix $X \geq 0$ and $AX \leq B$ that makes

$$\text{Reciprocal security} = CX$$

as large as possible. Here,

$$A = G = \begin{bmatrix} 6 & 1 \\ 2 & 7 \end{bmatrix} \qquad B = \begin{bmatrix} 1 \\ 1 \end{bmatrix} \qquad C = [1 \quad 1]$$

The corresponding simplex tableau is

$$\left[\begin{array}{cc|cc|c} 6 & 1 & 1 & 0 & 1 \\ 2 & 7 & 0 & 1 & 1 \\ \hline -1 & -1 & 0 & 0 & 0 \end{array} \right]$$

The numbers in X will appear in the circled area of the tableau after the simplex method is applied. But notice—this is exactly the same simplex tableau as the one we used to find P^* for the same matrix! Therefore, in order to solve this matrix game, it is sufficient to operate on a single tableau by the simplex method. The solution (P^*, Q_*) to the game can be calculated easily from the numbers in the last row and the numbers in the last column.

Next we apply the simplex method to the tableau above, going through the steps in some detail. The pivot column is the one with the largest negative value in the last row, so either of the first 2 columns will serve. We choose the first column. Dividing each of the positive numbers in the first column into the corresponding number in the last column gives us

$$\left[\begin{array}{cc|cc|c} 6 & 1 & 1 & 0 & 1 \\ 2 & 7 & 0 & 1 & 1 \\ \hline -1 & -1 & 0 & 0 & 0 \end{array} \right] \begin{array}{c} \dfrac{1}{6} \\[2mm] \dfrac{1}{2} \\[2mm] {} \end{array}$$

This tells us that the pivot element, which corresponds to the smaller quotient, is in the upper left-hand corner of the tableau. Since we wish the first column to look like

$$\begin{bmatrix} ① \\ 0 \\ \hline 0 \end{bmatrix}$$

we begin by dividing each number in the first row by 6:

$$\left[\begin{array}{cc|ccc|c} ① & \dfrac{1}{6} & \dfrac{1}{6} & 0 & \dfrac{1}{6} \\ 2 & 7 & 0 & 1 & 1 \\ \hline -1 & -1 & 0 & 0 & 0 \end{array}\right]$$

Adding appropriate multiples of the first row to the other 2 rows, as in Chapters 3 and 5, the tableau takes the form

$$\left[\begin{array}{cc|ccc|c} ① & \dfrac{1}{6} & \dfrac{1}{6} & 0 & \dfrac{1}{6} \\ 0 & \dfrac{20}{3} & -\dfrac{1}{3} & 1 & \dfrac{2}{3} \\ \hline 0 & -\dfrac{5}{6} & \dfrac{1}{6} & 0 & \dfrac{1}{6} \end{array}\right]$$

The pivot element is in the second column, because the only remaining negative number in the last row is in the second column. To identify the pivot element, we divide each positive number in the second column into the corresponding number in the last column and find

$$\left[\begin{array}{cc|ccc|c} ① & \dfrac{1}{6} & \dfrac{1}{6} & 0 & \dfrac{1}{6} \\ 0 & \dfrac{20}{3} & -\dfrac{1}{3} & 1 & \dfrac{2}{3} \\ \hline 0 & -\dfrac{5}{6} & \dfrac{1}{6} & 0 & \dfrac{1}{6} \end{array}\right] \quad \begin{array}{c} 1 \\[1em] \dfrac{1}{10} \end{array}$$

Then we convert the second column to the desired form:

$$\left[\begin{array}{cc|ccc|c} ① & 0 & \dfrac{21}{120} & -\dfrac{1}{40} & \dfrac{3}{20} \\ 0 & ① & -\dfrac{1}{20} & \dfrac{3}{20} & \dfrac{1}{10} \\ \hline 0 & 0 & \dfrac{1}{8} & \dfrac{1}{8} & \dfrac{1}{4} \end{array}\right]$$

All the numbers in the last row are greater than or equal to zero and the left-hand corner of the tableau is an identity matrix, so the simplex method has been completed.

The solution to the minimum problem is

$$Z = [z_1 \quad z_2] = \begin{bmatrix} \frac{1}{8} & \frac{1}{8} \end{bmatrix}$$

so

$$z_1 + z_2 = \frac{1}{8} + \frac{1}{8} = \frac{1}{4} = \frac{1}{v}$$

and thus $v = 4$. We conclude that

$$P^* = [p_1 \quad p_2] = [vz_1 \quad vz_2] = \begin{bmatrix} 4\left(\frac{1}{8}\right) & 4\left(\frac{1}{8}\right) \end{bmatrix} = \begin{bmatrix} \frac{1}{2} & \frac{1}{2} \end{bmatrix}$$

The solution to the maximum problem is

$$X = \begin{bmatrix} x_1 \\ x_2 \end{bmatrix} = \begin{bmatrix} \dfrac{3}{20} \\ \dfrac{1}{10} \end{bmatrix}$$

which implies that

$$Q_* = \begin{bmatrix} vx_1 \\ vx_2 \end{bmatrix} = \begin{bmatrix} 4\left(\dfrac{3}{20}\right) \\ 4\left(\dfrac{1}{10}\right) \end{bmatrix} = \begin{bmatrix} \dfrac{3}{5} \\ \dfrac{2}{5} \end{bmatrix}$$

As a final remark, we observe that

$$x_1 + x_2 = z_1 + z_2 = \frac{1}{v} = \frac{1}{4}$$

appears in the lower right-hand corner of the tableau after the method has been applied. Neither the particular numbers in the matrix

$$G = \begin{bmatrix} 6 & 1 \\ 2 & 7 \end{bmatrix}$$

nor even its size played an essential role in the analysis that has been performed in this section. The only property we used was the fact that G is positive. Though possibly involving more equations and more symbols, the same reasoning can be applied to any $k \times n$ matrix G where all the entries are positive. For the game with matrix G, we seek optimum mixed strategies

$$P^* = [p_1 \quad p_2 \cdots p_k]$$

and

$$Q_* = \begin{bmatrix} q_1 \\ q_2 \\ \cdot \\ \cdot \\ \cdot \\ q_m \end{bmatrix}$$

for players A and B, respectively. To calculate these strategies, it is sufficient to set up a simplex tableau of the following very special kind:

$$\begin{bmatrix} & & & & 1 \\ & G & & I & 1 \\ & & & & \cdot \\ & & & & \cdot \\ & & & & 1 \\ \hline -1 & -1 \cdots -1 & 0 \; 0 \cdots 0 & & 0 \end{bmatrix}$$

Here, I is the $k \times k$ identity matrix. After the simplex method has been applied, the tableau has the form

$$\begin{bmatrix} & & & x_1' \\ & & & \cdot \\ & & & \cdot \\ & & & x_k' \\ \hline & & z_1 \cdots z_k & \dfrac{1}{v} \end{bmatrix}$$

The value of the game is $v^* = v_* = v$. The optimum strategy P^* for player A is calculated by $p_1 = vz_1, p_2 = vz_2, \ldots, p_k = vz_k$. The maximum problem from which we will obtain Q_* may well be degenerate, so in general its solution $x_1, x_2, \ldots, x_m$ consists of some of the x_i' and some 0s. The optimum strategy Q_* for player B is then calculated by means of the equations $q_1 = vx_1, q_2 = vx_2, \ldots, q_m = vx_m$.

6-9 **Example** Solve the matrix game with matrix

$$\begin{bmatrix} 1 & 2 & 2 \\ 3 & 4 & 1 \end{bmatrix}$$

Solution We set up the simplex tableau,

$$\left[\begin{array}{ccc|cc|c} 1 & 2 & 2 & 1 & 0 & 1 \\ 3 & 4 & 1 & 0 & 1 & 1 \\ \hline -1 & -1 & -1 & 0 & 0 & 0 \end{array}\right]$$

and apply the simplex method (with some intermediate steps omitted):

$$\left[\begin{array}{ccc|cc|c} 1 & 2 & 2 & 1 & 0 & 1 \\ ③ & 4 & 1 & 0 & 1 & 1 \\ \hline -1 & -1 & -1 & 0 & 0 & 0 \end{array}\right] \quad \begin{array}{c} 1 \\ \dfrac{1}{3} \end{array}$$

$$\left[\begin{array}{ccc|cc|c} 0 & \dfrac{2}{3} & \dfrac{5}{3} & 1 & -\dfrac{1}{3} & \dfrac{2}{3} \\ ① & \dfrac{4}{3} & \dfrac{1}{3} & 0 & \dfrac{1}{3} & \dfrac{1}{3} \\ \hline 0 & \dfrac{1}{3} & -\dfrac{2}{3} & 0 & \dfrac{1}{3} & \dfrac{1}{3} \end{array}\right] \quad \begin{array}{c} \dfrac{2}{5} \\ 1 \end{array}$$

$$\left[\begin{array}{ccc|cc|c} 0 & \dfrac{2}{5} & ① & \dfrac{3}{5} & -\dfrac{1}{5} & \dfrac{2}{5} \\ ① & \dfrac{18}{15} & 0 & -\dfrac{1}{5} & \dfrac{6}{15} & \dfrac{1}{5} \\ \hline 0 & \dfrac{9}{15} & 0 & \dfrac{2}{5} & \dfrac{1}{5} & \dfrac{9}{15} \end{array}\right]$$

Therefore, $v = 15/9$ and

$$P^* = [vz_1 \quad vz_2] = \left[\left(\dfrac{15}{9}\right)\left(\dfrac{2}{5}\right) \quad \left(\dfrac{15}{9}\right)\left(\dfrac{1}{5}\right)\right] = \begin{bmatrix} \dfrac{2}{3} & \dfrac{1}{3} \end{bmatrix}$$

The maximum problem represented by the tableau is degenerate. There is a pivot element in the first column and the corresponding number in the last column is $1/5$, so $x_1 = 1/5$.

There is no pivot element in the second column, so $x_2 = 0$. The pivot element in the third column corresponds to 2/5 in the last column, so $x_3 = 2/5$. Thus, the solution to the maximum problem is

$$x_1 = \frac{1}{5} \qquad x_2 = 0 \qquad x_3 = \frac{2}{5}$$

and

$$Q_* = \begin{bmatrix} vx_1 \\ vx_2 \\ vx_3 \end{bmatrix} = \begin{bmatrix} \left(\dfrac{15}{9}\right)\left(\dfrac{1}{5}\right) \\ 0 \\ \left(\dfrac{15}{9}\right)\left(\dfrac{2}{5}\right) \end{bmatrix} = \begin{bmatrix} \dfrac{1}{3} \\ 0 \\ \dfrac{2}{3} \end{bmatrix}$$

As we promised at the beginning of this section, the extension of the technique from matrix games with positive matrices to all matrix games is very easy. Recall the matrix that was the principal example in Section 6-3:

$$\begin{bmatrix} 3 & -2 \\ -1 & 4 \end{bmatrix}$$

Let us add a fixed positive number to every number in this matrix, large enough so that all the sums are greater than zero. It does not matter what number we use, but since 3 is the smallest positive whole number that will work, we use it in hopes that the subsequent arithmetic will not get to be too messy. The result is

$$\begin{bmatrix} 6 & 1 \\ 2 & 7 \end{bmatrix}$$

This is the matrix we analyzed earlier in this section. It can be demonstrated that if we add a fixed number to every entry in the matrix, the new game has the same solution as the original game. Thus, the solution

$$P^* = \begin{bmatrix} \dfrac{1}{2} & \dfrac{1}{2} \end{bmatrix} \qquad Q_* = \begin{bmatrix} \dfrac{3}{5} \\ \dfrac{2}{5} \end{bmatrix}$$

which we calculated for the game with matrix

$$\begin{bmatrix} 6 & 1 \\ 2 & 7 \end{bmatrix}$$

is also the solution to the game with matrix

$$\begin{bmatrix} 3 & -2 \\ -1 & 4 \end{bmatrix}$$

In general, the value of a matrix game can be zero or negative as well as positive, while games with positive matrices always have positive value. Therefore, the operation of adding a fixed number to each entry in the matrix must have some effect on the value of the game. To get some idea of this effect, consider the strictly determined game

$$G = \begin{bmatrix} -1 & 2 & 0 \\ -2 & -2 & 2 \end{bmatrix}$$

The solution to this game is for player A to choose row 1 and for player B to choose column 1, and the value of the game is -1. Now, if we add 3 to every number in G, we obtain the payoff matrix

$$G' = \begin{bmatrix} 2 & 5 & 3 \\ 1 & 1 & 5 \end{bmatrix}$$

The number in the first row and first column is still the lowest in its row and the highest in its column, so the solution is the same as before. However, the value of the game has been increased by 3 from -1 to 2 because, like all other numbers in the matrix, the number in the first row and first column has been increased by that amount. In general, if we add a constant to every number in the matrix of a strictly determined game, then the location of a number lowest in its row and highest in its column does not change, only its value changes. Consequently, the value of the game increases by the amount of the constant.

If the game is not strictly determined, it is more difficult to see what happens to its value when a constant is added to every number in the matrix. We will not try to give an analysis of the problem here; we will only state the conclusion that our discussion of the strictly determined case should lead you to expect. If a constant is added to each number in the payoff matrix of any matrix game, then the value of the game increases by that amount.

We can check this statement readily for the game with matrix

$$G = \begin{bmatrix} 3 & -2 \\ -1 & 4 \end{bmatrix}$$

We showed before that the solution to the game with matrix

$$G' = \begin{bmatrix} 6 & 1 \\ 2 & 7 \end{bmatrix}$$

constructed by adding 3 to every entry in G was

$$P^* = \begin{bmatrix} \dfrac{1}{2} & \dfrac{1}{2} \end{bmatrix} \qquad Q_* = \begin{bmatrix} \dfrac{3}{5} \\ \dfrac{2}{5} \end{bmatrix}$$

We quoted the fact that a solution to the game with matrix G' is also a solution to the game with matrix G. This implies that the value of the game with matrix G is

$$E(P^*, Q_*) = P^*GQ_* = \begin{bmatrix} \dfrac{1}{2} & \dfrac{1}{2} \end{bmatrix} \begin{bmatrix} 3 & -2 \\ -1 & 4 \end{bmatrix} \begin{bmatrix} \dfrac{3}{5} \\ \dfrac{2}{5} \end{bmatrix} = 1$$

Thus the game with matrix G favors player A to the extent that if both players use their optimum strategies, then the expected value of the payoff is 1 unit for player A. On the other hand, we previously calculated that the value of the game whose matrix is G' is 4— and that is indeed 3 units greater than the value of the game with matrix G.

6-10   **Example**   Solve the game with matrix

$$\begin{bmatrix} 1 & -1 \\ 0 & 2 \\ 4 & 0 \\ -3 & -2 \end{bmatrix}$$

and determine which player is favored in this game.

Solution   In order to obtain a positive matrix, we add 4 to each number in the matrix:

$$\begin{bmatrix} 5 & 3 \\ 4 & 6 \\ 8 & 4 \\ 1 & 2 \end{bmatrix}$$

We form the tableau,

$$\begin{bmatrix} 5 & 3 & 1 & 0 & 0 & 0 & 1 \\ 4 & 6 & 0 & 1 & 0 & 0 & 1 \\ 8 & 4 & 0 & 0 & 1 & 0 & 1 \\ 1 & 2 & 0 & 0 & 0 & 1 & 1 \\ \hline -1 & -1 & 0 & 0 & 0 & 0 & 0 \end{bmatrix}$$

and apply the simplex method:

$$\begin{bmatrix} 0 & \frac{1}{2} & 1 & 0 & -\frac{5}{8} & 0 & \frac{3}{8} \\[2mm] 0 & ④ & 0 & 1 & -\frac{1}{2} & 0 & \frac{1}{2} \\[2mm] ① & \frac{1}{2} & 0 & 0 & \frac{1}{8} & 0 & \frac{1}{8} \\[2mm] 0 & \frac{3}{2} & 0 & 0 & -\frac{1}{8} & 1 & \frac{7}{8} \\[2mm] \hline 0 & -\frac{1}{2} & 0 & 0 & \frac{1}{8} & 0 & \frac{1}{8} \end{bmatrix}$$

$$\begin{bmatrix} 0 & 0 & 1 & -\frac{1}{8} & -\frac{9}{16} & 0 & \frac{5}{16} \\[2mm] 0 & ① & 0 & \frac{1}{4} & -\frac{1}{8} & 0 & \frac{1}{8} \\[2mm] ① & 0 & 0 & -\frac{1}{8} & \frac{3}{16} & 0 & \frac{1}{16} \\[2mm] 0 & 0 & 0 & -\frac{3}{8} & \frac{1}{16} & 1 & \frac{11}{16} \\[2mm] \hline 0 & 0 & 0 & \frac{1}{8} & \frac{1}{16} & 0 & \frac{3}{16} \end{bmatrix}$$

$$\begin{bmatrix} ① & 0 & 0 & -\frac{1}{8} & \frac{3}{16} & 0 & \frac{1}{16} \\[2mm] 0 & ① & 0 & \frac{1}{4} & -\frac{1}{8} & 0 & \frac{1}{8} \\[2mm] 0 & 0 & 1 & -\frac{1}{8} & -\frac{9}{16} & 0 & \frac{5}{16} \\[2mm] 0 & 0 & 0 & -\frac{3}{8} & \frac{1}{16} & 1 & \frac{11}{16} \\[2mm] \hline 0 & 0 & 0 & \frac{1}{8} & \frac{1}{16} & 0 & \frac{3}{16} \end{bmatrix}$$

We conclude that $v = 16/3$, and that

$$P^* = \left[0 \quad \left(\frac{16}{3}\right)\left(\frac{1}{8}\right) \quad \left(\frac{16}{3}\right)\left(\frac{1}{16}\right) \quad 0 \right] = \left[0 \quad \frac{2}{3} \quad \frac{1}{3} \quad 0 \right]$$

$$Q_* = \begin{bmatrix} \left(\frac{16}{3}\right)\left(\frac{1}{16}\right) \\ \left(\frac{16}{3}\right)\left(\frac{1}{8}\right) \end{bmatrix} = \begin{bmatrix} \frac{1}{3} \\ \frac{2}{3} \end{bmatrix}$$

The original game favors player A, because its value, namely $16/3 - 4 = 4/3$, is positive.

───

6-11   **Example**   A company makes three models of pop-up toaster: deluxe, regular, and economy. Company planners estimate the value of each model to the company in the year ahead by assigning profitability points as follows: If the economy is strong, the deluxe model will be worth 3 points to the company, the regular model 2 points, and the economy model 1 point. If the economy is weak, the deluxe model is valued at 0 points, the regular model is worth 2 points, and the economy model gets 5 points. What manufacturing strategy should the company follow?

Solution   The game can be described by the table below.

		Economy	
		Strong	Weak
	Deluxe	3	0
Company	Regular	2	2
	Economy	1	5

In order to obtain a positive matrix, we add 1 to each number in the table and form the simplex tableau:

$$\begin{bmatrix} 4 & 1 & 1 & 0 & 0 & 1 \\ 3 & 3 & 0 & 1 & 0 & 1 \\ 2 & 6 & 0 & 0 & 1 & 1 \\ -1 & -1 & 0 & 0 & 0 & 0 \end{bmatrix}$$

From this we calculate

$$\begin{bmatrix} \boxed{1} & \dfrac{1}{4} & \Big| & \dfrac{1}{4} & 0 & 0 & \Big| & \dfrac{1}{4} \\[2mm] 0 & \dfrac{9}{4} & \Big| & -\dfrac{3}{4} & 1 & 0 & \Big| & \dfrac{1}{4} \\[2mm] 0 & \dfrac{11}{2} & \Big| & -\dfrac{1}{2} & 0 & 1 & \Big| & \dfrac{1}{2} \\[2mm] \hline 0 & -\dfrac{3}{4} & \Big| & \dfrac{1}{4} & 0 & 0 & \Big| & \dfrac{1}{4} \end{bmatrix}$$

$$\begin{bmatrix} \boxed{1} & 0 & \Big| & \dfrac{3}{11} & 0 & -\dfrac{1}{22} & \Big| & \dfrac{5}{22} \\[2mm] 0 & \boxed{1} & \Big| & -\dfrac{1}{11} & 0 & \dfrac{2}{11} & \Big| & \dfrac{1}{11} \\[2mm] 0 & 0 & \Big| & -\dfrac{6}{11} & 1 & \dfrac{9}{11} & \Big| & \dfrac{1}{22} \\[2mm] \hline 0 & 0 & \Big| & \dfrac{2}{11} & 0 & \dfrac{3}{22} & \Big| & \dfrac{7}{22} \end{bmatrix}$$

Thus, $v = 22/7$ and the manufacturing strategy should be

$$P^* = \left[\left(\frac{22}{7}\right)\left(\frac{2}{11}\right) \quad 0 \quad \left(\frac{22}{7}\right)\left(\frac{3}{22}\right)\right] = \left[\frac{4}{7} \quad 0 \quad \frac{3}{7}\right]$$

so the company should only make the deluxe and economy models. In principle, the company should decide which model to manufacture by spinning a spinner, but since there is no need to fool the economy, the strategy can be intrepreted as recommending that the company devote 4/7 (about 57%) of its capacity to the deluxe model and the rest to the economy model.

Summary

Suppose we are given a matrix game with matrix G. Player A's pure strategies are the rows of the matrix, and player B's pure strategies are the columns. The optimum strategies P^* for player A and Q_* for player B can be calculated by means of the simplex method of linear programming. If the $k \times m$ matrix G is a **positive matrix** (every number in G is greater than 0), we form the simplex tableau

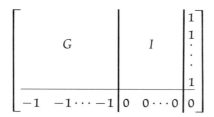

and apply the simplex method. Let $x_1, x_2, \ldots, x_m$ be the solution to the (possibly degenerate) maximum problem represented by the tableau, let $z_1, z_2, \ldots, z_k$ be the solution to the minimum problem represented by the same matrix, and let $1/v$ be the number in the lower right-hand corner of the tableau after the simplex method has been completed. Then

$$P^* = [vz_1, vz_2, \ldots, vz_k] \qquad Q_* = \begin{bmatrix} vx_1 \\ vx_2 \\ \vdots \\ vx_m \end{bmatrix}$$

and the positive number v is the value of the game. If G is not positive, let c be a positive number large enough so that $g_{ij} + c$ is greater than zero for every number g_{ij} in G. Define G' to be the positive matrix obtained by adding c to every number in G and apply the simplex method as above to G'. The solution (P^*, Q_*) to the matrix game with matrix G' is the same as the solution to the original game with matrix G. The value v of the original game with matrix G is related to the value v' of the new game with matrix G' by $v = v' - c$.

Exercises

- Note: Some of the games in Exercises 6-40 through 6-59 are strictly determined. Be sure to check for this.

- In Exercises 6-40 through 6-51, solve the matrix game with the matrix given and determine which of the two players is favored in the game by calculating its value.

6-40 $\begin{bmatrix} 3 & 1 \\ 1 & 2 \end{bmatrix}$

6-41 $\begin{bmatrix} 0 & 2 \\ 2 & -1 \end{bmatrix}$

6-42 $\begin{bmatrix} 1 & 2 & 1 \\ 2 & 1 & 2 \end{bmatrix}$

6-43 $\begin{bmatrix} 0 & 0 & -1 & 0 \\ -1 & 0 & 0 & 0 \end{bmatrix}$

6-44 $\begin{bmatrix} 1 & -3 \\ -1 & 0 \end{bmatrix}$

6-45 $\begin{bmatrix} 2 & -1 \\ 0 & 1 \\ -1 & -2 \end{bmatrix}$

6-46 $\begin{bmatrix} 1 & 2 & 1 & 1 \\ 1 & 1 & 1 & 2 \\ 2 & 1 & 1 & 2 \end{bmatrix}$

6-47 $\begin{bmatrix} 0 & 1 & -1 \\ -1 & 0 & 2 \end{bmatrix}$

6-48 $\begin{bmatrix} -2 & 2 & 3 \\ -1 & 0 & -1 \\ -2 & -3 & 3 \end{bmatrix}$

6-49 $\begin{bmatrix} 1 & 0 & 1 & 1 \\ 0 & 1 & 1 & 0 \end{bmatrix}$

6-50 $\begin{bmatrix} \dfrac{1}{2} & 0 \\ -\dfrac{1}{2} & \dfrac{1}{2} \\ 0 & 0 \end{bmatrix}$

6-51 $\begin{bmatrix} 1 & 0 & -\dfrac{1}{2} \\ 1 & 1 & 0 \end{bmatrix}$

6-52   Each of two players holds a playing card and, unseen by the other player, each may place
the card either face up or face down under a cloth. When the cloths are removed, if both
cards are face up or both face down, then player A wins \$1. If one card is face up and the
other face down, then player B wins \$1. Calculate the optimum strategy for each player
and the value of the game.

6-53   In planning a backpacking trip into the wilderness, a hiker has to decide whether to take
along a heavy jacket. If the weather is warm, she estimates the unpleasantness of carrying
the extra weight of the jacket will cause her to lose 1 point on her personal scale of happiness
resulting from the trip. If the weather is cool, the greater comfort from wearing the jacket
part of the time will just balance the bother of carrying it the rest of the time, with no
effect on total happiness. However, if the weather is cold, the satisfaction of having made
the correct decision together with the greater comfort will add 2 points to her happiness.
Alternatively, if it is warm, leaving the jacket behind will please the hiker 1 point worth,
and if it is cool, leaving the jacket will have no happiness effect. But in cold weather the
shivering hiker will give up 3 points of happiness if she has no jacket. The hiker will spin
a spinner once to decide whether to take the jacket. What is the optimum design for the
spinner? Also, determine whether she is likely to enjoy the trip by finding out whether
the value of the game is positive or negative.

6-54   Player A has a penny and a nickel and player B has a penny, a nickel, and a dime. Each
chooses a coin. If the sum of the values of the two coins chosen is an even number of cents,
then player B gives player A the coin which player B chose. For example, if A chose a penny
and B chose a nickel, then the sum (6¢) is even, so B gives A the nickel. If the sum is an
odd number of cents, then A gives B the coin which A chose. Thus, for example, if A
chose a penny and B chose a dime, the sum is odd (11¢), so A gives the penny to B. Calculate the optimum strategy for each player and the value of the game.

6-55   An investor has \$10,000 to invest in the stock market. If he buys conservative blue chip
stocks, then at the end of the year the expected value of the investment will be \$8000 if
the market prices decline, \$10,000 if the market holds generally steady, and \$12,000 if
that year the stock market averages show a substantial increase. However, if he chooses
to invest in speculative growth stocks, then the expected value of the investment is \$3000
at the end of the year if the stock market registers a decline in prices, \$9000 in a year of
steady market values, and \$16,000 if average stock prices increase to a considerable
extent. What is the optimum investment strategy for this investor? That is, how much
money should he invest in blue chip stocks and how much money should he invest in
growth stocks?

6-56   A football quarterback must decide whether to call a running play or a pass play. The
linebackers on the opposing team must decide as the play begins whether a running
play or a pass attempt is taking place. A gain of yardage for one team is a loss for the other,
and vice versa. Suppose that if the quarterback calls a running play, then the average
gain is only 2 yards if the opposing linebackers expect the running play, but the average
gain is 4 yards if the linebackers think the quarterback is about to attempt a pass. If the

quarterback calls a pass play and the linebackers expect a running play, then we suppose that, on the average, the outcome is an 8 yard gain for the quarterback's team. If the linebackers expect the quarterback's pass, the average result will be that the quarterback is thrown for a 3 yard loss. Calculate the optimum strategies for the quarterback and for the linebackers, and determine the expected outcome of a play by computing the value of this game.

6-57 A farmer must decide whether to plant potatoes, sugar beets, or barley. If there is little rain during the growing season, the potato crop will be small, earning the farmer only $20 an acre; the sugar beet crop will be poor so the farmer will make $40 an acre; and a somewhat reduced barley crop will bring in $30 an acre. In a year of moderate rain, the potatoes will do well, earning $200 an acre; the sugar beets will do fairly well, bringing in $100 an acre; and the barley will be worth $50 an acre. In a year of heavy rain, most of the potato crop will rot in the ground, so the little that is left will earn only $20 an acre; the sugar beets will flourish and earn $150 an acre; and the barley crop will be somewhat damaged by rain and earn $20 an acre. What is an optimum planting strategy for the farmer (which can be interpreted as the fraction of land devoted to each crop)?

6-58 In the children's game "paper, rock, scissors," each of the two players chooses one of these three words. The rules of the game are given by the chant "paper covers rock, rock breaks scissors, scissors cut paper." In other words, if a player chooses paper and the other chooses rock, the player who chose paper wins. Similarly, rock beats scissors and scissors beats paper. If both players choose the same word, the payoff is 0. Otherwise, the payoff to the winning player is 1 unit. Calculate the optimum strategy for each player and the value of the game.

6-59 In a military exercise, the soldiers are divided into two groups, the green army and the blue army. The green army must send reinforcements to a threatened city. There are three routes available and the green army commander must choose one: over a mountain pass, through a river valley, or over a plain. The opposing blue army has just enough soldiers available to set up an ambush at one location. If the blue army fails to intercept the green army reinforcements, the green army is awarded points and an equal number of points are taken from the blue army. Points are awarded to the green army depending on which route is used (based on the speed with which the reinforcements can reach the city). Since the mountain pass route is the fastest, the green army receives 1 point for choosing that route—provided the blue army ambush was set up on another route. Under the same circumstances, the valley route and the plain route are each worth 1/2 point to the green army. However, if the blue army ambush is set up on either the mountain pass or the river valley route and if the green army commander chose the same route, then the blue army receives 1 point and 1 point is taken from the green army. On the other hand, if the ambush is set up on the plain by the blue army and if the green army chooses that route, the chances of a successful ambush in open country are so uncertain that neither army is awarded any points. Determine the optimum strategies for the commanders of the armies and calculate the value of the game to find out which side is likely to benefit.

**Competition,
Cooperation,
and the Prisoner's
Dilemma
(Technical Essay)**

In the introductory essay, we described a psychological experiment that made use of the two person game outlined in the table.

		Partner	
		Blue	Red
Subject	Blue	$(+5\text{¢}, +5\text{¢})$	$(-4\text{¢}, +6\text{¢})$
	Red	$(+6\text{¢}, -4\text{¢})$	$(-3\text{¢}, -3\text{¢})$

Thus, for example, if the subject chooses the red button and the partner the blue button, then the subject receives 6¢, while the partner loses 4¢. Clearly, this game is not a zero–sum matrix game, because the subject's gain does not equal the partner's loss. Therefore, the analysis we performed in Sections 6-2 through 6-4 does not apply.

Although the game is not zero–sum, it is nevertheless instructive to look at the game first from the matrix game point of view. Specifically, suppose the subject looks at her payoffs without regard to the partner's. The subject will then see

Subject	Blue	$+5\text{¢}$	-4¢
	Red	$+6\text{¢}$	-3¢

as a matrix game in which she has to find an optimum row strategy. Noticing that -3¢ is both lowest in its row and highest in its column, she concludes that the matrix game

$$\begin{bmatrix} 5 & -4 \\ 6 & -3 \end{bmatrix}$$

is strictly determined and that her optimum strategy is the pure strategy "push the red button."

Since the partner's payoffs are not just the negatives of the subject's, the subject can also look at these payoffs from the matrix game point of view in order to obtain some idea of what the partner might do:

	Partner	
	Blue	Red
	$+5\text{¢}$	$+6\text{¢}$
	-4¢	-3¢

Let us rewrite this as the usual sort of matrix game, that is, from the point of view of the player whose pure strategies are the rows. Then the numbers take the opposite signs to those above, which give the payoffs from the partner (column) viewpoint:

$$\begin{bmatrix} -5 & -6 \\ 4 & 3 \end{bmatrix}$$

Again, the number in the lower right-hand corner is the lowest in its row and highest in its column, so the game is strictly determined and the partner's optimum strategy is also "push the red button." Suppose the subject concludes that the partner will choose the red button in order to obtain the lowest possible security level as the column player in the matrix game

$$\begin{bmatrix} -5 & -6 \\ 4 & 3 \end{bmatrix}$$

Then the subject has no temptation to push the blue button, because if the partner does choose the red button, then the subject's loss will be 4¢ instead of 3¢. Thus, we see that not only does our matrix game type analysis lead to a strategy for the subject, but it leads to one with considerable stability built into it.

The trouble with this analysis is that if both players respond to it by pushing the red button on the assumption that the other will do the same, then *both* lose 3¢—an unattractive conclusion all around and one that could never happen in a zero–sum game. Furthermore, the double loss is unnecessary in the sense that if both players could agree ahead of time to push the blue button, then *both* win. The trouble is that neither player knows what the other pushed when the game is played. If one player double-crossed the other by pushing the red button, then the double-crosser is even further ahead, 6¢ instead of 5¢, while the player who is double-crossed has a 4¢ loss instead

of the expected 5¢ gain. In a purely competitive situation, then, a player who seeks an optimum strategy from the security level point of view has no choice but to push the red button. In contrast, a subject who expects her partner to cooperate by pushing the blue button can herself act in a cooperative manner by pushing the blue button. Thus, the buttons represent more than just choices of different but essentially neutral actions, as they do in zero–sum games. Rather, the chosen button can be interpreted as the expression of an attitude: competition (red button) or cooperation (blue button).

If the game were played just once, like the matrix games we studied before, then it could be argued that the subject's proper strategy is the competitive one because she has no information on how the other player will behave. However, recall that the psychologist has each subject play the game 30 times and that the partner (who in fact was the experimenter) chose the cooperative blue button strategy in all but three plays of the game. The partner's strategy thus appears to be the mixed strategy

$$Q = \begin{bmatrix} \dfrac{3}{30} \\ \dfrac{27}{30} \end{bmatrix} = \begin{bmatrix} \dfrac{1}{10} \\ \dfrac{9}{10} \end{bmatrix}$$

which is certainly very cooperative. As the plays of the game take place, the subject must observe that most of the time the partner is making the cooperative choice. Thus, after a few plays it would seem that a subject with a positive attitude toward her partner would be encouraged to choose the blue button in expectation of mostly 5¢ gains for both players—even at the cost of an occasional 4¢ loss if the partner were to change to the red button. But a frequent choice of the red button by the subject would seem to indicate either that she did not trust her partner to continue to cooperate or that she deliberately chose a strategy that would give her a slightly larger gain (6¢ instead of 5¢) at considerable loss to the other player, if that player continued to try to cooperate by pushing the blue button. Since the experimenter found a significantly greater number of competitive red button choices when the subjects thought their partners were black than when they thought their partners were white, he concluded that the evidence indicated that the subjects had a negative attitude toward blacks in spite of the fact that they thought they were unprejudiced.

The game used in this psychological experiment is of a type of nonzero–sum game called the "Prisoner's Dilemma." The name comes from the following description of the game: Two people suspected of having together committed a major crime are taken into custody by the police and kept separated so they cannot communicate in any way. Each suspect has the choice of confessing or not. If neither confesses, then there is still sufficient evidence to convict both of a lesser crime for which each will receive a sentence of 1 year in prison. If both confess to the major crime, both will be sentenced to 8 years in prison. However, if one confesses and the other does not, then the one who confessed will get off with a 3 months' sentence in return for turning state's evidence against the other prisoner who will, as a result, be given the maximum sentence of 10 years. Since any time in prison is of negative benefit to the recipient, we express the game in years as shown in the table.

The reader will observe that confessing is the proper competitive strategy for both prisoners in the sense that it gives the most desirable security level. Yet if each prisoner can trust the other to cooperate so that neither confesses, then they are both better off.

| | | Prisoner B | |
		Not confess	Confess
Prisoner A	Not confess	$(-1, -1)$	$(-10, -1/4)$
	Confess	$(-1/4, -10)$	$(-8, -8)$

Concluding Essay

Game theory is a mathematical technique for the analysis of conflict first introduced by John von Neumann in 1927. It was brought to the attention of the mathematical community with the publication of von Neumann and Morgenstern's *Theory of Games and Economic Behavior* in 1944 (see References). At first, because of the formal nature of its presentation, game theory was of interest primarily to mathematicians.

However, people soon recognized the possibilities of game theory as a research tool in sociology and economics. It is used by economists and management scientists to analyze various complex business situations. The main example of this chapter shows a representative application to social psychology. Political scientists also make use of game theory. One example, Malvern Lumsden's article "The Cypress Conflict as a Prisoner's Dilemma Game," appeared in the same issue of *The Journal of Conflict Resolution* as Baxter's article on prejudice (see References). Games are also used to train officers in military strategy.

One of the most interesting and significant facts about game theory is that it is the first major mathematical development to arise from problems in the social sciences. And although it has not proved to be the magical solution to sociological and economic problems as was originally hoped, it has developed into a valuable research tool.

References

The example in the introductory essay of this chapter is based on George W. Baxter, Jr.'s article "Prejudiced Liberals? Race and Information Effects in a Two-Person Game," which appeared in *The Journal of Conflict Resolution,* Vol. 17 (March 1973). Many issues of *The Journal of Conflict Resolution* contain articles that use game theory to analyze various forms of conflict.

Martin Shubik has edited three anthologies dealing with the applications of game theory to various social sciences. They are *Game Theory and Related Approaches to Social Behavior* (New York: John Wiley, 1964), *Readings in Game Theory and Political Behavior* (Garden City, N.Y.: Doubleday, 1954), and *Strategy and Market Structure: Competition, Oligopoly and the Theory of Games* (New York: John Wiley, 1960).

Anatol Rapoport with Albert M. Chammah has written a book entitled *Prisoner's Dilemma* (Ann Arbor, Mich.: University of Michigan Press, 1966). Rapoport's article "Escape From Paradox," which appeared in *Scientific American* (July 1967) explains the work of Nigel Howard in solving the Prisoner's Dilemma game. Rapoport is also the author of *Two Person Game Theory: The Essential Ideas* (Ann Arbor, Mich.: University of Michigan Press, 1966), which is a not too technical exposition of the subject.

For a more complete and somewhat more mathematical treatment of game theory, see *Games and Decisions* by R. Duncan Luce and Howard Raiffa (New York: John Wiley, 1957).

The basic reference on the subject is John von Neumann and Oskar Morgenstern's *Theory of Games and Economic Behavior*, rev. ed. (Princeton, N.J.: Princeton University Press, 1947).

Chapter 7
Opinion Sampling:
An Introduction
to Statistics

Introductory Essay

Polling, polls, and pollsters arouse strong feelings among candidates and the public alike as each important election approaches. Many of us have been amazed or annoyed when the television networks declare a candidate elected after a very small percentage of the vote has been counted. The techniques the networks use to accomplish this are not new. The George Gallup organization, among others, has been predicting national elections with great accuracy for years. This accuracy is possible because of the development of scientific sampling based on statistical methods.

Before the development of sampling, the customary procedure for polling was the "straw vote." One group of people, such as the patrons of Joe's

433

Saloon, would be asked who they planned to vote for, and the pollster would count up the opinions that were expressed.

Before 1936, the *Literary Digest* had conducted straw votes with considerable success. This magazine mailed out millions of postcard ballots to persons whose names were found in telephone directories or on automobile registration lists. The *Literary Digest* prediction on the 1936 presidential election was that Alfred M. Landon would win with 57% of the vote and that Franklin D. Roosevelt would lose with 43%. Of course, that is not what happened. Roosevelt won with the large majority of 62.5% to Landon's 37.5%. Why did the *Literary Digest* come out with one of the largest errors—19%— in polling history? Their sample size was certainly large enough. They had mailed out 10,000,000 postcard ballots, enough to reach one out of three American families and a total of 2,376,523 people marked and returned the cards. However, the problem lay in the way in which the 10,000,000 people were picked. In 1936, the United States was in the midst of the great depression, and owning a telephone or an automobile meant that a person's income was at least average and probably somewhat above average. Before 1936, voters of all income levels were as likely to vote Democratic as Republican but, with the coming of the New Deal, the American electorate became sharply divided. The

majority of the "haves" tended to support the Republicans, while the "have-nots" supported the Democrats.

This 1936 election marked the beginning of scientific polling. George Gallup's, Elmo Roper's, and Archibald Crossley's polling organizations all predicted the outcome accurately. Not only were the predictions right, but Gallup, in a widely circulated newspaper article published on July 12, 1936, predicted the outcome of the *Literary Digest* poll before the *Digest* published its results. He also gave detailed reasons why the poll would be wrong. Gallup predicted the *Digest* poll by sending a random sample of 3000 postcard ballots to the same list of persons who received the *Literary Digest* ballot. As we will see later, this same sample should have provided virtually the same result as the *Digest's* huge sample. It did so, with an error of only 1%.

Gallup tries to come within a 4% range of the actual result in predicting election outcomes—and he generally succeeds. In fact, for 18 national elections from 1936 through 1970 his average error has been only 2.5%. Since his celebrated mistake of 1948 (when he was off by 5%), his average deviation has been only 1.6%.

The Gallup poll usually interviews five people in each of 362 sampling areas, a sample of approximately 1800 people. The Gallup office picks a random starting point within each area. Then the interviewer counts out households at a specified interval, which may be every third, fifth, or twelfth dwelling unit, depending on the population density of the area. Finally, one voter within each of the chosen households is interviewed.

One of the most important factors in the accuracy of the Gallup poll is the selection of truly representative areas. The organization separates all the precincts in the United States into seven categories by population: (1) cities of 1,000,000 or more, (2) cities of 250,000–1,000,000, (3) cities of 50,000–250,000, (4) urbanized areas around cities of 50,000 or more, (5) towns of 2500 or more, (6) villages of under 2500, and (7) open countryside. Then the 362 precincts are picked at random from these seven classifications in proportion to the percentage of the United States population (according to census data) in each classification.

Other polls use different techniques. For example, Sindlinger and Co. uses telephone interviews. The people called are chosen at random from telephone directories from a number of carefully selected counties. This organization feels that the telephone is common enough now so that, in contrast to 1936, few people are without one. They also believe that any error in results because of missing the very poor or those with unlisted numbers is more then compensated for by the speed with which the data is assembled and released.

One of the more controversial uses of polling techniques has been the television networks' use of special sample precincts to predict the result of an election minutes after the polls close. Vote Profile Analysis was developed after the 1960 elections by CBS News, Louis Harris Associates, and IBM. It was first used in the 1962 gubernatorial election in Michigan. At 10:05 p.m. Eastern Standard Time, CBS predicted Romney would win with 52% of the vote. When all the votes were counted his actual percentage was 51.4.

The reasoning behind Vote Profile Analysis, which, as we shall see, is a refined and computerized version of stratified sampling, was this: If a small sample could be drawn to represent all important groups in the electorate by proportionate voting strengths, if a method could be devised to keep track of this sample vote on election night, and if a technique could be developed for comparing those results with the past voting performances of the same group, then an accurate prediction could be made. A combination of careful research and the electronic computer has made this possible.

Vote Profile Analysis chooses representative precincts which reflect the distribution of population in a state and which in past elections have voted with the same percentages as the rest of the state. For example, in Missouri in 1964, Vote Profile Analysis used 40 precincts out of 4400 to predict the result of the presidential election.

Actually, Vote Profile Analysis usually predicts an election on the basis of incomplete information; often when half or even fewer of the representative precincts have reported. In Missouri in 1964, only 10 of the 40 precincts had reported when CBS posted Johnson as the winner. However, few elections are as one-sided as the Johnson–Goldwater contest of 1964, and the television broadcasters must exercise some caution so that in their desire to give the first announcement of the winner they do not speak prematurely and make a mistake.

Mistakes in polls and in predicting elections as they are happening have become very rare. This is because the pollsters use scientific sampling techniques and base their conclusions on statistical theory. This chapter will give you an introduction to statistics in general and, in particular, to the part of statistics that is applicable to opinion sampling.

7-1   Random Sampling

Sindlinger and Co. samples the opinions of the voters in one of its selected counties by calling telephone numbers at random. The phrase **at random** means each number is chosen without system or pattern, so any number is as likely to be called as any other.

One might imagine the Sindlinger employees flipping pages of the telephone book and then, eyes averted, sticking a pin in the page to select the next number to call. The act of flipping pages may not, however, lead to a truly random selection. A person asked to choose a page at random would not be likely to choose the first page or the last page of the book and yet, strictly speaking, numbers on these pages should have as good a chance of being selected as those near the middle of the book—if the selection is really random. In addition, employees might develop patterns of choice. For example, a lazy employee might tend to choose pages near the front of the book, because it takes less effort to reach them. Thus, voters with names near the beginning of the alphabet would be more likely to be called than other voters.

In order to select telephone numbers which are really chosen at random, the Sindlinger organization uses an important tool of the pollsters, a list of random numbers. The numbers on the list are entirely without pattern. The number at a particular location in the list does not depend in any way on the numbers which precede or follow it. At any location in the list, no number is more likely to appear than any other. A short table of random numbers is given in Table 7-1.

					Table 7-1								
					Random Numbers								
23	15	75	48	59	93	76	24	97	08	86	95	23	03
05	54	55	50	43	10	53	74	35	08	90	61	18	37
14	87	16	03	50	32	40	43	62	23	50	05	10	03
38	97	67	49	51	94	05	17	58	53	78	80	59	01
97	31	26	17	18	99	75	53	08	70	94	25	12	58
11	74	26	93	81	44	33	93	08	72	32	79	73	31
43	36	12	88	59	11	01	64	56	23	93	00	90	04
93	80	62	04	78	38	26	80	44	91	55	75	11	89
49	54	01	31	81	08	42	98	41	87	69	53	82	96
36	76	87	26	33	37	94	82	15	69	41	95	96	86
07	09	25	23	92	24	62	71	26	07	06	55	84	53
43	31	00	10	81	44	86	38	03	07	52	55	51	61
61	57	00	63	60	06	17	36	37	75	63	14	89	51
31	35	28	37	99	10	77	91	89	41	31	57	97	64
57	04	88	65	26	27	79	59	36	82	90	52	95	65
09	24	34	42	00	68	72	10	71	37	30	72	97	57
97	95	53	50	18	40	89	48	83	29	52	23	08	25
93	73	25	95	70	43	78	19	38	85	56	67	16	68
72	62	11	12	25	00	92	26	82	64	35	66	65	94
61	02	07	44	18	45	37	12	07	94	95	91	73	78
97	83	98	54	74	33	05	59	17	18	45	47	35	41
89	16	09	71	92	22	23	29	06	37	35	05	54	54
25	96	68	82	20	62	87	17	92	65	02	82	35	28
81	44	33	17	19	05	04	95	48	06	74	69	00	75

Let us suppose, as a simple example, that ten numbers are to be called from a telephone book 50 pages long. In order to decide from which pages the numbers are to be chosen, we will use the first ten different numbers between 01 and 50 that appear in the top two rows of Table 7-1. We will select the numbers as we read from left to right along the rows, starting with the top row and going down. We could have started anywhere in the table and gone in any direction. Thus, the pages to be consulted are

$$23, \quad 15, \quad 48, \quad 24, \quad 8, \quad 3, \quad 5, \quad 50, \quad 43, \quad 10$$

A similar procedure can be used to determine the particular telephone number to be called on each page.

In general, a random sample of a specified size is selected by writing down that many numbers (of the appropriate size) from a list of random numbers and then choosing the corresponding individuals to make up the sample. The numbers should be chosen in some systematic way from the list, but there is no restriction on how this is to be done. Since any number is as likely to appear as any other at any point in the list of random numbers, any individual is as likely to be chosen for the sample as any other.

7-1   **Example**   Use the Table 7-1 to take a random sample of size five from the following list of names:

Appel, Richard	Bleser, John	Coate, Roland
Arbas, Edmund	Borbals, Stanley	Cooling, Richard
Ayres, Don	Burton, Douglas	Day, Carl
Barsocchini, Michael	Cartozian, Richard	Delong, Roland
Bleaman, Jeffrey	Claeyssens, Pierre	

Solution   First number the names in order (we omit first names):

01	Appel	06	Bleser	11	Coate
02	Arbas	07	Borbals	12	Cooling
03	Ayres	08	Burton	13	Day
04	Barsocchini	09	Cartozian	14	Delong
05	Bleaman	10	Claeyssens		

Turning to the table, we will start in the upper left-hand corner and read along rows, noting the first five distinct numbers between 01 and 14 that we find. We could have started anywhere and gone in any direction.

23	15	75	48	59	93	76	24	97	08	86	95	23	03
05	54	55	50	43	10	53	74	35	08	90	61	18	37
14	87	16	03	50	32	40	43	62	23	50	05	10	03

Thus, the five chosen numbers are

$$03, \quad 05, \quad 08, \quad 10, \quad 14$$

and the corresponding random sample is

03 Ayres
05 Bleaman
08 Burton
10 Claeyssens
14 Delong

Summary

A **random sample** from a set is a subset whose members are selected in such a way that every member of the set has precisely the same probability of being in the subset as any other. A **table of random numbers** is a list of numbers in which, at every location, any number is as likely to appear as any other.

In order to select a random sample of size n from a set, first assign a different number to each member of the set. Then, in a table of random numbers, start at any position and go in any direction in a systematic manner taking the first n distinct numbers that appear which represent members of the set. The n members of the set with these numbers assigned to them constitute a random sample.

Exercises

7-1 Use Table 7-1, starting at the upper left-hand corner and reading down columns, to take a random sample of seven names from the list in Example 7-1.

7-2 Take a random sample of ten names from the following list:

Martin	Mike	Pauline	Robert
Marvin	Miles	Peggy	Roberta
Mary-Louise	Newton	Peter	Ronald
Maud	Olive	Philip	Ruby
Michael	Paul	Ray	Sharon

7-3 Take a random sample of six women's names from the list in Exercise 7-2. (Treat the women's names as a new list.)

7-4 Take a random sample of nine automobile makes from the list below, using Table 7-1, starting at the lower right-hand corner and reading from right to left along rows.

Austin-Healey	Bentley	BMW	Citroen	Datsun	Ferrari
Fiat	Honda	Jaguar	Mazda	Mercedes Benz	Opel
Peugeot	Porsche	Renault	Rolls Royce	Saab	Toyota
Triumph	Volkswagen	Volvo			

Take a random sample of eight cities from the following list:

Paso Robles	Gilroy	Petaluma
Pomona	Indio	Anaheim
Arcadia	Lompoc	Redding
Redlands	Merced	Bakersfield
Barstow	Modesto	Riverside
Sacramento	Monterey	Bermuda Dunes
Buellton	Morro Bay	Salinas
San Diego	Napa	Camarillo
Chico	Oakland	San Jose
San Luis Obispo	Oceanside	El Centro
El Monte	Ontario	Santa Barbara
Santa Maria	Palm Springs	Escondido
Fairfield	Palo Alto	Fresno

7-2   Histograms

Let us imagine a collection of data from which information is to be obtained. The collection, called the **population,** is too large to be examined in its entirety, so a random sample of the data is examined instead. For example, pollsters ask the opinions of a random sample of voters in order to get some idea of the opinions held by the population of all voters. In this section, we will take up the question of how one presents the results obtained by consulting the sample.

If an automobile is stolen, the insurance company pays its owner an amount of money equal to the value of the car. The company establishes the value by estimating the price of such a car in the local used-car market; the estimate is based on a sampling of what comparable cars are sold for. Let us suppose the insurance company's sample of sale prices for cars of a certain make and year gives the following data (in dollars):

998	993	1075
1191	1393	1488
1100	1250	1495
775	1150	1000
1050	750	1495

One method of summarizing these data is through the construction of a picture called a **histogram.** In order to construct the histogram, we group together numbers in the list that are close in value. For the data above, we use the following groups or **classes:**

$705–$905 $905–$1105 $1105–$1305 $1305–$1505

Our choice of classes was determined by three rules: the price range is the same for every class (here $200), every price can be put in some class (note that the prices range from $750 to $1495, while the classes can accommodate all numbers between $705 and $1505), and no price in the list appears as the extreme value of a class (called a **class boundary**). This last rule explains why we did not use the classes

$700–$900 $900–$1100 $1100–$1300 $1300–$1500

Had we done so, there would be no way to determine whether the car sale for $1100 should be placed in the $900–$1100 class or in the $1100–$1300 class. There are no rules about the size of the classes; we could have placed the class boundaries $100 apart or $250 apart if we had wished. Also, the actual location of the class boundaries does not matter, so long as the rule about avoiding numbers on the list is observed. Now we place the data in the classes listed in Table 7-2.

Table 7-2		
Class	Number of sales	Sales
$705–$905	2	775, 750
$905–$1105	6	998, 993, 1075, 1100, 1000, 1050
$1105–$1305	3	1191, 1250, 1150
$1305–$1505	4	1393, 1488, 1495, 1495

The histogram itself is constructed from the table by marking off the class boundaries on a number line and then erecting a rectangle over each portion of the line between class boundaries. The height of the rectangle depends on the **relative frequency** of the class, which is defined to be the number of items in the class divided by the **sample size** (the total number of items in the sample).

There are 15 sales in the insurance company's sample so, from the table, the relative frequency of the $705–$905 class is 2/15, of the $905–$1105 class is 6/15, and so on. The histogram of the car sales data is shown in Figure 7-1.

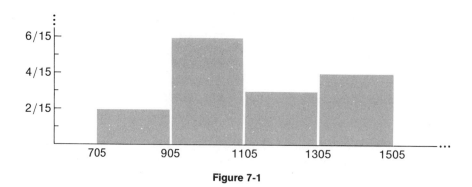

Figure 7-1

7-2   **Example**   A random sample of students finds their heights to be (in inches) 78, 70, 60, 68, 77, 64, 67. Construct a histogram of these data with class boundaries 3 inches apart.

Solution   We will use class boundaries involving 1/2 inches to ensure that no number in the list is a class boundary. We note that the numbers in the sample range from 60 to 78, so we begin our class boundaries with 59.5 and increase by 3 inch amounts until we get above 78. Using the term **frequency** to denote the number of items in the class, we construct Table 7-3. The histogram is shown in Figure 7-2.

Table 7-3	
Class	Frequency
59.5–62.5	1
62.5–65.5	1
65.5–68.5	2
68.5–71.5	1
71.5–74.5	0
74.5–77.5	1
77.5–80.5	1

You may be wondering at this point just what all this has to do with opinion sampling. The data discussed thus far has all been in numerical form, but a sample of voters, when asked for whom they will vote, does not appear to use numbers in its reply. How, then, can one hope to construct a histogram of such information?

We will illustrate the answer to this question by means of the following example: Fifteen people chosen at random are asked if their eyes are brown. The responses are

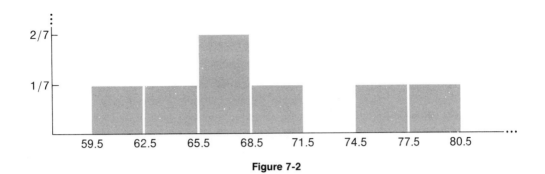

Figure 7-2

Not brown	Not brown	Not brown
Not brown	Brown	Brown
Not brown	Brown	Brown
Not brown	Brown	Not brown
Brown	Not brown	Not brown

The trick for converting these nonnumerical responses into numerical form is just to assign a number, either 0 or 1, to each possible response. It does not matter whether Brown or Not brown is represented by the 0. Our choice is

$$\text{Not brown} = 0 \qquad \text{Brown} = 1$$

Thus, the data on eye color become

0	0	0
0	1	1
0	1	1
0	1	0
1	0	0

For data in the form of 0s and 1s, the traditional classes are $-1/2$–$1/2$ and $1/2$–$3/2$. In this case, we have the following table:

Class	Frequency
$-1/2$–$1/2$	9
$1/2$–$3/2$	6

The histogram is shown in Figure 7-3.

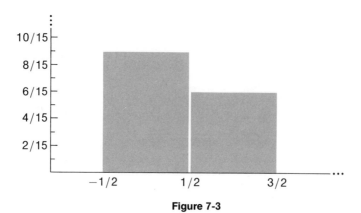

Figure 7-3

7-3   **Example**   Two students, Green and White, are running for class president. A sample of
ten student voters gives the following preferences: Green, Green, Green, Green, Green,
Green, Green, White, Green, White. Construct a histogram of the sample information.

Solution   Eight students support Green and two prefer White, so if we let a vote for Green
be represented by 0 and a vote for White by 1, the histogram is that shown in Figure 7-4.

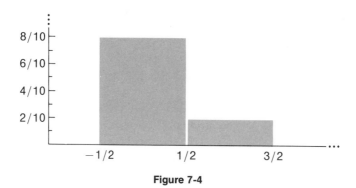

Figure 7-4

Summary

A random sample is taken from some larger collection of data, called the **population.**
Data from the sample are grouped into **classes,** ranges of numerical values all of the same
size. Every such piece of information must belong to some class. The extremes of the
classes, called the **class boundaries,** are chosen so that no number from the sample
coincides with a class boundary. The **frequency** of a class is the number of items from
the sample in it and the **relative frequency** of a class is its frequency divided by the size

of the sample. In order to construct a **histogram** of the data from a sample, the class boundaries are marked out on a number line and, over each interval on the line corresponding to a class, a rectangle is erected with height determined by the relative frequency of that class.

Suppose the members of a sample are questioned and there are two possible responses. The results of the examination can be converted to numerical form by assigning the number 0 to one response and 1 to the other. The histogram of such data uses the classes $-1/2–1/2$ and $1/2–3/2$.

Exercises

7-6 A sample of secretaries' monthly salaries in a certain city produces the following figures:

$$\begin{array}{ccccc} \$800 & \$850 & \$650 & \$700 & \$750 \\ \$675 & \$575 & \$650 & \$850 & \$800 \end{array}$$

Construct a histogram of these data with class boundaries $100 apart.

7-7 In baseball, a pitcher's earned run average (ERA) is a measure of his skill. A sample of ERAs of National League pitchers gave the following results:

$$\begin{array}{cccc} 3.40 & 3.93 & 3.09 & 4.33 \\ 3.34 & 2.77 & 3.81 & 3.75 \end{array}$$

Construct a histogram of these ERAs with class boundaries 0.5 apart.

7-8 A sample of twenty married couples was asked "Who usually balances the checkbook in your family?" The replies were:

Husband	Husband	Wife	Wife
Husband	Wife	Wife	Husband
Wife	Wife	Wife	Husband
Husband	Husband	Wife	Husband
Wife	Husband	Husband	Wife

Construct a histogram of these replies.

7-9 A sample of research mathematicians was found to have the following numbers of publications:

$$\begin{array}{cccccccc} 1 & 1 & 1 & 1 & 15 & 1 & 5 & 1 \\ 5 & 8 & 3 & 1 & 1 & 3 & 1 & 4 \end{array}$$

Construct a histogram of these data with class boundaries 2 units apart.

7-10   A sample of performances by women professional golfers in tournament play is taken
       to investigate the question of whether they tend to improve their score from the first
       round of the tournament to the second. The scores of the sample in the first and second
       rounds were:

First	Second		First	Second		First	Second
75	76		72	70		72	70
73	72		76	74		71	69
75	73		73	68		77	78
69	73		69	69		75	73
69	74		71	69		69	73
69	76		74	74		75	76

       Construct a histogram showing improvement (a lower score) or not in the second round.
       Do *not* make a histogram of the scores themselves.

7-11   A random sample of the prices of mutual fund shares (in dollars) on a particular day
       produced the following data:

7.76	3.93	3.17	2.09	3.90
2.10	11.70	6.21	5.82	8.41
6.12	4.22	7.91	8.01	10.07

       Construct a histogram of these data with class boundaries $1 apart.

7-12   A sample of baseball players' batting averages in the American League produced the
       figures:

.259	.233	.262	.247	.233	.291
.262	.300	.295	.272	.289	.251

       Construct a histogram of these figures with class boundaries .020 unit apart.

7-13   A sample of mathematics texts found the number of pages in the books to be 186, 300,
       347, 229, 149, 258, 247, 528, 237, 328, 298, 484, 374. Construct a histogram indicating
       whether the number of pages is even or odd. Do *not* construct a histogram of the actual
       number of pages of the books in the sample.

7-3 Mean
and Variance

In Section 7-2, we discussed an automobile insurance company that wished to determine the value of a certain automobile. The company took a sample of sale prices for cars of the same make and year in the area and obtained the following figures:

998	993	1075
1191	1393	1488
1100	1250	1495
775	1150	1000
1050	750	1495

A reasonable figure for the value of the automobile would be the **average** price being paid for such cars, since that is approximately what it would cost to replace that particular car with one like it. The insurance company does not know the average price being paid for this sort of car if by average we mean the average of *all* sales of such cars in that area. However, the company can calculate the average price in the sample and use that figure as an **estimate,** an educated guess, of the actual average selling price.

Recall that the average of a set of numbers is obtained by adding up all the numbers and then dividing the sum by how many numbers there are in the set. For the average sale price in the sample, we add up

$$998 + 993 + 1075 + 1191 + 1393 + 1488 + 1100 + 1250$$

$$+ 1495 + 775 + 1150 + 1000 + 1050 + 750 + 1495 = 17{,}203$$

Dividing by 15, the number of sales in the sample, gives the sample average $17{,}203/15 = 1146.87$ from which the insurance company would estimate the average of all such sales to be $1146.87. The technical name for the average in statistics is the **mean.** Therefore, the mean of the data from a sample is defined to be the sum of all the numbers obtained from the sample divided by the size of the sample.

7-4 **Example** Students selected in a random sample from a university were found to be the following heights (in inches): 78, 70, 60, 68, 77, 64, 67. Estimate the mean height of students at that university on the basis of this information.

Solution The mean height of the students in the sample is

$$\frac{78 + 70 + 60 + 68 + 77 + 64 + 67}{7} = \frac{484}{7} = 69 \quad \text{(approximately)}$$

so we estimate the mean height of students at that university is 69 inches.

Next let us examine what the mean tells us about questions of public opinion. In Example 7-3, two students named Green and White were running for class president. A voter sample produced the preferences Green, Green, Green, Green, Green, Green, Green, White, Green, White. We converted these results to numerical form by letting Green = 0 and White = 1. Then the sample data becomes 0, 0, 0, 0, 0, 0, 0, 1, 0, 1 and we calculate the mean to be

$$\frac{0 + 0 + 0 + 0 + 0 + 0 + 0 + 1 + 0 + 1}{10} = \frac{2}{10}$$

Notice that the mean in this case is just the number of 1s in the sample divided by the sample size. If the numerical data consists of just 0s and 1s, then the mean is always calculated in this same way: the number of 1 responses divided by the sample size. Another name for the mean, when the responses are 0s and 1s, is the **proportion** (of 1s).

In the sample of Example 7-3, the proportion of 1s, that is, the fraction of the sample favoring White, is 2/10. Thus, 20% of the sample favor White. Using the sample mean to estimate the mean of the population, here the population of all students voting in the election, we would predict that White will lose the election, receiving only 20% of the votes.

7-5   **Example**   In an election for governor, out of the 948 voters in a sample who expressed an opinion, 517 intended to vote for the Democratic candidate. Estimate the percentage of the total vote in the election the Democratic candidate will receive.

Solution   Counting each voter in the sample who supports the Democratic candidate as a 1, the proportion of voters in the sample in favor of that candidate is

$$\frac{517}{948} = .55 \quad \text{(approximately)}$$

so we estimate that 55% of the voters will choose the Democratic candidate in the election.

The calculation of the mean tells us something very useful about the data in a sample, but the information contained in this single number is necessarily rather limited. For example, if a sample of three measurements produces data 0, 5, 10, and another sample of size three gives measurements 5, 5, 5, then both have the same mean, 5. However, the samples are really quite different; in the first the measurements are spread out, but in the second they are all the same.

It is often important to measure the scattering of the numbers. After taking a sample of car sale prices to determine the value of a certain car, the insurance company might well want to know how much variation occurs in the prices it is using. Suppose there is little differences in price from sale to sale. Then if the car is stolen, the payment of the mean price to the policyholder will be just about the replacement cost. But suppose there is a great variation in the sale prices in the sample for that same make and year. Then it may be inappropriate to pay an amount equal to the mean of the sample. For example, the sample might include both desirable and undesirable models of that automobile. In that case, the variation in sale prices would reflect the fact that the car's value depends on its model type as well as on its make and year. The company may well wish to use only sale prices of the same year, make, and model in establishing the automobile's value.

One way to examine the spread of the numbers in the sample is to refer to the histogram. Certainly a histogram that looks like Figure 7-5 has much more spread than one that looks like Figure 7-6. But mere visual observation of the histogram is not sufficient, because it lacks precision. We can determine whether the data appear to be highly varied or not, but we cannot measure just how much spread occurs. In order to obtain this information, we will next discuss a measure of diversity in a collection of data. This measure is the **variance** of the sample.

> The variance s^2 of a sample of n numbers is calculated by first subtracting the mean of the sample from each number in the sample, next squaring each result, then adding up all the squares, and, finally, dividing the sum by $n - 1$.

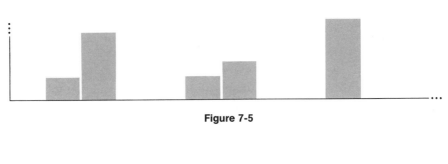

Figure 7-5

Figure 7-6

For example, suppose the sample consists of $n = 3$ measurements: 0, 5, 10. The mean of the sample is 5, so we subtract 5 from each number in the sample:

$$0 - 5 = -5 \qquad 5 - 5 = 0 \qquad 10 - 5 = 5$$

Next we square each difference:

$$(-5)^2 = 25 \qquad (0)^2 = 0 \qquad (5)^2 = 25$$

Then we add up the squares:

$$25 + 0 + 25 = 50$$

Dividing by $n - 1$ gives

$$s^2 = \frac{50}{3 - 1} = \frac{50}{2} = 25$$

7-6 **Example**　Calculate the variance of the sample of heights (in inches): 78, 70, 60, 68, 77, 64, 67.

Solution　By Example 7-4, the mean of the sample is approximately 69 inches. Therefore,

$$s^2 = \frac{(78 - 69)^2 + (70 - 69)^2 + (60 - 69)^2 + \cdots + (64 - 69)^2 + (67 - 69)^2}{7 - 1}$$

$$= \frac{81 + 1 + 81 + 1 + 64 + 25 + 4}{6} = \frac{257}{6} = 43 \qquad \text{(approximately)}$$

When we apply the definition of the variance of a sample to a sampling of opinion for which the responses can be represented by 0s and 1s, we obtain a very convenient formula. Recall that the mean of such a sample is called the "proportion" (of 1s). We use the symbol p for the proportion and define a symbol q by the formula $q = 1 - p$. We can attach a meaning to q as follows: The proportion p is calculated by dividing the number of 1s in the sample by the sample size n. Then, q is just the number of 0s in the sample divided by n. It can be shown that for a sample of 0s and 1s with proportion p, the definition of the variance reduces to the formula

$$s^2 = \frac{n}{n - 1} pq$$

We will not attempt to derive this formula here, but we will illustrate it by means of the sample of ten student voters of which eight preferred Green (denoted by 0) and the other two White, so that $p = 2/10$. According to the general definition of the variance,

$$s^2 = \frac{\overbrace{\left(0 - \frac{2}{10}\right)^2 + \cdots + \left(0 - \frac{2}{10}\right)^2}^{\text{Eight terms}} + \left(1 - \frac{2}{10}\right)^2 + \left(1 - \frac{2}{10}\right)^2}{10 - 1}$$

$$= \frac{8\left(-\frac{2}{10}\right)^2 + 2\left(\frac{8}{10}\right)^2}{9} = \frac{8\left(\frac{4}{100}\right) + 2\left(\frac{64}{100}\right)}{9} = \frac{160}{900} = \frac{8}{45}$$

On the other hand,

$$\frac{n}{n-1}pq = \left(\frac{10}{9}\right)\left(\frac{2}{10}\right)\left(\frac{8}{10}\right) = \frac{8}{45}$$

so the formula is seen to be correct in this case.

7-7 **Example** For the sample of 948 voters from Example 7-5, of which 517 favored the Democratic candidate, calculate the variance of the sample.

Solution According to Example 7-5, $p = 517/948$, so $q = 431/948$ and

$$s^2 = \frac{n}{n-1}pq = \left(\frac{948}{947}\right)\left(\frac{517}{948}\right)\left(\frac{431}{948}\right) = \frac{211{,}239{,}996}{851{,}072{,}688} = .2482 \quad \text{(approximately)}$$

When we are working with large samples, which is generally the case in opinion sampling, the fraction $n/(n-1)$ is very close to 1 (e.g., from Example 7-7, 948/947 is about 1.001). In that case, the exact formula $s^2 = [n/(n-1)]pq$ can be replaced by a simpler approximation, $s^2 = pq$. In Example 7-7, if we had used this approximation, we would have obtained

$$pq = \left(\frac{517}{948}\right)\left(\frac{431}{948}\right) = \frac{222{,}827}{898{,}704} = .2479 \quad \text{(approximately)}$$

which only differs from the answer we obtained by .0003. We will use the approximate formula $s^2 = pq$ whenever sample size n is at least 100.

Summary

The **mean** of a set of measurements is the sum of the measurements divided by the number of measurements. If a random sample is taken from some larger population, the mean of the sample is used as an **estimate** of the mean of the entire population. When the measurements are all 0s and 1s, the mean is called the **proportion** (of 1s), denoted by p, and p is computed by dividing the number of 1s in the sample by the size of the sample.

The **variance** s^2 of a sample of size n is calculated by first subtracting the sample mean from each measurement in the sample, next squaring each difference, then adding up the squares, and, finally, dividing the sum by $n - 1$. If the numbers in the sample are all 0s and 1s and the proportion is p, then

$$s^2 = \frac{n}{n - 1}pq$$

where $q = 1 - p$. When n is at least 100, we may use the approximate formula $s^2 = pq$.

Exercises

7-14   A sample of typing speed requirements for receptionist positions gave these figures (in words per minute):

45	50	50	55	50	50
55	45	55	55	40	

Estimate the mean typing speed required of receptionists and calculate the variance of the sample.

7-15   A sample of monthly rents for unfurnished one bedroom apartments in a certain city produced the following figures:

$160	$165	$170	$140
$160	$125	$165	$200

Estimate the mean monthly rent for such apartments in that city to the nearest dollar and use that figure to calculate the variance of the sample.

7-16   The volume of trading is a measure of stock market activity. A random sample of nine stocks listed with the American Stock Exchange produced the following figures for numbers of shares traded on a given day:

20,000	340,000	160,000
60,000	50,000	30,000
50,000	960,000	110,000

Estimate the mean number of shares traded on the American Stock Exchange that day to the nearest 100 shares.

7-17 Twelve National League baseball players chosen at random had the following batting averages:

.347	.270	.331	.277
.300	.294	.209	.274
.213	.257	.341	.298

Estimate the mean batting average of players in the National League.

7-18 In an election for state attorney general, out of a sample of 851 voters who had formed an opinion on the election, 539 planned to vote for the incumbent. Estimate the percentage of the vote that the incumbent will receive in the election and calculate the variance of the sample.

7-19 The mean increase or decrease in stock market prices over the previous day is a measure of stock market behavior. Price increases are denoted by positive numbers, price decreases by negative numbers, and 0 indicates no change from the previous day's price. The units are fractions of a dollar. Use the sample data below, based on prices of over-the-counter stocks traded on a certain day, to estimate the mean change in price for all over-the-counter stocks on that day:

$$-\frac{1}{2} \quad -\frac{1}{8} \quad +\frac{1}{8} \quad 0 \quad 0 \quad 0 \quad 0 \quad 0 \quad 0$$

$$+\frac{1}{8} \quad -1 \quad -\frac{9}{8} \quad -2 \quad 0 \quad +\frac{1}{8} \quad -\frac{1}{4} \quad +\frac{1}{2} \quad 0$$

7-20 A sample of the weights of professional football linemen gives the following figures (in pounds):

248	240	250	252
245	269	270	238

Estimate the mean weight of professional football linemen and calculate the variance of the sample.

7-21 In a poll of 3000 American women conducted a few years ago, 1800 women said they considered divorce an acceptable solution to a bad marriage. Estimate the proportion of all American women who share this opinion and calculate the variance of the sample.

7-22   In nine American cities chosen at random, the highest temperatures (in °F) recorded in 1971 were:

$$98 \quad 94 \quad 99$$
$$95 \quad 93 \quad 98$$
$$94 \quad 89 \quad 75$$

Estimate the mean of these temperatures to the nearest whole degree and use that figure to calculate the variance of the sample.

7-23   A sample of scores achieved by the most successful male professional golfers produced the following results:

$$67 \quad 68 \quad 71 \quad 72 \quad 72 \quad 69 \quad 69 \quad 69 \quad 73$$
$$73 \quad 67 \quad 69 \quad 73 \quad 69 \quad 73 \quad 69 \quad 71 \quad 70$$

Estimate the mean score that such golfers get in a round of golf.

Essay on Errors in Sampling

A short time before the New York gubernatorial election of 1942, the Office of Public Opinion Research (OPOR) selected a sample of 200 New York State voters and asked them who they would vote for in the coming election. Thomas Dewey was named by 58% of the voters (116 of the 200 people in the sample).

If we were asked who we would expect to become governor of New York in 1942 and we based our answer on the sample, then the answer would certainly have to be Dewey. We would, however, wish to qualify our response by saying that Dewey would *probably* win. Clearly, 58% is an impressive margin. Nevertheless, the sample probably would not predict the results

exactly and, in particular, it might be off by enough so that Dewey would actually lose the election. Therefore, we also want to know how likely it is that the prediction based on the sample will be correct, or at least very close to the actual outcome.

Let us list some of the reasons the sample might give an incorrect prediction. One obvious reason is that, since the sample was taken some time before the election, many people might have changed their minds as a result of the political campaign or because news events influenced their decision. Some respondents might have given incorrect information. For example, some might have said they were going to vote even though they had no such intention, because they did not want the interviewer to get the impression that they were not good citizens. The enumerators may also have been a source of error. They might, through carelessness, have checked a response in the wrong column of the tabulation sheet. They might have shown their own opinion in the way they questioned the voters, thus encouraging uncertain voters to name the same candidate.

You can probably think of many other difficulties that might plague the pollster. All the possible sources of error we have mentioned so far could arise no matter whether OPOR took a sample of only four voters, 200 voters, or even if they queried every voter in New York. The "inaccuracy" of the data is the extent to which the sample data would be in error even if every member of the population were included. To put the matter in more positive terms, we speak of the "accuracy" of the data used in the sample, and we seek to make that data as accurate as possible.

There are many techniques for improving the accuracy of the data. Opinion shifts just before an election can be dealt with to some extent by taking the sample of opinion as close to the date of the election as possible. The great advantage of the Sindlinger telephone poll is that it takes its final sample just before the election. Thus, in this way at least, it uses very accurate data. The problems arising both from dishonest answers and from enumerator error can be reduced through the careful selection, training, and supervision of the employees who collect the data. However, no system, whether of sampling or of census of the entire population, can produce completely accurate information; people still change their minds in the voting booth. At best, we can hope that the inaccuracies are small and that they tend to cancel each other out. There is no known way to measure the accuracy of data of this sort and so, in effect, pollsters act as if inaccuracy does not exist. As long as the polling organizations continue to make correct predictions, as the record of the Gallup organization described in the introduction to this chapter suggests they do, it is reasonable to believe that inaccuracy of data is not a severe problem in predicting elections.

Notice that the techniques pollsters use to produce reasonably accurate data can be applied only because they take small samples. It may be desirable for pollsters to take larger samples, as we shall see in Section 7-6, but we should observe that large samples require more time to collect, and even with the aid of high-speed computers, they take more time to analyze. Also, because a larger staff is needed to take a large sample, the economic and administrative personnel problems become greater. Thus, we should keep in mind that to some extent data gathered from a complete census of the population, or even from a large sample, tends to be less accurate than the data from a small sample.

We turn now to the source of sampling error which probably occurred to you first. If there were millions of voters in New York State in 1942, it is likely that the sample of 200 voters in the OPOR sample did not exactly reflect the opinions of the millions who were not included in the sample. The fact that 116 out of the 200 voters in the sample said they planned to vote for Dewey did not necessarily mean that the exact same percentage, 58%, of all New York voters preferred Dewey. Perhaps, at the time the sample was taken, 50% of all New York voters planned to vote for Dewey, or possible 60%, or some other percentage planned to do so. The extent to which a sample measures the responses of a larger population is called the "precision" of the sample. We will assume, as the pollsters do, that the data used is entirely accurate. That is, we will assume the responses from the sample are exactly what the members of the sample will do in the election. Thus, we proceed as if none of the problems with the collection of data that we mentioned earlier exist. Then the original question, "How much faith can we have in the prediction based on a sample?" is answered by determining the precision of the sample.

In contrast to the problem of estimating the accuracy of data, it is possible to obtain useful information on the precision of a sample provided the sample was selected in a scientifically sound manner.

In the next sections, we will assume all samples are the result of random selection. In particular, then, we will assume OPOR chose the 200 voters in its sample from New York State voter registration lists with the aid of a table of random numbers in a manner like that described in Section 7-1.

7-4 The Normal Distribution

Section 7-6 will be devoted to the question of how much faith one may have in the results of a random sample. To answer that question, we will need to be able to use one of the most important concepts in statistics—the **normal distribution.** Therefore, in this section, we take up the aspects of the normal distribution that we will use later.

If we construct a histogram of the heights of all the students at some large university, the histogram will look something like Figure 7-7. Most students would be of about average height, while relatively few would be either very short or very tall. A histogram of scores on a nationally administered test would have a similar shape. Most students would achieve approximately average scores, while a few would do very well or very poorly on the examination. The fact that the histograms have this appearance can be supported by observation; histograms with this shape are a familiar sight.

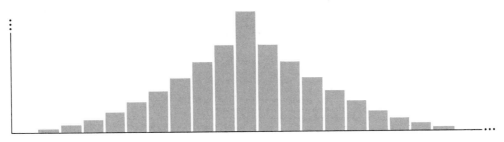

Figure 7-7

Smoothing out a histogram that bunches up in the middle and tapers off at the extremes produces a shape like that shown in Figure 7-8. We can think of this curve as a histogram if we imagine that it is made up of rectangles so thin they seem to flow into each other. Generally, histograms with this shape are constructed from measurements taken from some large population (all students at the university, all people who took the national examination). In order to get an accurate picture of so much data, a histogram of it will use many classes, with class boundaries close together, so the rectangles really will be very thin. The shape is called a "bell-shaped curve" or, more formally, a **normal distribution.**

There are infinitely many different normal distributions, but they all have certain characteristics in common. For one thing, they all peak in the middle; the highest point of the curve occurs at the mean. In other words, if we think of a normal distribution as the histogram of measurements from some population, then the class with the largest relative frequency is the one containing the mean of the population (traditionally denoted by μ, the Greek letter mu). Furthermore, a normal distribution is symmetric. By looking at Figure 7-9, it is evident that half the measurements lie to the right or above the mean and half to the left or below it, but even more is true. Suppose that x is any positive number, and points x units above and below the mean are marked off on the normal distribution, as in Figure 7-10. The symmetry of the normal distribution is such that the shape of the curve in the region below the mean is a mirror image of what lies above, so the area under

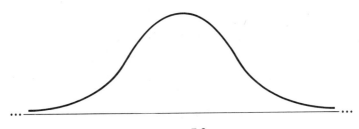

Figure 7-8

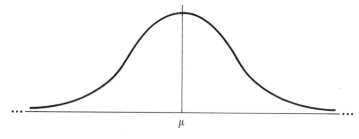

Figure 7-9

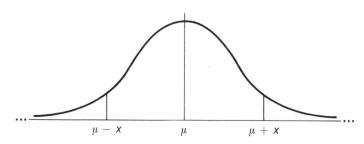

Figure 7-10

the curve lying in the portion between $\mu - x$ and μ is exactly the same as the corresponding area between the lines through μ and $\mu + x$. In terms of histograms, this means there are as many members of the population with measurements between $\mu - x$ and μ as there are members with measurements between μ and $\mu + x$, and this is true for any positive number x.

For example, if the mean height of the students in the university is 68 inches and 40% of all students are between 62 and 68 inches tall, then, assuming the histogram of heights looks like a normal distribution, we can conclude that another 40% of the students are between 68 and 74 inches in height, because $\mu = 68$ and $62 = \mu - 6$ so $\mu + 6 = 74$ (see Figure 7-11).

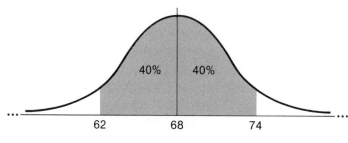

Figure 7-11

7-8   **Example**   Suppose the mean score on some portion of the Graduate Record Examination is 500 points and 45% of the people who take the examination receive between 500 and 720 points on that portion. If the histogram of scores is shaped like a normal distribution, what percentage receive a score between 280 and 500?

Solution   Between $\mu = 500$ and $720 = \mu + 220$ there are 45% of all measurements, so the symmetry of the normal distribution implies that another 45% of all scores lie between $\mu - 220 = 280$ and 500.

There is more that can be said of the common characteristics of all normal distributions, but in order to discuss it we need first to examine the ways in which they differ. Of course, different normal distributions might have different means; the mean can be any number at all. But the two distributions shown in Figure 7-12 are quite different, even though they have the same mean. The difference can be described by saying that in the first distribution the measurements are clustered around the mean, while in the second the measurements are much more spread out. The reader will recall that we have already discussed a measure of the spread of a histogram. In Section 7-3, we defined the variance of the sample. The variance of a sample of size n was computed by subtracting the mean of the sample from each measurement in the sample, squaring the result, adding up the squares, and then dividing by $n - 1$. Since we think of a normal distribution as arising from the histogram of data from an entire population rather than a sample, we need a definition for the **variance of the population.** The definition is almost the same as before. If we

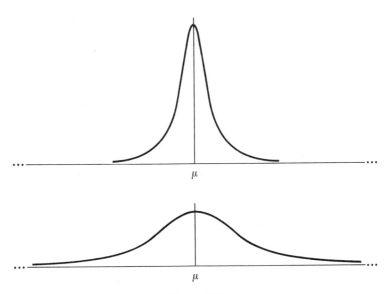

Figure 7-12

have a population of N measurements with mean μ, the variance of the population is denoted by σ^2 (σ is the Greek letter sigma) and calculated by first subtracting μ from each measurement in the population, next squaring the result, then adding up all the squares, and, finally, dividing the sum by N. The exact analogy with the variance of the sample would occur if we had divided by $N - 1$ instead of N. Since N is generally very large, there is little arithmetic significance to whether N or $N - 1$ is used, so the distinction occurs only at a theoretical level. We can certainly think of the variance of the population and the variance of the sample as corresponding concepts, just as we think of the mean of the population and the mean of the sample as corresponding. Now we can describe the difference between the two normal distributions in Figure 7-12. The first has a small variance, while the second has a much larger variance. The normal distributions are now considered to be histograms of population data and variance means variance of the population.

As a measure of spread, it is convenient to use not the variance of the population, but rather its (positive) square root, called the **standard deviation of the population** and denoted by the symbol σ. Since the variance is represented by a squared symbol σ^2, it is natural that its square root just deletes the exponent. Now we can describe the crucial similarity among all normal distributions. We are interested again in the area under the normal curve between the mean μ and $\mu + x$ for some positive number x. Thinking of the normal distribution as the histogram of measurements from some population, it has a standard deviation σ. We can express x in terms of σ thus:

$$x = z\sigma$$

that is, x is some number z of standard deviations. For example, if the normal distribution has standard deviation $\sigma = 6$ and we wish to consider the portion of the distribution between μ and $\mu + 9$, then we have

$$9 = x = (1.5)(6) = 1.5\sigma$$

so $z = 1.5$. Thus, we are considering the part of the normal distribution between the mean and the mean plus 1.5 standard deviations. The important characteristic that all normal distributions have in common is this: The area under the curve between μ and $\mu + z\sigma$ depends only on z and not at all on μ and σ. For example, in any normal distribution, 43% of the area under the curve lies between μ and $\mu + 1.5\sigma$, as shown in Figure 7-13. To put it another way, if the histogram of measurements from some population is in the shape of a normal distribution, then 43% of all the measurements will be between the mean and 1.5 standard deviations greater than the mean.

We illustrate these ideas by supposing that the heights of the students at a university are normally distributed with mean 69 inches and standard deviation 3.5 inches. If we wish

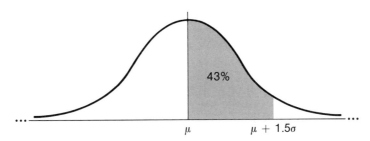

Figure 7-13

to know what percentage of the students are between 69 and 76 inches tall, we proceed as follows: To write 76 in the form $\mu + z\sigma$, we know that $\mu = 69$ and $\sigma = 3.5$, so we solve the equation

$$76 = 69 + z(3.5)$$

for z. Subtracting 69 from both sides and dividing the quotient by 3.5 produces

$$z = \frac{76 - 69}{3.5} = \frac{7}{3.5} = 2$$

So 76 is 2 standard deviations above the mean. (You may have observed this without going through the bother of solving the equation.) Now turning to Table 7-4, we look for $z = 2.0$. The figure for area just below $z = 2.0$ is .48; this means that 48% of all measurements in a normal distribution lie between the mean and 2.0 standard deviations above the mean. We conclude that 48% of all the students at that university are between 69 and 76 inches tall.

Table 7-4
Areas Under the Normal Curve

z	.1	.2	.3	.4	.5	.6	.7	.8	.9	1.0
Area	.04	.08	.12	.16	.19	.23	.26	.29	.32	.34
z	1.1	1.2	1.3	1.4	1.5	1.6	1.7	1.8	1.9	2.0
Area	.36	.38	.40	.42	.43	.45	.46	.46	.47	.48
z		2.1		2.2		2.3		2.4		2.5	2.6 or more
Area		.48		.49		.49		.49		.49	.49+

In general, if we wish to determine the percentage of all measurements in a normally distributed population that lie between the mean and some number y, we have to solve the equation

$$y = \mu + z\sigma$$

for z, where σ is the standard deviation of the population. Subtracting μ from both sides gives

$$z\sigma = y - \mu$$

Now dividing by σ solves the equation

$$z = \frac{y - \mu}{\sigma}$$

The area in Table 7-4 which corresponds to this value of z is the percentage we seek.

7-9   **Example**   The scores on a portion of the Graduate Record Examination are normally distributed with mean 500 and standard deviation 120. What percentage of the people taking the examination receive scores between 500 and 560?

Solution   We have been told that $\mu = 500$ and $\sigma = 120$. We must find out how many standard deviations $y = 560$ is above the mean, so we calculate

$$z = \frac{y - \mu}{\sigma} = \frac{560 - 500}{120} = \frac{60}{120} = .5$$

According to the table, $z = .5$ corresponds to area $= .19$, so 19% of all people taking the test receive a score between 500 and 560.

7-10   **Example**   Suppose the histogram of the operating lives of flashlight batteries of a certain type looks like a normal distribution, the mean is 5 hours, and the standard deviation is 1.4 hours. What percentage of all batteries of this type will last for at least 4 hours of operation?

Solution   We consider a normal distribution with $\mu = 5$ and $\sigma = 1.4$. We are trying to determine the percentage of measurements (battery lives) indicated by the shaded region in Figure 7-14. The procedure is to break the problem up into two parts. First recall that half of all measurements in a normal distribution are above the mean, so 50% of all the battery lives are greater than 5 hours. To determine the percentage of the batteries that last between 4 and 5 hours, we just go through the formula above with $y = 4$ and get

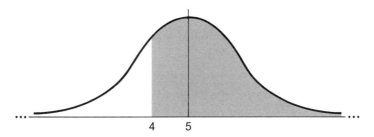

Figure 7-14

$$z = \frac{y - \mu}{\sigma} = \frac{4 - 5}{1.4} = -.7 \quad \text{(approximately)}$$

This answer makes sense because $y = 4$ is less than the mean and we have established that

$$4 = \mu + (-.7)\sigma = \mu - .7\sigma$$

Recall from our discussion of the symmetry of the normal distribution that the area under the curve between $\mu - .7\sigma$ and μ must be the same as that between μ and $\mu + .7\sigma$. Consequently, the area corresponding to $z = -.7$ is the same as that for $z = .7$, which, according to the table is .26. We conclude that 26% of the batteries last between 4 and 5 hours, while 50% last at least 5 hours and thus, all together, $26\% + 50\% = 76\%$ of the batteries last at least 4 hours.

7-11 **Example** Suppose the mean IQ is 100, the standard deviation of IQs is 12, and IQs are normally distributed. What percentage of the population has an IQ between 115 and 125?

Solution The problem is to find the area of the shaded region in Figure 7-15. The method above lets us determine the percentage of IQ scores between the mean 100 and 125. But that percentage is too large, because it includes people whose IQ is between 100 and 115.

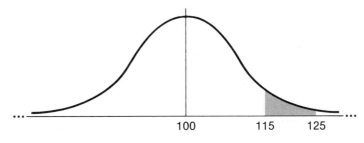

Figure 7-15

However, the same technique will determine the percentage of IQs between 100 and 115, so if we subtract that figure from the group of IQs between 100 and 125, we will get the number we want. To find out what percentage of IQs lie between 100 and 125, we recall that $\mu = 100$ and $\sigma = 12$, so

$$z = \frac{y - \mu}{\sigma} = \frac{125 - 100}{12} = \frac{25}{12} = 2.1 \quad \text{(approximately)}$$

and Table 7-4 informs us that 48% of all IQs are in this range. For the scores between 100 and 115, the corresponding equation is

$$z = \frac{y - \mu}{\sigma} = \frac{115 - 100}{12} = \frac{15}{12} = 1.3 \quad \text{(approximately)}$$

which, according to the same table, represents 40% of all IQs. Thus, 48% $-$ 40% = 8% of the population has an IQ between 115 and 125.

7-12   **Example**   The mean wind velocity throughout the year in San Francisco is 10.5 miles per hour. If the wind velocity in San Francisco behaves like a normal distribution with standard deviation 3 miles per hour, what percentage of the time is the wind blowing at less than 5 miles per hour?

Solution   This time, Figure 7-16 shows the area that concerns us. The trick is to calculate the percentage of the time the wind is blowing at a velocity between 5 miles per hour and 10.5 miles per hour. For this, the usual sort of calculation gives

$$z = \frac{y - \mu}{\sigma} = \frac{5 - 10.5}{3} = -\frac{5.5}{3} = -1.8 \quad \text{(approximately)}$$

So, from Table 7-4, 46% of the time the wind velocity is between 5 and 10.5 miles per hour. Now we know the wind is below mean velocity 50% of the time, because we assume a normal distribution. Since the velocity is between 5 and 10.5 miles per hour 46% of the time, during the other 4% the wind dies down below 5 miles per hour.

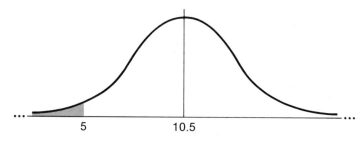

5   10.5

Figure 7-16

Summary

The **normal distribution** is a curve with the shape shown in Figure 7-17. This may be thought of as representing the histogram of measurements from some population, where μ is the mean of the population. The area under the curve from μ to $\mu + x$, for any positive number x, is the same as that between $\mu - x$ and μ.

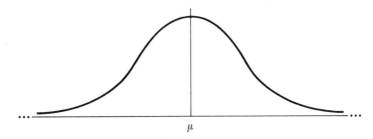

μ

Figure 7-17

The **variance of the population** σ^2 is defined by first subtracting the mean μ of the population from each measurement, next squaring the result, then adding up the squares, and, finally, dividing the sum by the number N of members of the population. The **standard deviation of the population,** written σ, is the square root of the variance of the population. The percentage of all measurements in a normally distributed population between μ and $\mu + z\sigma$ (z positive or negative) depends only on z. To find the percentage of measurements between the mean and some number y, calculate

$$z = \frac{y - \mu}{\sigma}$$

Then the value of area corresponding to z (or the same-sized positive number if z is negative) in Table 7-4 (p. 461) is the required percentage.

Exercises

7-24 The durations of long-distance telephone calls between two cities are normally distributed with mean 6 minutes and standard deviation 1.5 minutes. What percentage of all calls are between 5 and 6 minutes long?

7-25 The distribution of salaries in a certain occupation is normal with mean $20,000 a year and standard deviation $4000. What percentage of people with this occupation earn at least $14,000 a year?

7-26   Suppose the life expectancy of American women can be described by a normal distribution with mean 75 years and standard deviation 7 years. What percentage of American women live between 70 and 80 years?

7-27   The mean temperature in Fairbanks, Alaska, in January is $-11°F$. Suppose the temperature in Fairbanks in January is normally distributed with standard deviation $14°F$. During what percentage of the month of January is the temperature of Fairbanks below $-20°F$?

7-28   Suppose the heights of 10 year old American boys form a normal distribution with mean 54 inches and standard deviation 3 inches. What percentage of 10 year old American boys are between 56 and 60 inches tall?

7-29   If the life of a television picture tube is normally distributed with mean 7 years and standard deviation .9 year and the manufacturer guarantees to replace the tube if it burns out within the first 5 years, what percentage of picture tubes must be replaced by the manufacturer under the guarantee?

7-30   A metal rod will pass inspection if its diameter is within .002 inch of the desired diameter of .110 inch. A lathe produces rods in such a way that the diameters form a normal distribution with mean .109 inch and standard deviation .001 inch. What percentage of the rods produced by this lathe will pass inspection?

7-31   The mean discharge of water from the mouth of the Missouri River is 70,000 cubic feet per second. Suppose the rate of discharge is normally distributed with standard deviation 7000 cubic feet per second. What percentage of the time does the Missouri River discharge at the rate of at least 65,000 cubic feet per second?

7-32   Suppose the mean age of American women college graduates at the time of their graduation is 23 years and 3 months and that the ages at graduation are normally distributed with standard deviation 15 months. What percentage of American women college graduates are between 21 and 22 years of age when they graduate from college?

7-33   A box of dry breakfast cereal holds 272 cubic inches. The machine that fills the boxes discharges amounts of breakfast cereal that form a normal distribution with mean 270 and standard deviation .8 cubic inch. What percentage of the time does the machine put so much breakfast cereal in the box that it overflows?

7-5   A Central Limit Theorem

Let us imagine there are nine people in a room and we are interested in estimating how many of them have brown eyes, using a sample of size three. The population size of nine is so small that we can actually list all the things that could happen. To turn the problem into a numerical one, we use 0 to denote that a person's eyes are not brown and a 1 to denote that they are brown. We assign numbers to the individuals in the room and we suppose for illustration that four of them have brown eyes, as listed in Table 7-5.

					Table 7-5				
Individual	1	2	3	4	5	6	7	8	9
Eye color	0	0	0	1	0	1	0	1	1

The reader who has studied combinations in Section 2-3 will remember that there are $C_3^9 = 84$ different samples of size $n = 3$ out of a population of size $N = 9$. The mean of a sample is, in this case, the proportion p of brown-eyed people in it, that is, the number of people with brown eyes divided by 3. All 84 samples and the corresponding proportions are listed in Table 7-6. We thus have 84 measurements, the sample proportions, for which we construct a histogram as indicated in Table 7-7 and Figure 7-18.

				Table 7-6					
Individual	1	2	3	4	5	6	7	8	9
Eye color	0	0	0	1	0	1	0	1	1

Sample	p	Sample	p	Sample	p	Sample	p
1,2,3	0	1,5,9	1/3	2,5,9	1/3	3,8,9	2/3
1,2,4	1/3	1,6,7	1/3	2,6,7	1/3	4,5,6	2/3
1,2,5	0	1,6,8	2/3	2,6,8	2/3	4,5,7	1/3
1,2,6	1/3	1,6,9	2/3	2,6,9	2/3	4,5,8	2/3
1,2,7	0	1,7,8	1/3	2,7,8	1/3	4,5,9	2/3
1,2,8	1/3	1,7,9	1/3	2,7,9	1/3	4,6,7	2/3
1,2,9	1/3	1,8,9	2/3	2,8,9	2/3	4,6,8	1
1,3,4	1/3	2,3,4	1/3	3,4,5	1/3	4,6,9	1
1,3,5	0	2,3,5	0	3,4,6	2/3	4,7,8	2/3
1,3,6	1/3	2,3,6	1/3	3,4,7	1/3	4,7,9	2/3
1,3,7	0	2,3,7	0	3,4,8	2/3	4,8,9	1
1,3,8	1/3	2,3,8	1/3	3,4,9	2/3	5,6,7	1/3
1,3,9	1/3	2,3,9	1/3	3,5,6	1/3	5,6,8	2/3
1,4,5	1/3	2,4,5	1/3	3,5,7	0	5,6,9	2/3
1,4,6	2/3	2,4,6	2/3	3,5,8	1/3	5,7,8	1/3
1,4,7	1/3	2,4,7	1/3	3,5,9	1/3	5,7,9	1/3
1,4,8	2/3	2,4,8	2/3	3,6,7	1/3	5,8,9	2/3
1,4,9	2/3	2,4,9	2/3	3,6,8	2/3	6,7,8	2/3
1,5,6	1/3	2,5,6	1/3	3,6,9	2/3	6,7,9	2/3
1,5,7	0	2,5,7	0	3,7,8	1/3	6,8,9	1
1,5,8	1/3	2,5,8	1/3	3,7,9	1/3	7,8,9	2/3

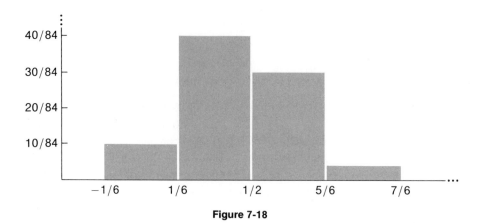

Figure 7-18

	Table 7-7	
Value of p	Frequency	Relative frequency
0	10	10/84
1/3	40	40/84
2/3	30	30/84
1	4	4/84

As a second example, consider a population of eight high school basketball players whose heights in inches are given in Table 7-8. If we try to estimate the mean height of the eight players by taking a sample of four players, we will get one of the 70 possible samples which are listed, along with the resulting sample means, in Table 7-9. We construct a histogram of the 70 sample means as indicated in Table 7-10 and Figure 7-19, replacing the relative frequencies by decimal approximations because it makes it easier to construct the picture.

For a final example, we have a class of fifteen students; seven of these students are left-handed. We wish to take a sample of five students in order to estimate the proportion of left-handed students in the class. We let 1 represent left-handed and 0 represent right-handed students. The possible values of p, the proportion of left-handed students in a

	Table 7-8							
Player	1	2	3	4	5	6	7	8
Height	72	74	72	73	77	74	71	70

	Table 7-9							
Player	1	2	3	4	5	6	7	8
Height	72	74	72	73	77	74	71	70

Sample	Mean	Sample	Mean	Sample	Mean
1,2,3,4	72.75	1,3,7,8	71.25	2,4,5,8	73.5
1,2,3,5	73.75	1,4,5,6	74	2,4,6,7	73
1,2,3,6	73	1,4,5,7	73.25	2,4,6,8	72.75
1,2,3,7	72.25	1,4,5,8	73	2,4,7,8	72
1,2,3,8	72	1,4,6,7	72.5	2,5,6,7	74
1,2,4,5	74	1,4,6,8	72.25	2,5,6,8	73.75
1,2,4,6	73.25	1,4,7,8	71.5	2,5,7,8	73
1,2,4,7	72.5	1,5,6,7	73.5	2,6,7,8	72.25
1,2,4,8	72.25	1,5,6,8	73.25	3,4,5,6	74
1,2,5,6	74.25	1,5,7,8	72.5	3,4,5,7	73.25
1,2,5,7	73.5	1,6,7,8	71.75	3,4,5,8	73
1,2,5,8	73.25	2,3,4,5	74	3,4,6,7	72.5
1,2,6,7	72.75	2,3,4,6	73.25	3,4,6,8	72.25
1,2,6,8	72.5	2,3,4,7	72.5	3,4,7,8	71.5
1,2,7,8	71.75	2,3,4,8	72.25	3,5,6,7	73.5
1,3,4,5	73.5	2,3,5,6	74.25	3,5,6,8	73.25
1,3,4,6	72.75	2,3,5,7	73.5	3,5,7,8	72.5
1,3,4,7	72	2,3,5,8	73.25	3,6,7,8	71.75
1,3,4,8	71.75	2,3,6,7	72.75	4,5,6,7	73.75
1,3,5,6	73.75	2,3,6,8	72.5	4,5,6,8	73.5
1,3,5,7	73	2,3,7,8	71.75	4,5,7,8	72.25
1,3,5,8	72.75	2,4,5,6	74.5	4,6,7,8	72
1,3,6,7	72.25	2,4,5,7	73.75	5,6,7,8	73
1,3,6,8	72				

sample of size five, are in decimal form, 0, .2, .4, .6, .8, and 1. Since there are 3003 different samples of five individuals out of fifteen, it is not practical to list them all as we did in the previous examples. We ask you to believe, or verify by the methods of Section 2-3, that Table 7-11 is correct. The histogram of sample means (proportions) looks like Figure 7-20.

The purpose of all these examples is an observation that we very much hope has by now occurred to you: The histogram of sample means (or proportions) is shaped rather like

Table 7-10		
Class	Frequency	Relative frequency (approximate)
71.1–71.6	3	.04
71.6–72.1	10	.14
72.1–72.6	16	.23
72.6–73.1	13	.19
73.1–73.6	15	.21
73.6–74.1	10	.14
74.1–74.6	3	.04

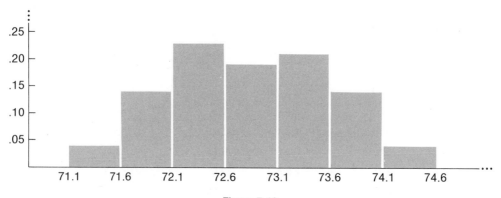

Figure 7-19

Table 7-11		
p	Frequency	Relative frequency (approximate)
0	56	.02
.2	490	.16
.4	1176	.39
.6	980	.33
.8	280	.09
1	21	.01

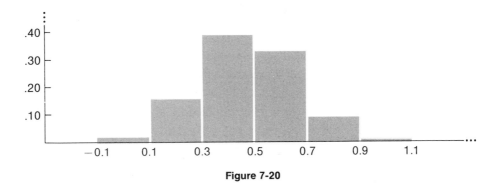

Figure 7-20

a normal distribution. The resemblance is not just an observation from experience but rather a mathematically verifiable fact, called the **central limit theorem.**

The central limit theorem states that when certain conditions are met, the histogram of means of all samples of size n from a population is approximately a normal distribution. We will not attempt to describe the general conditions. We will instead state conditions below which assure that the theorem is valid for the kind of opinion sampling situations we are interested in. We will also leave imprecise the meaning of an approximation to a normal distribution, because the mathematical concepts we need to go into greater detail are beyond the scope of this book.

The central limit theorem also tells us what the mean and the standard deviation of the normal distribution are. To get an idea of what the mean should be, let us calculate the mean of the data in Table 7-6, that is, the mean of the proportions of brown-eyed people in all samples of size three. Out of the 84 different samples, $p = 0$ in 10 of them, $p = 1/3$ in 40 samples, $p = 2/3$ in 30 samples, and $p = 1$ in the other 4 samples. Therefore,

$$\text{Mean} = \frac{(10)(0) + (40)\left(\frac{1}{3}\right) + (30)\left(\frac{2}{3}\right) + (4)(1)}{84} = \frac{112}{252} = \frac{4}{9}$$

But the total population consisted of nine people of whom four were brown-eyed, so in this case, the mean of all the sample proportions is precisely the number we were trying to estimate by taking a sample.

This example is not an isolated instance, but rather it illustrates the general fact: The mean μ of the population from which the samples were drawn is also the mean of the normal distribution which approximates the histogram of sample means. Thus, according to the central limit theorem, the means of the various samples tend to cluster about the population mean—the number we estimate by the sample mean. This explains why statistics generally "works." The estimate of the population mean based on the evidence from a sample is usually pretty close to the mean we would get if we investigated the entire population.

As we mentioned above, the central limit theorem includes a formula for the standard deviation of the normal distribution that approximates the histogram of sample means. We will not discuss that standard deviation, called the **standard deviation of the mean,** in general here. Rather, we will restrict ourselves to a form of the theorem that is sufficient for our opinion sampling questions.

When the data consists of 0s and 1s, the population mean is called the **population propor-tion** and is represented by the letter P. We will assume that P is between .20 and .80, thus excluding very one-sided elections. Further, we require that the sample size n is at least 25. Under these circumstances, the central limit theorem is valid: For all samples of size n from the population, the histogram of sample proportions is approximately a normal distribution with mean P and standard deviation

$$\sqrt{\frac{PQ}{n}} \qquad \text{where } Q = 1 - P$$

In Section 7-6, we will show how this form of the central limit theorem can be used to evaluate the worth (precision) of an estimate of the population proportion based on a sample.

Summary

The **central limit theorem** states that when certain conditions are met, the histogram of the means of all samples of size n from a population is approximately a normal distribution. The mean μ of the population from which the samples are drawn is also the mean of this normal distribution.

If the data consists of 0s and 1s, the population mean is called the **population proportion** and is represented by the letter P. The central limit theorem for such data states that if P is between .20 and .80 and if sample size n is at least 25, then for all samples of size n from the population, the histogram of sample proportions is approximately a normal distribution with mean P and standard deviation $\sqrt{PQ/n}$, where $Q = 1 - P$.

7-6 Applying the Central Limit Theorem

Imagine that all possible samples of a fairly large size n are taken from a population of voters in an election that is not extremely one-sided. According to the central limit theorem, the proportion p of the sample voting for a particular candidate, varies in a reasonably simple, predictable manner. To illustrate how we can obtain useful information from this theorem, we consider the 1952 presidential election. Since Eisenhower actually received 55.7% of the popular vote, we will assume that, at some time shortly before the election, 56% of the voters in the United States intended to vote for Eisenhower. In the symbols of the previous section, we are assuming that $P = .56$. If a sample of 1000 voters were chosen at random from among all voters in the United States, what are the chances that the sample would have correctly predicted an overwhelming victory for Eisenhower? Specifically, how likely is it that the proportion p of Eisenhower voters in the random sample would be within .03 of the correct proportion P, that is, that the sample will contain between 53% and 59% Eisenhower voters?

The question is a probability question in the sense of Chapter 2, though we do not really need any information from Chapter 2 in order to discuss it. We are asking: If we had taken a random sample of 1000 voters in 1952, what is the probability that the proportion p of Eisenhower voters in the sample would be between .53 and .59? Think of all possible samples of 1000 voters from the population of American voters in 1952, just as we thought of all samples of four basketball players from a group of eight in the previous section. Each sample contains a proportion p of Eisenhower voters that varies from sample to sample. Since each sample is as likely to be chosen as any other, if we know what percentage of all samples have p between .53 and .59, then that percentage represents the probability that a sample chosen at random will have this property.

The central limit theorem tells us that the histogram of all sample proportions for samples of size 1000 is approximately a normal distribution with mean P, here equal to .56, and standard deviation $\sqrt{PQ/n}$. This standard deviation is what we called the "standard deviation of the mean" in the last section. We will denote it by the symbol S_p. For American voters in 1952 with respect to their support of Eisenhower,

$$S_p = \sqrt{\frac{PQ}{n}} = \sqrt{\frac{(.56)(.44)}{1000}} = \sqrt{.0002464} = .016 \quad \text{(approximately)}$$

Therefore, as in Section 7-4, we want to calculate the area of the shaded region in Figure 7-21 for mean .56 and standard deviation .016. To find this area, we first determine the area from .56 to .59. That is, we compute

$$z = \frac{.59 - .56}{.016} = \frac{.03}{.016} = 1.9 \quad \text{(approximately)}$$

The corresponding area, according to Table 7-4 (p. 461), is .47. The symmetry of the normal distribution tells us that the area between .53 and .56 is also .47. We conclude that $47\% + 47\% = 94\%$ of all samples of size 1000 from the American electorate in 1952 would have predicted an overwhelming victory for Eisenhower; between 53% and 59% of the popular vote.

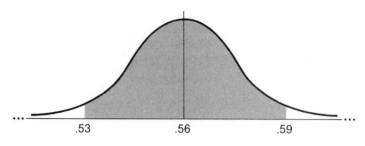

.53   .56   .59

Figure 7-21

7-13   **Example**   In the 1972 race for United States Senator from Nebraska, the incumbent Carl Curtis received 52.5% of the vote. Suppose that shortly before the election, 52% of Nebraska's voters preferred Senator Curtis. What is the probability that a sample of 400 Nebraska voters chosen at random would have contained a proportion of Curtis supporters within 2% of the correct figure of 52%

Solution   The central limit theorem tells us that the picture is approximately that shown in Figure 7-22. The standard deviation is

$$S_p = \sqrt{\frac{PQ}{n}} = \sqrt{\frac{(.52)(.48)}{400}} = \sqrt{.000624} = .025 \qquad \text{(approximately)}$$

To find the area from .52 to .54, we calculate

$$z = \frac{.54 - .52}{.025} = \frac{.02}{.025} = .8$$

which corresponds to .29 in Table 7-4. By the symmetry of the normal distribution, the area from .50 to .52 is also .29, so the probability is .29 + .29 = .58 that a sample of 400 voters would have been within 2% of the correct answer.

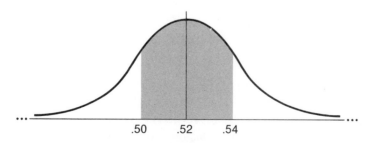

.50   .52   .54

Figure 7-22

Let us now put ourselves in the position of a pollster in order to see what bearing the central limit theorem has on the pollster's work. We will use the historical example we described in the essay on errors in sampling (p. 454), the OPOR sample of 200 New York State voters in 1942, of whom 116 planned to vote for Dewey for governor. All the information we have is the sample size $n = 200$ and the proportion $p = 116/200 = .58$ of voters for Dewey in the sample. The central limit theorem informs us that if we were to construct the histogram of proportions of voters for Dewey out of all samples of size $n = 200$ from the voters of New York in 1942, we would obtain approximately a normal distribution with mean P, the proportion of New York State voters who preferred Dewey, and standard deviation $S_p = \sqrt{PQ/200}$. An apparent difficulty in using the theorem is that it seems to require us to know the value of P. But the purpose of taking the sample was to get some information on how large P is!

The way out of this difficulty is to observe that, in calculating S_p, we are using not just P but rather the product PQ, where $Q = 1 - P$. The important property of PQ is demonstrated by the following table:

P	.1	.2	.3	.4	.5	.6	.7	.8	.9
PQ	.09	.16	.21	.24	.25	.24	.21	.16	.09

As P varies between, say, .3 and .7, there is very little change in the product PQ. Thus, even a very rough guess as to the value of P will produce about the right value for PQ, so long as P is neither very large nor very small. In the formula for S_p, the product PQ is divided by the rather large sample size, so errors in estimating P have little influence indeed on the value of S_p. If P is very large or very small, this reasoning does not apply, but then the election is so lopsided there is little need to take the sample anyway. Therefore, we replace the population proportion P we do not know by the sample proportion p from the sample we do know (in our example, the OPOR sample), to obtain a formula

$$\sqrt{\frac{pq}{n}} \qquad \text{where } q = 1 - p$$

which estimates S_p. If the election is at all close, the estimate will be very near the correct value of the standard deviation of the mean S_p, even if p is quite far from the correct proportion P.

In the historical example we have been using, $p = .58$ and $n = 200$, so

$$\sqrt{\frac{pq}{n}} = \sqrt{\frac{(.58)(.42)}{200}} = \sqrt{.001218} = .035 \qquad \text{(approximately)}$$

Even though we do not know what the value of P is, we claim the central limit theorem tells us that 96% of all samples of size 200 of New York voters in 1942 would have contained a proportion p of Dewey supporters within .07 of the population proportion P, whatever that may have been. We support this claim as follows: The histogram of sample proportions for samples of size 200 from that population has mean P and a standard deviation we estimated to be .035. Thus, the picture is shown in Figure 7-23. Now, .07 is 2 standard deviations, so looking up $z = 2.0$ in Table 7-4 gives us area $= .48$, which means the percentage of samples with proportion p between P and $P + .07$ is 48%. By the symmetry of the normal distribution, there is another 48% of the samples with p between $P - .07$ and P, so altogether 96% of the samples have proportions within .07 of P, as we claimed.

If the sample proportion p (here, .58) is within .07 of P, then, equivalently, P is within .07 of the proportion p from the OPOR sample. Thus, if the OPOR sample were one of those 96%, then the actual proportion of voters for Dewey in the entire population would be within .07 of the figure from that sample. There is, therefore, a 96% chance that P is between $.58 - .07 = .51$ and $.58 + .07 = .65$. There is no way of determining, however, whether the OPOR sample was one of the 96% which came within .07 of the correct proportion or whether it was one of the unfortunate 4% which missed the population proportion P by more than .07. In the latter case, the figure of $p = .58$ might have been off from P by, say, .10, and in fact Dewey might be headed for defeat (assuming a two person race) with merely 48% of the vote.

Since the OPOR sample was chosen at random, it is as likely to be any one of all the possible samples of 200 New York State voters as any other. We know that 96% of these samples contain a proportion of Dewey voters within .07 of the proportion in the total population, so the OPOR sample had a chance of 96 out of 100 of being a sample of this type. Thus, we cannot state what the precision of a sample is, only what it probably is. In the case of the OPOR sample, we say there is a "96% probability" that the sample is precise to within .07, or there is a 96% probability that the precision of the OPOR sample is .07.

To define precision in general, we suppose we are given a certain probability A. Let z correspond to the area equal to half of A in Table 7-4. The **precision** d of a sample, with probability A, is defined by the equation $d = zS_p$, where S_p is the standard deviation of the mean.

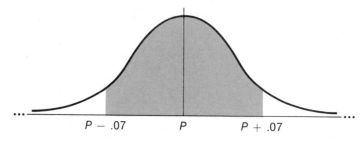

Figure 7-23

7-14 **Example** Suppose a sample of 500 voters in Texas in 1972 contained a proportion $p = .55$ who favored John Tower to retain his United States Senate seat. Use this information to estimate the precision which could be attained, with probability .94, by samples of size 500 with respect to this election.

Solution We estimate S_p by

$$\sqrt{\frac{pq}{n}} = \sqrt{\frac{(.55)(.45)}{500}} = \sqrt{.000495} = .022 \quad \text{(approximately)}$$

According to Table 7-4, the value of z corresponding to an area of $.94/2 = .47$ is $z = 1.9$. Therefore, with 94% probability,

$$\text{Precision} = d = zS_p = (1.9)(.022) = .04 \quad \text{(approximately)}$$

Now we return to the 1942 New York gubernatorial election to answer the question, "How much faith can we have in the prediction based on the OPOR sample?"

Since the percentage of Dewey voters in the sample was 58%, we observe that Dewey would have the support of a majority of New York voters if this estimate, $p = .58$, were within .08 of the actual proportion P of Dewey voters. Thus, our question is: What is the probability that p based on a sample of 200 voters is within .08 of P? In other words, if the precision d is .08, what is the probability A that such precision can be obtained? Using the information $n = 200$ and $p = .58$, we calculated an estimate of .035 above for S_p. We substitute into the equation

$$d = zS_p$$

and get

$$.08 = z(.035)$$

We solve the equation for z:

$$z = \frac{.08}{.035} = 2.3 \quad \text{(approximately)}$$

and recall that z corresponds to half of A in Table 7-4. Turning to the table, we find that $z = 2.3$ corresponds to .49. Therefore, the probability is $2(.49) = .98$ that the proportion in the OPOR sample is within .08 of the correct proportion of supporters of Dewey. It follows that the probability is at least .98 that Dewey would receive a majority of votes in the election.

In fact, Dewey did win the election—with 53% of the vote. A Gallup poll of 2800 voters conducted at the same time as the OPOR sample predicted the outcome exactly, while a more informal poll run by the *New York Daily News* predicted on the basis of 48,000 responses that Dewey would receive 57% of the votes. Thus, the OPOR sample of only 200 voters chosen in a scientific manner came almost as close to the correct outcome as did the huge *New York Daily News* sample.

Any time we wish to calculate the probability A of attaining a given precision d with samples of a certain size, we copy the procedure we just followed. That is, we estimate S_p and substitute it into the equation

$$z = \frac{d}{S_p}$$

Then the probability A is twice as large as the area in Table 7-4 corresponding to this value of z.

7-15 **Example**   A sample of 150 voters in 1944 was asked who they would vote for in the presidential election. Eighty named Roosevelt. What is the probability that the sample proportion is within .05 of the proportion of Roosevelt voters in the American population in 1944?

Solution   Since $p = 80/150 = .53$, we estimate S_p to be

$$\sqrt{\frac{pq}{n}} = \sqrt{\frac{(.53)(.47)}{150}} = \sqrt{.0017} = .04 \qquad \text{(approximately)}$$

We calculate

$$z = \frac{d}{S_p} = \frac{.05}{.041} = 1.2 \qquad \text{(approximately)}$$

In Table 7-4, $z = 1.2$ corresponds to area $= .38$, which implies there is a probability of $A = 2(.38) = .76$ that the proportion in the sample is within .05 of the proportion of Roosevelt voters in the population.

Summary

Given a probability A, let z correspond to the area in Table 7-4 (p. 461) equal to half of A. The **precision** d of samples of size n, with probability A, is defined by the equation $d = zS_p$, where S_p is the standard deviation of the mean for samples of that size.

To calculate the probability A of attaining a certain precision d, compute

$$z = \frac{d}{S_p}$$

and find the area in Table 7-4 corresponding to that value of z. Then the probability A is twice that area.

Suppose a sample of voters of size n is taken from some population and a proportion p of them are in favor of a candidate. Then S_p is estimated by the formula $\sqrt{pq/n}$, where $q = 1 - p$.

Exercises

● *For Exercises 7-34 through 7-37, we use the fact that Lyndon Johnson received 61% of the popular vote in the 1964 presidential election to justify letting $P = .61$.*

7-34 What percentage of the samples of size 200 of voters in 1964 would have p, the proportion for Johnson, between .55 and .61?

7-35 In what percentage of 1964 voter samples of size 100 would p have been between .53 and .69?

7-36 What percentage of 1964 voter samples of size 30 would contain a proportion of Johnson supporters within .11 of P?

7-37 What percentage of samples of 1000 voters in 1964 would contain a proportion of Johnson supporters within .01 of P?

7-38 Suppose a sample of 500 voters in 1940 contained 255 who favored Roosevelt for president. Estimate the standard deviation of the mean for all American voters in 1940 from this information. Determine the percentage of all such samples of 500 voters which will produce a proportion p favoring Roosevelt within 1% of the actual proportion of all voters favoring Roosevelt.

7-39 Suppose 60 out of a sample of 100 voters in 1956 favored Eisenhower for president. Estimate S_p and the percentage of samples of size 100 which contain a proportion p for Eisenhower within 10% of P.

● *For Exercises 7-40 through 7-44, we use the fact that in the 1956 presidential election, Eisenhower received 55% of the popular vote to justify letting $P = .55$.*

7-40 For samples of size 200 of 1956 presidential voters, what was the probability that the sample proportion p was within .035 of P?

7-41   For samples of size 2000 of voters in the 1956 presidential election, what was the probability that the sample proportion p was within .035 of P?

7-42   For samples of 250 voters in the 1956 presidential election, what precision could have been obtained with probability .90?

7-43   For samples of size 1500 of voters in the 1956 presidential election, what precision could have been obtained with probability .90?

7-44   For samples of size 1500 of voters in the 1956 presidential election, what precision could have been obtained with probability .80?

7-45   In the 1970 congressional races, Gallup predicted that 53% of American voters would vote for Democratic candidates. Assuming Gallup used a random sample of 1500 voters, what was the probability that this figure is within .03 of the proportion P of all voters who would vote for a Democratic congressional candidate in that election?

7-46   In the 1950 congressional races, Gallup predicted that 51% of American voters would vote for Democratic candidates. Assuming Gallup used a random sample of 3000 voters, what was the probability that this figure is within .01 of the proportion P of all voters who would vote for a Democratic congressional candidate in that election?

7-7   Sample Size

Now let us put ourselves in the position of a pollster and ask a new question. How large a sample should be taken in order to get results precise enough so that the public will come to have faith in the pollster's predictions?

Recall from Section 7-6 that precision d with probability A is defined by $d = zS_p$, where S_p is the standard deviation of the mean for samples of a certain size from the population and z depends on A. To determine sample size, the pollster must decide both the precision required and how often the error can be permitted to exceed that amount. That is, the pollster must specify both d and A.

To explain the procedure for determining sample size, we will use the following example: Let us assume the pollster decides to accept a proportion p for the candidate (from his sample) which misses the correct proportion P (from the entire population of voters) by more than .02 no more than 6% of the time. In other words, we suppose the pollster wishes to obtain precision equal to .02 with probability .94. By definition, precision is zS_p, where z corresponds to half of A. Since $A = .94$ in this example and $.94/2 = .47$, then we find in Table 7-4 (p. 461) that $z = 1.9$. Thus, the pollster requires that

$$\text{Precision} = d = .02 = (1.9)S_p$$

We do not know the value of S_p, but assuming the pollster is taking a random sample and the total population size N is large, we may use the estimate $\sqrt{pq/n}$. The pollster does not know what p is, because the sample has not yet been taken; the pollster is trying to decide how large a sample to take. Since only close elections are difficult to predict, the pollster had best assume the election will be close, i.e., that p will be near .50. But in that case, pq will be .25 or a bit less. The larger the value of pq, the larger the sample size will be, so there is no great harm in assuming that $pq = .25$. Now, the pollster's problem is to find the sample size n such that $1.9S_p = .02$. This is estimated by solving

$$1.9\sqrt{\frac{.25}{n}} = .02$$

for n. Squaring both sides of the equation gives

$$(1.9)^2\left(\frac{.25}{n}\right) = (.02)^2$$

Dividing both sides by $(1.9)^2(.25)$, we have

$$\frac{1}{n} = \frac{(.02)^2}{(1.9)^2(.25)}$$

and taking the reciprocal of both sides, we compute,

$$n = \frac{(1.9)^2(.25)}{(.02)^2} = 2257$$

(Of course, we round decimal answers up to the next largest whole voter.)

We turn now to the general question of sample size. Suppose d stands for the precision desired by the pollster and the pollster requires probability A that this precision will be obtained. Letting z be the number in Table 7-4 corresponding to $A/2$, the pollster requires

$$zS_p = d.$$

If we estimate S_p by $\sqrt{pq/n}$ and take $pq = .25$ (the largest value possible) as before, then the equation becomes

$$z\sqrt{\frac{.25}{n}} = d$$

We square both sides of the equation,

$$z^2\left(\frac{.25}{n}\right) = d^2$$

and divide through by $(z^2)(.25)$ to get

$$\frac{1}{n} = \frac{d^2}{(z^2)(.25)}$$

The reciprocal is

$$n = \frac{(z^2)(.25)}{d^2}$$

and since $.25 = 1/4$, the equation for sample size is

$$n = \frac{z^2}{4d^2}$$

where z corresponds to half the probability A, and d is the required precision.

7-16   **Example**   Make a table of sample sizes for precisions .04, .03, .02, .01, and .005; each is to be attained with 94% probability and 97% probability.

Solution   For probability 94%, we compute

$$n = \frac{(1.9)^2}{4d^2}$$

for each value of d, because, as we saw above, $z = 1.9$ corresponds to area $= .47 = .94/2$ in Table 7-4. Thus,

$$n = \frac{(1.9)^2}{4(.04)^2} = 565$$

$$n = \frac{(1.9)^2}{4(.03)^2} = 1003$$

and so on. In the case of 97% probability, $.97/2 = .485$ so, according to the table, z must lie between 2.1 and 2.2. The larger value gives slightly larger sample sizes than we need to attain the desired precision, and the pollster will then err on the side of caution. So the formula is

$$n = \frac{(2.2)^2}{4d^2}$$

which gives, for instance,

$$n = \frac{(2.2)^2}{4(.04)^2} = 757$$

Table 7-12 is the required table.

Table 7-12					
d	.04	.03	.02	.01	.005
Probability 94%	565	1,003	2,257	9,025	36,100
Probability 97%	757	1,345	3,025	12,100	48,400

Notice that the size of the sample required to obtain a given level of precision does not appear to depend on the size of the total population. A random sample of 2257 will provide precision to within .02 in 94% of all elections no matter whether the total population consists of the voters in a city of a few hundred thousands or of the millions of voters throughout the United States. We have come to this conclusion because the formula $S_p = \sqrt{PQ/n}$ ignores the total population size N. So N does not appear in the formula for the sample size n. Our procedure is justified provided the sample size we end up with is small compared to the total population size. With a small sample, $\sqrt{pq/n}$ really is a reasonable estimate of the standard deviation of the mean. Otherwise, the determination of sample size we have given will be incorrect (e.g, you cannot take a sample of $n = 2257$ individuals from a population of $N = 2000$). For example, if we wish to obtain precision of .02 in 94% of elections for mayor of a city of 10,000, the formula above which requires a sample size of 2257 should not be used. A more complicated formula for sample size, given in the exercises, will show that in this case, a sample of 1842 will be sufficient.

Summary

To attain precision to within d with probability A, a sample of size

$$n = \frac{z^2}{4d^2}$$

is required, where z corresponds to $A/2$ in Table 7-4 (p. 461). Answers should be rounded up to the next highest whole number.

Exercises

- In Exercises 7-47 through 7-49, make a table of sample sizes for precisions .04, .03, .02, .01, and .005; each is to be attained with the indicated probability A.

7-47 $A = .90$

7-48 $A = .99$ (use $z = 2.6$)

7-49 $A = .96$ (use $z = 2.1$)

- In Exercises 7-50 through 7-54, calculate the sample size required to attain a precision of .05 with the indicated probability A.

7-50 $A = .80$

7-51 $A = .90$

7-52 $A = .94$

7-53 $A = .96$ (use $z = 2.1$)

7-54 $A = .99$ (use $z = 2.6$)

- In Exercises 7-55 through 7-59, calculate the sample size required to attain a precision of .015 with the indicated probability A.

7-55 $A = .80$

7-56 $A = .90$

7-57 $A = .94$

7-58 $A = .96$ (use $z = 2.1$)

7-59 $A = .99$ (use $z = 2.6$)

- If the formula for sample size n gives a value n greater than 1/20 of the population size N, compute

$$\frac{n}{1 + \dfrac{n}{N}}$$

and use this new, smaller, number for the sample size. In Exercises 7-60 through 7-65, compute the sample size for the indicated precision, probability, and population size.

7-60 $d = .02; A = .94; N = 10,000$

7-61 $d = .01; A = .94; N = 500,000$

7-62 $d = .01; A = .96; N = 5000$ (use $z = 2.1$)

7-63 $d = .02; A = .96; N = 20{,}000$ (use $z = 2.1$)

7-64 $d = .01; A = .90; N = 1000$

7-65 $d = .02; A = .90; N = 35{,}000$

Essay on Stratified Sampling

Voters generally vote for the candidate who seems to understand their needs, agrees with their opinions, and will support their interests. People who have a great deal in common tend to have similar needs, opinions, and interests. Therefore, it is hardly surprising that experience extending over many elections supports the principle that voters who share important characteristics tend to vote for the same candidate. Pollsters have exploited this principle in order to improve substantially the precision of their predictions.

One sampling technique which takes advantage of the tendency of homogeneous populations to vote as a block is called "stratified sampling." Instead of just taking a random sample of the entire population of voters, the population is first divided up into smaller groups and then a random sample is taken separately from each group. Each of the smaller groups, called "strata," is selected in such a way that the members of the group have common characteristics of the sort that will tend to make them vote alike. Once it has been decided how large the total sample will be, a sample is taken from each stratum so the total number of individuals sampled in all strata adds up to the desired total sample size. Generally, the size of the stratum sample is proportional to the fraction of the total population which belongs to that stratum.

For example, suppose a pollster wishes to predict the outcome of an election for mayor of a city using a sample of 150 voters. The pollster could, of course, take a random sample of 150. However, the pollster observes that about 2/3 of the voters live in inner city precincts where the population is relatively poor and of minority ethnic background. Living in outlying precincts are the remaining 1/3 of the population who tend to be well-educated, middle-class, and white, like the population of the suburban communities just across the city line. If the pollster employs stratified sampling, then 2/3 of the sample of 150 is chosen at random from the voter registration lists of the inner city precincts, since the inner city represents 2/3 of the total population of the city. The remaining 1/3 is selected at random from the outlying precincts.

Voters from similar population areas have a decided tendency to vote alike. For example, voters from very large cities tend to agree on national candidates, even though the cities are widely separated from one another geographically. This big city vote, referring to voters who live in cities with over 1 million inhabitants is one of the seven strata used by the Gallup organization. Similarly, Gallup treats the voters who live in open countryside as a single group. (Gallup's stratification was described in detail on p. 434.) Thus, when Gallup selects his sample of 1800 voters, some are chosen from among the populations of large cities, while a separate choice is made from among rural voters. Since many more people live in large cities than in open countryside, the size of the sample from the big city vote is much larger than that from the open countryside. Strictly speaking, Gallup's sample is not the kind of stratified sample we have been discussing, because he does not use the kind of random samples within each stratum that we have been considering, but the principle of stratification is the same.

The Vote Profile Analysis sample of 40 precincts in Missouri mentioned on p. 435 depended primarily on the type of geographical area stratification, as Gallup does, to select precincts (rather than voters) to be sampled. Thus, since about 1/3 of the population of Missouri in 1964 lived in the St. Louis area, about 1/3 (13 out of 40) of the precincts were selected from that area. However, Vote Profile Analysis also takes other factors into account; three precincts with predominantly black populations were used to represent the 8% of the population of Missouri who are black. Similarly, the 487 counties in the Sindlinger poll are chosen to represent various strata of the total voter population.

Pollsters prefer stratified sampling to simple random sampling, because the sample results are more precise (in the technical sense of Section 7-6) for the same sample size. Consequently, the pollsters are correct more often.

Concluding Essay

Polling of public opinion is only one of the many ways in which sampling is used. The United States Census Bureau takes a sample of 25,000 out of the 60,000,000 American families to obtain current estimates of unemployment and other population characteristics. The Air Force has developed a sampling procedure which it uses in auditing the shipping operations at Air Force bases and materiel centers. The Internal Revenue Service uses sampling techniques to determine which tax returns to audit and also to tabulate statistics on income and taxes.

Businesses employ sampling procedures in many different ways such as market research and quality control of manufactured goods. The work time of employees can be analyzed by sampling methods to find out what percentage of working time an employee spends on various aspects of a job. The information obtained in this way helps a company make the best possible use of its employees. There are many more applications of sampling to more specialized business problems. For example, railroads and airlines use sampling techniques to settle intercompany accounts for passenger tickets and freight handling.

Among the most dramatic uses of sampling have been its applications to medical research. Before any new medicine can be put into general use, it must pass elaborate tests, based on sampling, to establish its safety and effectiveness. Also, statistical tests have proved there is a correlation between smoking cigarettes and contracting lung cancer.

Sampling is only one aspect of the branch of applied mathematics called statistics. Statistics involves both designing sampling procedures and drawing conclusions from the data. The statistician is concerned with developing and using techniques that will provide the most useful and dependable information possible. The development of a sophisticated statistical theory during the present century has made this subject one of the most effective, as well as most widely used, branches of applied mathematics.

References

For a concise explanation of polls and how they work, see the article entitled "Political Pulse Taking: How the Pollsters Do It" in the October 16, 1972 issue of *U.S. News and World Report*. Robert A. Skedgell's article, "How Computers Pick an Election Winner," from the November 1966 issue of *Trans-Action* tells how Vote Profile Analysis was developed and how it works. George Gallup, in *The Sophisticated Poll Watcher's Guide* (Princeton, N.J., 1972), tells a lot about the polling process from Gallup's own viewpoint.

Sampling in all its aspects is discussed in *Sampling*, by Morris James Slonim (New York: Simon and Schuster, 1966). [The original version was published as *Sampling in a Nutshell* (New York: Simon and Schuster, 1960).] An amusing but nevertheless informative book on the general subject of statistics is Darell Huff's *How to Lie with Statistics* (New York: Norton, 1954).

For a more mathematical treatment of statistics, see *Introduction to Probability and Statistics* by William Mendenhall (North Scituate, Mass.: Duxbury Press, 1975).

Chapter 8
Computers

Computers have become virtually indispensable in applying mathematics to practical problems. They have not changed basic mathematical theory, but they have profoundly affected the way mathematics can be used. Because computers work at such great speed and can manipulate such vast quantities of data, it is now practical to use sophisticated mathematical techniques to solve difficult problems arising in our complicated technological society.

A computer is a machine that receives information (letters and numbers) from an external source in the form of punched cards or magnetic tape. It stores this information in its memory, transfers information from one location to another, and retrieves the informa-

489

tion from its memory when needed. The computer performs the basic arithmetic operations: addition, subtraction, multiplication, and division. It can compare two pieces of information to determine, for example, whether two words are identical or which of two numbers is greater. It has the capacity to perform long sequences of arithmetic, comparative, or manipulative operations. These operations are carried out according to a pre-assigned set of directions, called a "program," stored in its memory. The results of carrying out the program are communicated through printing or some other form of visual display.

One of the chief advantages of the computer over older means of computation (paper and pencil, the abacus, the mechanical calculator) is its memory capacity. For example, some models of the IBM 370 series can store over 16 million binary digits in their main memory—and retrieve any one of them in less than 1 microsecond (1 microsecond is one-millionth of a second). Another advantage of the computer is its ability to be pro-

grammed. Even in the most sophisticated mechanical calculator, each single operation must be punched into the machine by the operator. In a computer, programs involving thousands of individual steps are common, but once programmed and started, the computer will carry out the entire routine without more outside direction. The speed of the computer is one of its most obvious advantages. A CDC 7600 can add two numbers in 1/10 microsecond. Thus, it is not surprising that a project that formerly would have required months of effort on the part of many people operating mechanical calculators takes only minutes of computer time.

Throughout this book, we have been concerned with problems whose solution would be at best difficult, and at worst impossible, without the use of computers. In order to understand how a practical problem becomes a computer problem—and to supply some content to the generalizations about computers stated above—we will next present a case study, the calculation of monthly interest payments on a long-term loan.

8-1   The Loan Problem

Let us suppose a home buyer obtains a mortgage from a bank for $30,000 at an annual interest rate of 8%, to be repaid in equal monthly installments over a period of 25 years. Each monthly payment consists of the interest due that month on the unpaid balance of the loan plus an amount credited against the principal. The amount of the monthly payment is such that, at the end of the 25 years, all $30,000 will have been repaid. The amount of the monthly payment can be computed by means of a formula. In practice, the bank's loan officer just looks it up in a book of tables. The monthly payment on a $30,000 mortgage for 25 years at 8% interest is $231.54.

To perform calculations involving interest rates, it is necessary to be able to express them as decimal numbers as well as percentages. For example, 7% corresponds to 0.07 and $10\frac{1}{2}$ to 0.105. In our example, the annual interest rate of 8% is thus 0.08 in decimal form.

The monthly interest rate on a loan is 1/12 of the annual rate. The interest due as part of the first month's payment then is the principal times the monthly interest rate, i.e.,

$$\$30,000\left(\frac{0.08}{12}\right) = \$200$$

Thus, out of the first payment of $231.54, an amount of $200 is paid as interest, while the remaining $31.54 is applied to the principal. It follows that

$$\$30,000.00 - \$31.54 = \$29,968.46$$

remains to be paid on the principal. The next month, the interest owed to the bank is

$$\$29,968.46\left(\frac{0.08}{12}\right) = \$199.79$$

while the remaining

$$\$231.54 - \$199.79 = \$31.75$$

reduces the principal, and so on for 300 months (25 years).

This brings us to the loan problem: Determine the amount of interest paid to the bank in each of the 300 installment payments. This information is needed by the bank for its records. The home buyer also wants this information, because money paid out as interest on loans is tax-deductible. When itemizing deductions, the amount of interest paid during the year is subtracted from the home buyer's gross income in computing taxable income. If the mortgage payments begin in January, then the amount of interest in the first year on the loan described above will total $2385.81. That amount could have a significant effect on the taxes paid by the buyer.

We next describe in detail the general procedure for computing monthly interest. The home buyer must pay interest each month equal to a proportion of the unpaid balance (principal) of the loan; the proportion depends on the interest rate. The amount of interest due is computed by multiplying the amount of principal remaining by 1/12 of the interest rate. The formula is

$$\text{Interest} = \text{Balance} * (\text{Rate}/12)$$

where the symbols *, meaning "times" (multiplication), and /, meaning "divided by," are the ones used in the computer languages described in Sections 8-2 through 8-4.

Earlier in this section, we found the interest to be paid the first month of the sample loan by performing the calculation

$$(\$30,000) * (0.08/12) = \$200$$

The next month, the unpaid balance turned out to be $29,968.46. So that month,

$$\text{Interest} = (\$29,968.46) * (0.08/12) = \$199.79$$

We also must compute the new balance, after the payment, so the process can be repeated for the next month's computation of interest to be paid. The amount of the payment used to reduce the amount of the loan outstanding is the remainder of the payment after the interest has been paid. In the second month of the example, we just saw that $199.79 out of the payment of $231.54 went for interest, so

$$\$231.54 - \$199.79 = \$31.75$$

was used to reduce the balance of the loan. In symbols, the old balance is reduced by an amount equal to

$$\text{Payment} - \text{Interest}$$

(The familiar symbol − represents subtraction in the computer languages.) The balance after the first month of the example loan was $29,968.46. We saw that $31.75 of the payment would go toward reducing the balance. Therefore, the amount still unpaid of the $30,000 loan after two payments is

$$\$29,968.46 - \$31.75 = \$29,936.71$$

The new balance in general, then, is obtained by reducing the old balance by the amount Payment − Interest. In symbols, the new balance, that is, the amount of the principal remaining after the payment, will be

$$\text{New balance} = \text{Old balance} - (\text{Payment} - \text{Interest})$$

For each month, we repeat the calculations using the same payment and interest rate, but a different balance, to determine the interest paid and the new balance for the following month's calculation. Starting with a balance equal to the amount of the loan (for example, $30,000), and repeating the process above as many times as there are months in the life of the loan (e.g., 300 months), solves the loan problem.

8-1   **Example**   On October 1, a homeowner obtains a $4000 home-improvement loan from the bank. The loan is for 5 years at $10\frac{1}{2}\%$ annual interest. If the monthly payment on the loan is $85.96, determine the total interest paid on the loan during the first calendar year (i.e., the 3 months of October, November, and December).

Solution   The annual interest rate is $10\frac{1}{2}\%$, which in decimal form is 0.105. On October 1, the balance is $4000, the original amount of the loan, so

$$\text{Interest} = \text{Balance} * (\text{Rate}/12) = (\$4000) * (0.105/12) = \$35$$

The new balance on November 1 then is

$$\text{Old balance} - (\text{Payment} - \text{Interest}) = \$4000 - (\$85.96 - \$35)$$
$$= \$3949.04$$

Consequently, the interest due on November 1 is

$$\text{Balance} * (\text{Rate}/12) = (\$3949.04)(0.105/12) = \$34.55$$

The new balance as of December 1 will then be

$$\text{Old balance} - (\text{Payment} - \text{Interest}) = \$3949.04 - (\$85.96 - \$34.55)$$
$$= \$3897.63$$

The interest due on the December 1 balance is

$$\text{Interest} = \text{Balance} * (\text{Rate}/12) = (\$3897.63) * (0.105/12) = \$34.10$$

Adding together the three interest payments, we get

Date	Interest paid
October 1	$35.00
November 1	34.55
December 1	34.10
	$103.65

which is the total interest paid to the bank during the first calendar year of the loan.

The calculations involved in the loan problem are not difficult, especially if an electronic calculator is available to carry out the arithmetic. However, if one starts out to calculate a specific example with 300 months in the life of the loan, one quickly discovers that, even with the aid of the calculator, this is a tedious and time-consuming problem. It is not difficult to imagine what a bank is up against if it has to solve this problem for thousands of loans. On the other hand, we shall see that this problem is very well suited to the peculiar abilities of the computer.

Summary

Payment: monthly payment to the bank.

Rate: annual interest rate, in decimal form.

Balance: amount owed on the loan at any given time.

Interest: interest paid as part of the monthly payment.

Arithmetic Operations

$-$: minus.

$*$: times.

$/$: divided by.

Interest $=$ Balance $*$ (Rate$/12$).

New balance $=$ Old balance $-$ (Payment $-$ Interest).

Exercises

● *For Exercises 8-1 through 8-3, find the monthly interest.*

8-1   Balance $=$ \$6000; Rate $=$ 9%

8-2   Balance $=$ \$29,607.34; Rate $=$ 8%

8-3   Balance $=$ \$1214.20; Rate $=$ 10½%

● *For Exercises 8-4 and 8-5, a store advertises: "Buy in time for Christmas; no payments until February." The store charges 1½% per month interest. Although no payments are due until February 1, interest is charged from the date of purchase.*

8-4   If an item is bought for \$200 on December 1, how much does the customer owe the store on February 1?

8-5   If an item is bought for \$300 on November 1, how much does the customer, owe the store on February 1?

- *In Exercises 8-6 through 8-8, find the new balance.*

8-6   Old balance = $10,000.00; Payment = $327.39; Interest = $91.67

8-7   Old balance = $2420.00; Payment = $40.00; Interest = $26.15

8-8   Old balance = $29,708.46; Payment = $231.54; Interest = $198.50

- *Table 8-1 can be used to find the approximate amount of monthly payments for loans at various annual interest rates. For example, according to Table 8-1, the monthly payment per $1000 for a 25 year loan at 8% interest is $7.72. Therefore, the monthly payment on a $30,000 mortgage is approximately*

$$(30)(\$7.72) = \$231.60$$

This is a good approximation to the correct payment of $231.54 that would be found in a complete book of tables and that we used in Section 8-1. In Exercises 8-9 through 8-11, use Table 8-1 to compute the monthly payments.

Table 8-1

Monthly Payment per $1000 (in Dollars)

Percent / Year	7	7½	8	8½	9	9½	10	10½	11
2	44.77	45.00	45.23	45.46	45.68	45.91	46.14	46.38	46.61
3	30.88	31.11	31.34	31.57	31.80	32.03	32.27	32.50	32.74
5	19.80	20.04	20.28	20.52	20.76	21.00	21.25	21.49	21.74
10	11.61	11.87	12.13	12.40	12.67	12.94	13.22	13.49	13.78
20	7.75	8.06	8.36	8.68	9.00	9.32	9.65	9.98	10.32
25	7.07	7.39	7.72	8.05	8.39	8.74	9.09	9.44	9.80
30	6.65	6.99	7.34	7.69	8.05	8.41	8.78	9.15	9.45

8-9 $4000 for 2 years at 10½%

8-10 $35,000 for 30 years at 8%

8-11 $7500 for 5 years at 9%

- *For Exercises 8-12 through 8-15, suppose loan payments began in November. With the aid of Table 8-1, compute the total interest paid during the first calendar year (i.e., the first 2 months only).*

8-12 $2000 for 2 years at 9½%

8-13 $50,000 for 30 years at 7%

8-14 $8500 for 10 years at 8½%

8-15 $500 for 3 years at 10½%

● *On many loans for such items as automobiles and boats, as well as some home-improvement loans, the interest rate is advertised as what is known as "add-on" interest. The total interest owed during the life of the loan is computed by multiplying the amount of the loan times the life in years times the interest rate. For example, a $3000 3 year automobile loan at 5% add-on interest charges a total of*

$$(\$3000)(3)(0.05) = \$450$$

in interest. In order to calculate the total owed to the lender for the loan, we add on this interest to the principal and get

$$\$3000 + \$450 = \$3450$$

This amount is paid off in equal monthly payments. In the example, since there are 36 months in 3 years, each payment is

$$\frac{\$3450}{36} = \$95.83$$

To determine how add-on interest rates compare to the rates quoted for mortgages, first compute the total add-on interest for a $1000 loan. In the example of 5% add-on interest for 3 years, that amount will be

$$(\$1000)(3)(0.05) = \$150$$

Next calculate the monthly payment on such a $1000 loan. Thus, the total amount due in the example is

$$\$1000 + \$150 = \$1150$$

so the monthly payment is

$$\frac{\$1150}{36} = \$31.94$$

Finally, find the figure in Table 8-1 for loans of the same duration which is closest to the monthly add-on payment on $1000. In the example, the figure in the row of the table corresponding to 3 year loans which is closest to $31.94 is $32.03. This is the monthly payment per $1000 on a 3 year loan at $9\frac{1}{2}\%$ mortgage-type interest. We conclude that, for a 3 year loan, 5% add-on interest amounts to slightly less than $9\frac{1}{2}\%$ interest on the unpaid balance. In Exercises 8-16 through 8-19, convert the given add-on interest rates to mortgage-type interest rates.

8-16   5%: 5 year loan

8-17   5%: 10 year loan

8-18   6%: 3 year loan

8-19   6%: 10 year loan

8-20   On a 1 year deposit, bank A pays simple interest of $6\frac{1}{4}\%$, that is, $6\frac{1}{4}\%$ of the amount invested is paid as interest at the end of the year. Bank B pays interest of 6% compounded quarterly, so after 3 months you receive as interest 1/4 of 6% of the original investment. After 6 months you receive 1/4 of 6% of the sum of the amount invested and the interest paid at the end of the first 3 months, and so on. If you plan to leave $1000 in the bank for 1 year, which bank will pay you more interest?

8-2   Programming the Loan Problem in BASIC

In order to solve the loan problem described in Section 8-1, the computer must be told how to deal with problems of this type. The computer has no built-in problem-solving ability, so a list of the steps the computer will go through in solving the problem, called a **program,** must be prepared. The program, furthermore, must be in a form the computer can "understand."

Each type of computer has a language of its own. A single word of the language describes only a very simple step, so that even a quite uncomplicated program may require hundreds of words in the computer language. The words of the computer language are in the form of strings of numbers. A program written in this language would look something like this:

```
01531      21      01550      02444
01573      01      01135      00120
01578      43      01352      02418
```

and so on. Each line (of seventeen digits in this case) represents a single word, and thus a very short step toward carrying out the procedure. Clearly, computer language is not well suited to human communication, especially when one considers the likelihood of making an error in writing a long program in this language. Since even a single incorrect digit will change the meaning of the program, and thus cause the machine either to stop or to obtain an incorrect answer, the difficulties of communicating with a computer in its own language are apparent.

Ideally, one would like to describe the method of solution to the problem in English—just as we did in Section 8-1—submit the information to the computer in that form, and have the computer translate the message into its own language in order to perform the calculations. At present, only a compromise is possible. There are languages which are comprehensible to a properly equipped computer and, at the same time, are close enough to

English so that a person can learn them without excessive effort and can then write programs in them accurately. We will now describe a program for solving the loan problem, written in such a language. The language we have chosen to use in this section is called **BASIC.**

We begin by printing the entire program so that the reader can refer back to it as we analyze it in detail below.

```
10    REMARK LOAN PROGRAM
20    REMARK B MEANS BALANCE
30    REMARK R MEANS INTEREST RATE
40    REMARK P MEANS PAYMENT
50    REMARK I MEANS INTEREST PAID
60    INPUT B,R,P
70    IF B<=0 THEN 120
80    LET I=B*(R/12)
90    LET B=B-(P-I)
100   PRINT USING "######.## ###.##",B,I
110   GOTO 70
120   END
```

Each line of the BASIC program, called a **statement,** must be numbered. The computer will execute the statements in order unless instructed by the program to do otherwise. A BASIC program usually begins with some information for the human user of the program. This information is in the form of **REMARK** statements. The computer takes no action in response to a **REMARK** statement other than to include it in the program listing. Thus, the program begins

```
10    REMARK LOAN PROGRAM
20    REMARK B MEANS BALANCE
30    REMARK R MEANS INTEREST RATE
40    REMARK P MEANS PAYMENT
50    REMARK I MEANS INTEREST PAID
```

These statements inform the human user of the program of its purpose (**LOAN**) and the meaning of the symbols **B, R, P,** and **I** which will appear in the statements that follow. The symbols represent locations in the memory portion of the computer in which the numerical values of these quantities will be stored during the running of the program.

The computer begins its activity in response to the next statement

```
60    INPUT B,R,P
```

Since BASIC programs are primarily for use with a computer terminal, we imagine that the human user has typed in the entire program and then instructed the computer to run the program. In response to the statement

```
60   INPUT B,R,P
```

the computer prints a question mark indicating to the user that data is required. The user types in the values for the amount of the loan, the interest rate, and the monthly payment, in that order, separated by commas. Thus, for the problem of Section 8-1, this line would look like

```
?   30000.00,.08,231.54
```

where the question mark was printed by the computer and the numerical data by the user.

The next five statements of the program form a **loop,** that is, a sequence of operations which is to be repeated many times. The entire set of instructions for the loop is

```
70   IF B<=0 THEN 120
80   LET I=B*(R/12)
90   LET B=B-(P-I)
100  PRINT USING "######.## ###.##",B,I
110  GOTO 70
```

The idea is that B, the unpaid balance of the loan, changes (gets smaller) each time the computer runs through the operations that make up the loop. Then the new balance is used for the next repetition of the loop, and so on. The first time the loop is performed, B has the value 30000.00 typed into the machine by the user.

The statement

```
70   IF B<=0 THEN 120
```

tells the machine to check whether the contents of the B location in its memory is less than or equal to zero. If so, the computer next executes statement number 120. On the other hand, if the current value of B is positive, the computer goes on to statement 80 and through the remaining steps of the loop. At the beginning of each repetition of the loop, the computer will check the value of B to determine whether to perform the steps of the loop. It will continue to do so as long as the value of B is greater than zero. Since the first time through the loop the value of B is the number typed into the machine and therefore is positive, the computer follows the instructions in the next statement.

The next statement in the program is

```
80   LET I=B*(R/12)
```

On the one hand, the statement is the equation of Section 8-1 for computing the interest to be paid as part of the monthly payment. On the other hand, as an instruction to a computer it has a precise operational meaning. The symbol = in BASIC does not mean "equals" in the mathematical sense, although, in this case, such as interpretation is reasonable. The symbol = in programming language means that a copy of the result of the calculation described in the expression to the right of = is to be stored (placed) in the memory location named to the left of =. Thus, in this statement, the computer is instructed to take the current value of R, divide it by 12, and multiply the result by the current value of B. Then the product is to be placed in the memory location called I.

The meaning of the symbol = in BASIC is made more evident in the next instruction:

$$90 \quad LET \quad B=B-(P-I)$$

This time, the statement does not make sense as a mathematical equation. The purpose of the command is to change the value of the B memory location as a result of making a payment. The value of the I location, which was computed in statement 80, is to be subtracted from the (constant) value of the P location. The resulting difference is then subtracted from the value presently in the B memory location. Finally, that difference replaces the present value in B. Another way to describe this command is to say that the value of the B memory location is reduced by the amount P−I. The new value in B is the one that will be used in the next cycle through the loop.

The computer prints out the new balance and the amount of interest paid in response to the command

$$100 \quad PRINT \quad USING \quad "\#\#\#\#\#\#.\#\# \quad \#\#\#.\#\#",B,I$$

In the example of Section 8-1, the terminal output after the first run through of the loop will look like

```
?      30000.00,.08,231.54
    29968.46              200.00
```

The statement

$$110 \quad GOTO \quad 70$$

informs the computer that it should go back to the beginning of the loop, where it will check the new B (i.e., 29968.46) to see whether it is still greater than zero. If so, the loop will be repeated using the new value of B. Thus, after the second run through the loop in the example, the printout looks like this:

```
?      30000.00,.08,231.54
    29968.46              200.00
    29936.71              199.79
```

The computer repeats the loop over and over until it checks B and finds that the current value of that location is negative or zero. In that event,

$$70 \quad \text{IF } B<=0 \text{ THEN } 120$$

means, since $B<=0$ is true, that the computer should omit the loop and go directly to statement 120 of the program,

$$120 \quad \text{END}$$

which instructs the computer that the program has ended.

The program we have presented is very primitive. For example, a good program would identify payments by month and year and then add up the total interest paid at the end of the year so the home buyer would have the information readily available for tax purposes. A more satisfactory program would be several times the length of the one we used, and would involve a large number of different types of instructions in addition to the ones in this section. Nevertheless, the underlying principles would be the same.

Summary

P: monthly payment to the bank.

R: annual interest rate.

B: amount owed on the loan.

I: interest paid as part of the monthly payment.

REMARK: information for a human reader of the program, not executed by the computer.

INPUT . . . : request numerical values from the user.

IF . . . THEN . . . : test the truth of the expression after **IF** and, if it is true, next execute the statement whose number follows **THEN**; otherwise, just go to the next step in the program.

LET . . . : perform the calculation indicated.

PRINT: print out.

GOTO . . . : the next statement to be executed is the one whose number is given.

END: end of program.

=: place the number or the result of the calculation written to the right of this symbol in the memory location named to the left of the symbol.

Arithmetic Operations

+ : plus.

− : minus.

* : times.

/ : divided by.

Exercises

- *In Exercises 8-21 through 8-28, translate the BASIC statements into ordinary language.*

8-21 `LET A=5*X`

8-22 `LET C=(A*B)/(X+3.5)`

8-23 `LET A=A+2.31`

8-24 `LET B=B*(1+(R/52))`

8-25 `LET H=40`

8-26 `INPUT S,R`

8-27 `PRINT USING "####.## #.### ###.##",X,Y,Z`

8-28 `IF M<12 THEN 80`

- *Instead of requiring the user to supply the data for the problem, this information can be included in the program by means of* **DATA** *and* **READ** *statements which would replace* **INPUT** *statements. For example,*

```
40    DATA 45,5.2,-10
50    READ A,B,C
```

informs the computer that it should place the number 45 *into memory location* **A**, 5.2 *into* B *and* −10 *into* C. *There must be the same number of numbers in the* **DATA** *statement as there are symbols in the* **READ** *statement, and the ordering corresponds one to the other.*

8-29 Explain what the following program does and how it does it:

```
10    REMARK BANK PROGRAM
20    REMARK B MEANS BALANCE
30    REMARK R MEANS ANNUAL INTEREST RATE
40    REMARK W MEANS WEEK
50    DATA 100,.05,1
```

```
60    READ B,R,W
70    IF W>52 THEN 120
80    LET B=B*(1+(R/52))
90    PRINT USING "## #####.##",W,B
100   LET W=W+1
110   GOTO 70
120   END
```

- *The computer will respond to the command* PRINT ". . ." *by printing the message within the quotation marks, for example,*

PRINT "THANK YOU"

instructs the computer to print out the message THANK YOU.

8-30   Explain what the following program does and how it does it:

```
10    REMARK WINE PROGRAM
20    REMARK B MEANS NUMBER OF BOTTLES
      OF WINE PURCHASED
30    REMARK P MEANS PRICE PER BOTTLE
40    REMARK C MEANS COST
50    INPUT B,P
60    LET C=0
70    IF B<12 THEN 110
80    LET C=C+(12*(P*.9))
90    LET B=B-12
100   GOTO 70
110   LET C=C+(B*P)
120   PRINT "PLEASE PAY"
130   PRINT USING "#####.##",C
140   PRINT "THANK YOU"
150   END
```

8-31   Explain what the following program does and how it does it:

```
10    REMARK PAYROLL PROGRAM
20    REMARK H MEANS NUMBER OF HOURS WORKED
30    REMARK R MEANS HOURLY RATE OF PAY
40    REMARK P MEANS PAYMENT FOR HOURS WORKED
50    INPUT H,R
60    LET P=0
70    IF H<=40 THEN 100
80    LET P=(H-40)*(R*1.5)
90    LET H=40
```

```
100    LET P=P+(H*R)
110    PRINT USING "####.##",P
120    END
```

- *Although the* **IF . . . THEN . . .** *statement can test only the truth of a single expression, more complicated alternatives can be dealt with by the use of several* **IF . . . THEN . . .** *statements.*

8-32   Explain what the following program does and how it does it:

```
10     REMARK THIS PROGRAM IS DESIGNED TO MAKE CHANGE
20     REMARK P MEANS THE AMOUNT PAID BY THE CUSTOMER
30     REMARK C MEANS THE COST OF THE ITEMS PURCHASED
40     REMARK D DENOTES A DIFFERENCE
50     INPUT P,C
60     IF P=C THEN 170
70     IF P>C THEN 120
80     LET D=C-P
90     PRINT "UNDERPAYMENT: PLEASE PAY"
100    PRINT USING "####.##",D
110    GOTO 180
120    LET D=P-C
130    PRINT "YOUR CHANGE IS"
140    PRINT USING "####.##",D
150    PRINT "THANK YOU"
160    GOTO 180
170    PRINT "PAYMENT IS CORRECT, THANK YOU"
180    END
```

**8-3   Programming the
Loan Problem
in FORTRAN**

In order to solve the loan problem described in Section 8-1, the computer must be told how to deal with problems of this type. The computer has no built-in problem-solving ability, so a list of the steps the computer will go through in solving the problem, called a **program,** must be prepared. The program, furthermore, must be in a form the computer can "understand."

Each type of computer has a language of its own. A single word of the language describes only a very simple step, so that even a quite uncomplicated program may require hundreds of words in the computer language. The words of the computer language are in the form of strings of numbers. A program written in this language would look something like this:

```
01531      21    01550    02444
01573      01    01135    00120
01578      43    01352    02418
```

and so on. Each line (of seventeen digits in this case) represents a single word, and thus a very short step toward carrying out the procedure. Clearly, computer language is not well suited to human communication, especially when one considers the likelihood of making an error in writing a long program in this language. Since even a single incorrect digit will change the meaning of the program, and thus cause the machine either to stop or to obtain an incorrect answer, the difficulties of communicating with a computer in its own language are apparent.

Ideally, one would like to describe the method of solution to the problem in English— just as we did in Section 8-1—submit the information to the computer in that form, and have the computer translate the message into its own language in order to perform the calculations. At present, only a compromise is possible. There are languages which are comprehensible to a properly equipped computer and, at the same time, are close enough to English so that a person can learn them without excessive effort and can then write programs in them accurately. We will now describe a program for solving the loan problem, written in such a language. The language we have chosen to use in this section is called **FORTRAN.**

We begin by printing the entire program so that the reader can refer back to it as we analyze it in detail below.

```
 1      10    FORMAT(1H1,30X,12HLOAN PROGRAM)
 2      20    FORMAT(F10.2,F8.7,F10.2)
 3      30    FORMAT(1H0,F10.2,15X,F8.7,10X,F10.2)
 4      40    FORMAT(1H0,F10.2,8X,F10.2)
 5            WRITE(6,10)
 6            READ(5,20)BAL,RATE,PAY
 7            WRITE(6,30)BAL,RATE,PAY
 8     100    IF(BAL.LE.0)GO TO 110
 9            PDINT=BAL*(RATE/12)
10            BAL=BAL-(PAY-PDINT)
11            WRITE(6,40)BAL,PDINT
12            GO TO 100
13     110    .STOP
14            END
```

The numbers 1, 2, and so on to the far left of the program are just for reference in the following discussion and are not part of the program. On the other hand, the two and three digit numbers 10, 20, . . . , 100, 110 closer to the **statements** are an integral part of the program, as we shall see.

The computer does not execute **FORMAT** statements except in conjunction with another instruction, so it skips the first four steps of the program and the first instruction to which it will respond is

5   WRITE(6,10)

The part of the instruction **WRITE(6,** tells the computer that something is to be printed out on the high-speed printer. The **10** that follows means the directions for the printout can be found by the computer in the statement of the program labeled **10,** which is

1   10   FORMAT(1H1,30X,12HLOAN PROGRAM)

The symbol **1H1** tells the printer to begin on a fresh sheet of paper, while **30X** instructs it to start printing 30 spaces from the left-hand margin. The message to be printed will be twelve symbols long, including spaces, as indicated by the direction **12H,** and the message to be printed follows immediately: **LOAN PROGRAM.** Thus, steps 5 and 1 together cause the computer to print out

LOAN PROGRAM

on a fresh sheet of paper starting 30 spaces from the left-hand margin.

The next instruction executed by the computer will be

6   READ(5,20)BAL,RATE,PAY

The purpose of this instruction is to have the computer receive numerical data from a punched card and store it in memory locations within the computer. The symbol **5** in **READ(5,20)** tells the computer that it should read the contents of a punched card, while **BAL,RATE,PAY** indicates that three numbers will come into the computer from the card. Throughout the program, **BAL** will refer to the unpaid balance of the loan, **RATE** to the annual interest rate, and **PAY** to the amount of the monthly payment. The **20** in the phrase **READ(5,20)** tells the computer that the format information it needs is contained in the program statement labeled **20,** which is

2   20   FORMAT(F10.2,F8.7,F10.2)

The three expressions **F10.2,F8.7,F10.2** refer to the expected format of **BAL, RATE,** and **PAY,** in that order, on the input card. The **FORMAT** item **F10.2** corresponding to **BAL** indicates the number to be read into that memory location will occupy the first ten columns of the card. Furthermore, the **.2** means that if no decimal point appears in the value punched in the card, the decimal point is assumed to be between the eighth and ninth columns. Similarly, the **FORMAT** item **F8.7** tells the computer that the value of **RATE** will occupy columns 11–18 of the card and, unless a decimal

point is explicity used, the decimal point occurs between columns 11 and 12. The final
F10.2 informs the computer that the value of **PAY** is in columns 19–28 of the card
with a decimal point understood between columns 26 and 27 unless one is punched some-
where in these columns.

The computer next performs the instruction

$$7 \qquad\qquad \texttt{WRITE(b,30)BAL,RATE,PAY}$$

The symbol **b** in **WRITE(b,30)** means that the high-speed printer is to be used.
BAL,RATE,PAY in this instruction means the contents of the memory locations with
these names are to be copied to the printer. This action will not affect the numbers in
these locations; rather, it is as if the computer were taking a picture of these numbers to
display elsewhere. The computer still needs to know the form in which to display these
numbers (the format), and for this the instruction **WRITE(b,30)** sends the computer
to the instruction labeled **30,** that is,

$$3 \qquad \texttt{30} \quad \texttt{FORMAT(1HO,F10.2,15X,F8.7,10X,F10.2)}$$

The **FORMAT** item **1HO** causes the printer to double space before it begins to operate.
It starts at the left-hand margin of the page and prints out the contents of the **BAL** memory
location just as it is in the memory, because the **FORMAT** item **F10.2** is the same as
in the previous **FORMAT** instruction for storing the information. The next **FORMAT**
item, **15X,** causes the printer to move fifteen spaces to the right before printing out the
contents of the **RATE** memory location in response to the **F8.7** **FORMAT** item. Moving
ten spaces more to the right as directed by the **10X** command, the printer copies the
contents of the **PAY** location. Thus, for the example of Section 8-1, the computer will at
this point have printed out the message

```
                                   LOAN PROGRAM
30000.00                  .08        231.54
```

These are the figures it received from the card for balance, interest rate, and payment,
in that order. This information appears at the top of the sheet of paper that will contain
the answer to the problem. In this way, the bank will have a record of the loan conditions
to which the schedule of interest payments refers.

Statements **8–12** of the program form a **loop,** that is, a sequence of operations which
is to be repeated many times. The entire set of instructions for the loop is

```
 8      100   IF(BAL.LE.0)GO TO 110
 9            PDINT=BAL*(RATE/12)
10            BAL=BAL-(PAY-PDINT)
11            WRITE(b,40)BAL,PDINT
12            GO TO 100
```

The idea is that **BAL**, the unpaid balance of the loan, changes (gets smaller) each time the computer runs through the operations that make up the loop. Then the new balance is used for the next repetition of the loop, and so on. The first time the loop is performed, the **BAL** is the figure ЗОООО.ОО read by the machine in response to the **READ** command of instruction Ь. The statement

Ƀ   ٦ОО   IF(BAL.LE.O)GO TO ٦٦О

tells the machine to check whether the contents of the **BAL** location in its memory is less than or equal to (abbreviated as .LE.) zero. If so, the computer next executes statement number ٦٦О as required by the phrase GO TO ٦٦О. On the other hand, if the current value of the **BAL** location is positive, the computer goes through the remaining steps of the loop. At the beginning of each repetition of the loop, the computer will check the contents of the **BAL** location in its memory to determine whether to perform the steps of the loop. It will go through the loop provided the value of **BAL** is greater than zero. Since the first time through the loop the value of **BAL** is the number read into the machine and therefore is positive, the computer follows the instructions in the next statement.

The next statement in the program is

٩   PDINT=BAL*(RATE/٦੨)

On the one hand, the statement is the equation of Section 8-1 for computing the interest to be paid as part of the monthly payment. On the other hand, as an instruction to a computer, it has a precise operational meaning. The symbol = in FORTRAN does not mean "equals" in the mathematical sense, although, in this case, such an interpretation is reasonable. The symbol = in programming language means that a copy of the result of the calculation described in the expression to the right of = is to be stored (placed) in the memory location named to the left of =. Thus, in this statement, the computer is instructed to take the number in the memory location called **RATE**, divide it by 12, and multiply the result by the number in the memory location **BAL**. Then the product is to be placed in a new memory location called **PDINT**.

The meaning of the symbol = in FORTRAN is made more evident in the next instruction:

٦О   BAL=BAL—(PAY—PDINT)

This time, the statement does not make sense as a mathematical equation. The purpose of the command is to change the contents of the **BAL** memory location as a result of making a payment. The number in the **PDINT** location, which was computed in the previous statement, is to be subtracted from the (constant) number in the **PAY** location. The number which results is subtracted from the number presently in the **BAL** memory location. Finally, that difference replaces the present value in **BAL**. Another way to describe this command is to say that the number in the **BAL** memory location is reduced by the amount **PAY—PDINT**. The new number in **BAL** is the one that will be used in the next cycle through the loop.

The computer prints out the new balance and the amount of interest paid in response to the command

11   WRITE(6,40)BAL,PDINT

In the example of Section 8-1, the answer sheet after the first run through of the loop will look like

                                    LOAN PROGRAM
30000.00   .08   231.54
29968.46   200.00

The reason the printout has this form is that instruction 11 requires the printing of the contents of the **BAL** and **PDINT** memory locations according to the instructions contained in the **FORMAT** statement labeled 40. That statement is

4   40   FORMAT(1H0,F10.2,8X,F10.2)

which indicates that the printer should double space, print the contents of the **BAL** location at the left-hand margin, move eight spaces to the right, then print the contents of **PDINT**.

The statement

12   GO TO 100

sends the computer next to the statement labeled 100, that is,

8   100 IF(BAL.LE.0) GO TO 110

which is the beginning of the loop. Thus, the computer will check whether the contents of **BAL** (e.g., 29968.46) are less than or equal to zero and, if the contents are still positive, repeat the process using the new value of **BAL**. Thus, after the second run through the loop in the example, the printout looks like this:

                                    LOAN PROGRAM
30000.00   .08   231.54
29968.46   200.00
29936.71   199.79

The computer repeats the loop over and over until it checks **BAL** and finds that the number in that location is negative or zero. In that event,

8   100   IF(BAL.LE.0)GO TO 110

means, since **BAL.LE.0** is true, the computer should go directly to the statement of the program labeled **110**,

13   110   STOP

which instructs the computer to stop. The final statement

14   END

just indicates the end of the program.

The computer program is put onto punched cards and these are fed into the computer. Other cards containing the initial balance and the interest rate and payment figures are also put into the machine. The operator starts the program, and the machine does the rest. The program can easily be modified so that many cards, containing other values for **BAL** and so on, can be put in and the computer will work out as many interest schedules as the bank requires. The punched cards containing the program are retained and used again whenever the bank needs to compute more interest schedules.

The program we have presented is very primitive. For example, a good program would identify payments by month and year and then add up the total interest paid at the end of the year so the home buyer would have the information readily available for tax purposes. A more satisfactory program would be several times the length of the one we used, and would involve a large number of different types of instructions in addition to the ones in this section. Nevertheless, the underlying principles would be the same.

Summary

PAY: monthly payment to the bank.

RATE: annual interest rate.

BAL: amount owed on the loan.

PDINT: interest paid as part of the monthly payment.

FORMAT: a statement which gives the form in which data will be received in the computer or output will appear from the printer.

WRITE: print out.

READ: read in information.

IF(...)GO TO ...: test the truth of the expression in the parentheses and, if it is true, next execute the statement whose number follows GO TO; otherwise, just perform the next step in the program.

GO TO: the next statement to be executed is the one whose number is given.

STOP: stop.

END: end of program.

=: place the number or the result of the calculation written to the right of this symbol in the memory location named to the left of the symbol.

Arithmetic Operations

+: plus.

—: minus.

*: times.

/: divided by.

Exercises

● *In Exercises 8-33 through 8-38, translate the FORTRAN statements into ordinary language.*

8-33   `A=5*X`

8-34   `C=(A*B)/(X+3.5)`

8-35   `A=A+2.31`

8-36   `BOTTL=BOTTL—12`

8-37   `BAL=BAL*(1+(RATE/52))`

8-38   `HOURS=40`

● *In Exercises 8-39 and 8-40, write the printout that results from the instructions given.*

8-39   `10   FORMAT(1H1,25X,19HFORTRAN PROGRAMMING)`
       `     WRITE(6,10)`

8-40   `30   FORMAT (1H1,10X,5HMONTH,10X,7HBALANCE,`
       `     6X,13HINTEREST PAID)`
       `     WRITE(6,30)`

● *It frequently happens that the numbers we have to work with are integers (whole numbers) rather than decimals. In FORTRAN, the names of memory locations whose values are to be integers must begin with one of the letters* **I, J, K, L, M, N.** *The names of memory locations whose values are decimal numbers must begin with one of the other letters of the alphabet. Thus, for example, the memory location containing the year will be called* **IYEAR** *so that it begins with* **I** *and can therefore take integer values. The* **FORMAT** *items that refer to memory locations whose values are integers must consist of the letter* **I** *followed by a number. The number indicates which columns of the input card will contain the value for that memory location or gives corresponding information to the printer. In Exercises 8-41 and 8-42, write a possible printout for the instructions given, making up your own numbers (but be sure they agree with the* **FORMAT** *requirements).*

8-41 ```
10   FORMAT(1H0,I2,10X,I2,10X,I4)
      WRITE(6,10)MONTH,IDAY,IYEAR
```

8-42  ```
40    FORMAT(1H0,15X,I4,20X,F5.2)
      WRITE(6,40)NBOTTL,PRICE
```

● *Rather than placing numerical information into the computer by means of punched cards and a* **READ** *statement, such information can be included as part of the program. For example, the command*

```
4              BAL=100.00
```

causes the computer to set aside a memory location to be labeled **BAL** *and places the number* **100.00** *in that location. The symbol* **.GT.** *in the parentheses of an* **IF** *statement (see step* 7 *below) means greater than.*

8-43 Explain what the following program does and how it does it:

```
1      10    FORMAT(1H1,30X,12HBANK PROGRAM)
2      20    FORMAT(1H0,10X,I2,5X,F6.2)
3            WRITE(6,10)
4            BAL=100.00
5            RATE=.05
6            IWEEK=1
7      100   IF(WEEK.GT.52)GO TO 110
8            BAL=BAL*(1+(RATE/52))
9            WRITE(6,20)IWEEK,BAL
10           IWEEK=IWEEK+1
11           GO TO 100
12     110   STOP
13           END
```

8-44   Explain what the following program does and how it does it:

```
 1      10   FORMAT(1H1,30X,12HWINE PROGRAM)
 2      20   FORMAT(I4,F5.2)
 3      30   FORMAT(1H0,I4,1X,7HBOTTLES,7X,16HPRICE
             PER BOTTLE,3X,F5.2)
 4      40   FORMAT(1H0,10HPLEASE PAY,
             5X,F7.2,10X,9HTHANK YOU)
 5           WRITE(6,10)
 6           READ(5,20)NBOTTL,PRICE
 7           WRITE(6,30)NBOTTL,PRICE
 8           COST=0
 9     100   IF(NBOTTL.LE.12)GO TO 110
10           COST=COST+(12*(PRICE*.9))
11           NBOTTL=NBOTTL-12
12           GO TO 100
13     110   COST=COST+(NBOTTL*PRICE)
14           WRITE(6,40)COST
15           STOP
16           END
```

● *The symbol* **2F4.2** *in step* **2** *below is an abbreviated form of* **F4.2,F4.2**. *The symbol* **1H+** *in step* **4** *commands the printer to continue printing on the same line rather than moving on to the next line.*

8-45   Explain what the following program does and how it does it:

```
 1      10   FORMAT(1H1,30X,15HPAYROLL PROGRAM)
 2      20   FORMAT(2F4.2)
 3      30   FORMAT(1H0,F4.2,5X,F4.2)
 4      40   FORMAT(1H+,25X,F6.2)
 5           WRITE(6,10)
 6           READ(5,20)HOURS,RATE
 7           WRITE(6,30)HOURS,RATE
 8           PAY=0
 9           IF(HOURS.LE.40)GO TO 100
10           PAY=(HOURS-40)*(RATE*1.5)
11           HOURS=40
12     100   PAY=PAY+(HOURS*RATE)
13           WRITE(6,40)PAY
14           STOP
15           END
```

● *Another type of* **IF** *statement is of the form* **IF (. . .) . . . , . . . , . . .** *which tests the quantity in the parentheses and sends the program to the statements labeled with the numbers that follow the parentheses, depending on whether the quantity is negative, zero, or positive. Thus,*

8   IF(COST−PAID)100,110,120

means that if the value of the **PAID** *memory location is subtracted from the value of* **COST** *and the difference is negative, then the program will next execute the statement labeled* **100**. *On the other hand, if* **COST−PAID** *is zero, then the program next performs statement* **110**. *Finally, if* **COST−PAID** *is positive, the next command carried out by the computer is the one labeled* **120**.

8-46   Explain what the following program does and how it does it:

```
 1      10    FORMAT(1H1,30X,14HCHANGE PROGRAM)
 2      20    FORMAT(2F8.2)
 3      30    FORMAT(1H0,14HYOUR CHANGE
              IS,5X,F8.2,10X,9HTHANK YOU)
 4      40    FORMAT(1H0,29HPAYMENT IS CORRECT,
              THANK YOU)
 5      50    FORMAT(1H0,23HUNDERPAYMENT:PLEASE
              PAY,5X,F8.2)
 6            WRITE(6,10)
 7            READ(5,20)COST,PAID
 8            IF(COST−PAID)100,110,120
 9     100    DIFF=PAID−COST
10            WRITE(6,30)DIFF
11            GO TO 130
12     110    WRITE(6,40)
13            GO TO 130
14     120    DIFF=COST−PAID
15            WRITE(6,50)DIFF
16     130    STOP
17            END
```

8-4 Programming the Loan Problem in PL/I

In order to solve the loan problem described in Section 8-1, the computer must be told how to deal with problems of this type. The computer has no built-in problem-solving ability, so a list of the steps the computer will go through in solving the problem, called a **program,** must be prepared. The program, furthermore, must be in a form the computer can "understand."

Each type of computer has a language of its own. A single word of the language describes only a very simple step, so that even a quite uncomplicated program may require hundreds of words in the computer language. The words of the computer language are in the form of strings of numbers. A program written in this language would look something like this:

```
01531     21     01550     02444
01573     01     01135     00120
01578     43     01352     02418
```

and so on. Each line (of seventeen digits in this case) represents a single word, and thus a very short step toward carrying out the procedure. Clearly, computer language is not well suited to human communication, especially when one considers the likelihood of making an error in writing a long program in this language. Since even a single incorrect digit will change the meaning of the program, and thus cause the machine either to stop or to obtain an incorrect answer, the difficulties of communicating with a computer in its own language are apparent.

Ideally, one would like to describe the method of solution to the problem in English—just as we did in Section 8-1—submit the information to the computer in that form, and have the computer translate the message into its own language in order to perform the calculations. At present, only a compromise is possible. There are languages which are comprehensible to a properly equipped computer and, at the same time, are close enough to English so that a person can learn them without excessive effort and can then write programs in them accurately. We will now describe a program for solving the loan problem, written in such a language. The language we have chosen to use in this section is called **PL/I.**

We begin by printing the entire program so that the reader can refer back to it as we analyze it in detail below.

```
1     LOAN: PROCEDURE OPTIONS (MAIN);
2        DCL (BALANCE, PAYMENT, INT_PAID) DEC FIXED
            (10,2), INT_RATE DEC FIXED (8,8);
3        GET DATA (BALANCE, INT_RATE, PAYMENT) COPY;
4        LOOP: DO WHILE (BALANCE > 0);
5          INT_PAID = BALANCE * (INT_RATE/12);
6          BALANCE = BALANCE - (PAYMENT - INT_PAID);
7          PUT SKIP LIST (BALANCE, INT_PAID);
8          END LOOP;
9        END LOAN;
```

Each PL/I program begins with the name of the program (we call ours **LOAN**), followed by a standard phrase, **PROCEDURE OPTIONS (MAIN)**, which informs the computer that a program is beginning. Thus, the first line of the program is

1   LOAN: PROCEDURE OPTIONS (MAIN);

The semicolon is an important part of the line, because it signals the computer that the end of the **statement** has been reached.

The next statement, a **DECLARE** statement (abbreviated **DCL**), informs the computer of the kind of numbers with which it will be working in solving the problem. In this case, the balance, payment, and interest paid are all expressed as money, that is, as dollars and cents. We will employ figures like 30000.00, 231.54 and 200.00 with several digits to the left of the decimal point and exactly two digits to the right. The interest rate, on the other hand, is a decimal like .08. The computer instruction is

2   DCL (BALANCE, PAYMENT, INT_PAID) DEC FIXED
(10,2), INT_RATE DEC FIXED (8,8);

In response to this the computer makes internal adjustments to deal with figures that have up to ten digits in all and exactly two digits to the right of the decimal point [thus, (10,2)] for the amounts of balance, payment, and interest paid. Furthermore, the machine will now automatically cut off the results of all arithmetic operations two places after the decimal point. It also makes room in its memory for the interest rate, a number with up to eight digits, all to the right of the decimal point. The word **DEC** in the instruction indicates that all numbers will be fed into the machine in the usual decimal form. Finally, **FIXED** concerns the manner in which the computer handles the decimal point—we need not go into detail on this here.

In following the rest of the program, it is useful to keep in mind that the essential things a computer does in solving a problem are (1) to receive information, (2) to manipulate the information in some way, and (3) to print out the result of the manipulations. The key words to look for in a PL/I program are **GET**, which signals the computer to read in information; **DO**, which tells it to manipulate information; and **PUT**, which orders it to print an answer. Thus, the next statement in the program,

3   GET DATA (BALANCE, INT_RATE, PAYMENT) COPY;

means the computer should read the figures for the total amount of the loan (**BALANCE**), the annual interest rate, and the monthly payment. This information will have been put into the machine by the computer operator by means of punched cards before the computer begins the problem. The word **COPY** causes the machine to print out a copy of that information immediately. For the example of Section 8-1, the computer would print out

30000.00   .08   231.54

which are the figures it received for

```
        BALANCE          INT_RATE          PAYMENT
```

in that order. These numbers are printed at the top of the sheet of paper that will contain the answer to the problem. In this way, the bank will have a record of the loan conditions to which the schedule of interest payments refers.

The next PL/I statement sets up the actual computation. It reads

```
4           LOOP: DO WHILE (BALANCE > 0);
```

The word **LOOP**, which has no effect on what the computer does, is for the purpose of informing a human reader of the program that the instructions in the lines below (in our case, lines 4–8) will form a **loop**, that is, a sequence of operations which is to be repeated many times. The entire set of instructions for the loop is

```
4           LOOP: DO WHILE (BALANCE > 0);
5               INT_PAID = BALANCE * (INT_RATE/12);
6               BALANCE = BALANCE - (PAYMENT - INT_PAID);
7               PUT SKIP LIST (BALANCE, INT_PAID);
8               END LOOP;
```

The idea is that the **BALANCE**, the unpaid balance of the loan, changes (gets smaller) each time the computer runs through the operations that make up the loop. Then the new balance is used for the next repetition of the loop, and so on. The first time the loop is performed, the **BALANCE** is the figure, e.g., 30000.00, read by the machine in the previous (**GET**) step. The phrase DO WHILE (BALANCE > 0) tells the machine to check whether the current **BALANCE** is greater than zero and, if so, to perform the operations (note the key word **DO**) that follow. At the beginning of each repetition of the loop, the computer will check the value of the **BALANCE** location in its memory to determine whether to perform the steps of the loop. It will go through the loop provided the value of **BALANCE** is greater than zero. Since the first time, **BALANCE** > 0 is certainly true, the computer follows the instruction in the next statement.

The next statement in the program is

```
5               INT_PAID = BALANCE * (INT_RATE/12);
```

On the one hand, the statement is the equation of Section 8-1 for computing the interest to be paid as part of the monthly payment. On the other hand, as an instruction to a computer, it has a precise operational meaning. The symbol = in PL/I does not mean "equals" in the mathematical sense, although, in this case, such an interpretation is reasonable. The symbol = in programming language means that a copy of the result of the calculation described in the expression to the right of = is to be stored (placed) in the memory location named to the left of =. Thus, in this statement, the computer is instructed to take the value of the memory location called **INT_RATE**, divide it by 12, and multiply the result by the value of the memory location **BALANCE**. Then the product is to be placed in the memory location called **INT_PAID**.

The meaning of the symbol = in PL/I is made more evident in the next instruction:

```
6          BALANCE = BALANCE - (PAYMENT - INT_PAID);
```

This time, the statement does not make sense as a mathematical equation. The purpose of the command is to change the contents of the **BALANCE** memory location as a result of making a payment. The current value of the **INT_PAID** location, which was computed in the previous statement, is to be subtracted from the (constant) value of the **PAYMENT** location. The number which results is subtracted from the present value of the **BALANCE** memory location. Finally, that difference replaces the present value in **BALANCE**. Another way to describe this command is to say that the value of the **BALANCE** memory location is reduced by the amount **PAYMENT - INT_PAID**. The new value of **BALANCE** is the one that will be used in the next cycle through the loop.

The computer prints out the new values of **BALANCE** and **INT_PAID** in response to the command

```
7          PUT SKIP LIST (BALANCE, INT_PAID);
```

In the example of Section 8-1, the answer sheet after the first run through of the loop will look like

```
30000.00          .08              231.54
29968.46          200.00
```

The **SKIP** command caused the printer to print on the next line.

The statement

```
8              END LOOP;
```

informs the computer that it should go back to the beginning of the loop. The computer will then check the new **BALANCE** (e.g. 29968.46) to see whether it is still greater than zero and, if so, repeat the process using the new value of **BALANCE**. Thus, after the second run through the loop in the example, the printout looks like this:

```
30000.00          .08              231.54
29968.46          200.00
29936.71          199.79
```

The computer repeats the loop over and over until it checks a **BALANCE** and finds that it is negative or zero. In that event, **DO WHILE (BALANCE > 0)** means, since **BALANCE > 0** is false, that the computer should omit the loop and go directly to the step of the program immediately following **END LOOP**. But that statement is

9   END LOAN;

which instructs the computer that the program has ended.

The computer program is put onto punched cards and these are fed into the computer. Other cards containing the initial **BALANCE** and the **INT_RATE** and **PAYMENT** figures are also put into the machine. The operator starts the program, and the machine does the rest. More cards, containing other values for **BALANCE** and so on, can be put in and the computer will work out as many interest schedules as the bank requires. The punched cards containing the program are retained and used again whenever the bank needs to compute more interest schedules.

The program we have presented is very primitive. For example, a good program would identify payments by month and year and then add up the total interest paid at the end of the year so the home buyer would have the information readily available for tax purposes. A more satisfactory program would be several times the length of the one we used, and it would involve a large number of different types of instructions in addition to the ones in this section. Nevertheless, the underlying principles would be the same.

Summary

PAYMENT: monthly payment to the bank.

INT_RATE: annual interest rate.

BALANCE: amount owed on the loan.

INT_PAID: interest paid as part of the monthly payment.

COPY: print out information read in.

DCL: declare.

DO WHILE (...): check the condition in parentheses and, if it is true, follow the instructions below; if not, proceed to the step after **END**.

END: end of (part of) program.

GET: read in information.

PROCEDURE OPTIONS (MAIN): beginning of program.

PUT: print out.

SKIP: go to the next line in printing out.

; : end of a statement.

= : place the number or the result of the calculation written to the right of this symbol in the memory location named to the left of the symbol.

Arithmetic Operations

+ : plus.

− : minus.

∗ : times.

/ : divided by.

Exercises

- *In Exercises 8-47 through 8-52, translate the PL/I statements into ordinary language.*

8-47 `A = 5 * X;`

8-48 `C = (A * B)/(X + 3.5);`

8-49 `A = A + 2.31;`

8-50 `BOTTLES = BOTTLES - 12;`

8-51 `BALANCE = BALANCE * (1 + (INT_RATE/52));`

8-52 `HOURS = 40;`

- *In Exercises 8-53 through 8-55, give numerical examples which illustrate the* **DECLARE** *statements.*

8-53 `DCL VALUE DEC FIXED (7,2);`
`VALUE =`

8-54 `DCL (MONTH,DAY) DEC FIXED (2,0), YEAR DEC FIXED (4,0);`
`MONTH =        DAY =        YEAR =`

8-55 `DCL BOTTLES DEC FIXED (4,0), PRICE DEC FIXED (5,2);`
`BOTTLES =        PRICE =`

- *In Exercises 8-56 through 8-58, translate the PL/I statements into ordinary language.*

8-56 `GET DATA (COST, AMT_PD) COPY;`

8-57 `PUT SKIP LIST (MONTH, BALANCE);`

8-58 `DO WHILE (MONTH < 12);`

● *The word* **INITIAL** *in the* **DECLARE** *statement places starting data directly into the program. For example,*

BALANCE DEC FIXED (6,2) INITIAL (100.00)

means that the **BALANCE** *starts with a value of* **100.00** *when the program begins.*

8-59 Explain what the following program does and how it does it:

```
1      BANK: PROCEDURE OPTIONS (MAIN);
2         DCL BALANCE DEC FIXED (6,2) INITIAL (100.00),
             INT_RATE DEC FIXED (8,8) INITIAL (.05),
             WEEK DEC FIXED (2,0) INITIAL (1);
3         LOOP: DO WHILE (WEEK < 53);
4            BALANCE = BALANCE * (1 + (INT_RATE/52));
5            WEEK = WEEK + 1;
6            PUT SKIP LIST (WEEK, BALANCE);
7            END LOOP;
8         END BANK;
```

● *The command* **PUT LIST '...'** *tells the computer to print out the message between the single quotation marks, e.g.,*

PUT LIST 'THANK YOU'

instructs the computer to print out the message **THANK YOU.**

8-60 Explain what the following program does and how it does it:

```
1        WINE: PROCEDURE OPTIONS (MAIN);
2           DCL BOTTLES DEC FIXED (4,0), PRICE
               DEC FIXED (5,2), COST DEC FIXED
               (7,2) INITIAL (0);
3           GET DATA (BOTTLES, PRICE) COPY;
4           LOOP: DO WHILE (BOTTLES > 11);
5              COST = COST + (12 * (PRICE * .9));
6              BOTTLES = BOTTLES - 12;
7              END LOOP;
8           COST = COST + (BOTTLES * PRICE);
9           PUT SKIP LIST ('PLEASE PAY', COST,
               'THANK YOU');
10          END WINE;
```

● *The command* **IF ... THEN DO** *works like* **DO WHILE (...)** *but without the loop feature. If the statement* **...** *is true, then the program executes the steps that follow. On reaching*

END, *it proceeds to the next step in the program rather than going back to the* **IF ... THEN DO** *step. If the statement ... is false, then the computer jumps to the statement following* **END**. *Thus, in the program below, if* **HOURS** *is greater than 40, steps* 5–10 *are each executed once in order while if* **HOURS** *is less than or equal to 40, the program jumps to step* 8 *and goes on to the end.*

8-61   Explain what the following program does and how it does it:

```
1       PAYROLL: PROCEDURE OPTIONS (MAIN);
2         DCL (HOURS, RATE) DEC FIXED (4,2), PAY DEC
            FIXED (6,2) INITIAL (0);
3         GET DATA (HOURS, RATE) COPY;
4         OVERTIME: IF HOURS > 40 THEN DO;
5           PAY = (HOURS - 40) * (RATE * 1.5);
6           HOURS = 40;
7           END;
8         PAY = PAY + (HOURS * RATE);
9         PUT SKIP LIST (PAY);
10        END PAYROLL;
```

● *In order to prevent the program from carrying out the steps immediately following a subroutine headed by* **IF ... THEN DO**, *the programmer can head the next group of commands by* **ELSE**. *In the program below, if the condition of step* 4 *is true, then after carrying out that step, the computer jumps to the end of the program rather than going on to step* 5. *If the condition in step* 5 *is true, then steps* 6 *and* 7 *are executed and the computer then jumps to the end. Otherwise, the program moves down to step* 8.

8-62   Explain what the following program does and how it does it:

```
1       CHANGE: PROCEDURE OPTIONS (MAIN);
2         DCL (AMT_PD, DIFF, COST) DEC FIXED (8,2);
3         GET DATA (COST, AMT_PD);
4         IF COST = AMT_PD THEN PUT LIST ('PAYMENT
            IS CORRECT, THANK YOU');
5           ELSE IF COST > AMT_PD THEN DO;
6             DIFF = COST - AMT_PD;
7             PUT LIST ('UNDERPAYMENT: PLEASE PAY',
                DIFF);
8             ELSE DO;
9               DIFF = AMT_PD - COST;
10              PUT LIST ('YOUR CHANGE IS', DIFF,
                  'THANK YOU');
11        END CHANGE;
```

Historical Essay

Counting objects and computing sums are necessary but dull tasks. Throughout history, people have searched for mechanical aids to make these jobs easier. One such aid, the abacus with its beads representing digits and wires showing place, has been in use in China since the sixth century B.C.

Until the seventeenth century, there was no further significant progress in the simplification of computation. Then in 1604, Sir John Napier invented logarithms. Actually, logarithms had been used by the Arabs, but Napier discovered them independently. It was through his efforts that they became known in Europe. Logarithms are tables of numbers whose usefulness depends on a mathematical principle that turns multiplication problems into addition problems. For a very simple example, consider the table of base-2 logarithms shown below. To find 4 times 8, take the numbers below 4 and 8, i.e., 2 and 3, respectively; add them, and look above their sum (5) to find the answer, 32. Logarithms must have been particularly useful in the seventeenth century when knowledge of the multiplication table was not common.

In 1622, another Englishman, William Oughtred, invented the slide rule, a refinement of an earlier device called "Napier's rods." Across the channel, Pascal in 1640 devised a mechanical device that performed addition and subtraction. Shortly thereafter, Leibniz invented another that would also handle multiplication and division. These devices are the ancestors of the electromechanical calculator.

Charles Babbage (1792–1871) gave up a professorship at Cambridge University and sacrificed his personal fortune in an attempt to develop an "engine" that would perform mathematical computations of much greater complexity than those done on the devices of Pascal and Leibniz. His first project, partly supported by the British government, was a "difference engine" that was expected to carry out arithmetic manipulations with numbers of up to 26 digits. He finished a working model of part of this machine, which now resides in the science museum in South Kensington, London. Before work on the difference engine could be completed, however, Babbage had a dispute with his chief engineer, who closed the workshop and claimed ownership of crucial diagrams and working papers.

Babbage could not bear the tedious chore of rebuilding what was lost through the dispute, because he was now full of ideas for a superior "analytic

Number	$\frac{1}{8}$	$\frac{1}{4}$	$\frac{1}{2}$	1	2	4	8	16	32	64	128
Log$_2$	-3	-2	-1	0	1	2	3	4	5	6	7

engine" designed to perform long sequences of complicated arithmetic operations. This project failed, because the technology available at the time was not sufficiently developed to build Babbage's design. The elements of this design—the introduction of data and instructions into the machine by means of punched cards and the machine's ability to alter its own instructions—were so fundamentally sound that Howard Aiken, one of the pioneers of the modern digital computer, once said that if Babbage had been born 75 years later he, Aiken, would have been out of a job.

Although Babbage used punched cards, it was Herman Hollerith who first patented both the punched cards and an electromechanical device that read the cards and tabulated the data they contained. Using Hollerith's equipment, the United States census of 1890 was performed in one-third the time of the previous census and with increased accuracy.

About 1937, George Slibitz and Howard Aiken began to work independently on a sequentially operated digital computer. The first digital computer was demonstrated in 1940 at the American Mathematical Society meeting in Hanover, New Hampshire. Meanwhile, on a contract from the Navy, IBM worked in conjunction with Harvard University to produce the Automatic Sequence Controlled Calculator, or Mark I, which operated by means of electromechanical relays. The next stage of computer development, represented by Mark IV and ENIAC,

used vacuum tubes and electronic circuitry.

During the late 1940s, several different machines were developed and, more importantly, many advances in computer theory were made. John von Neumann, an eminent mathematician from the Institute for Advanced Study in Princeton, became interested in computers after seeing a demonstration of ENIAC. In a series of articles written with Herman Goldstine and A. S. Burks, he introduced the idea of storing operating instructions (the program) as well as data in the computer memory, and also suggested the use of binary numbers because of the economy of operation possible through the employment of two state devices. It was during this period—with MANIAC at Princeton, UNIVAC at the Bureau of Standards, and Mark IV at Harvard—that an early market forecast stated that the entire United States "might" eventually need as many as *nine* computers.

By 1954, the situation had changed completely. UNIVAC I and the IBM 650 had become the first commercially successful stored-program, vacuum-tube, drum-memory machines. Almost 1000 IBM 650s were installed. The "second generation" of computers, such as the RCA 501, which used transistors to replace vacuum tubes and high-speed magnetic core elements to replace the drum memory, were developed in the late 1950s. The most recent "third generation" of computers use monolithic integrated circuitry (circuits as complicated as transistor radios can be contained within a piece of silicon 1/10 inch square). Computers of this type are represented by the IBM 360 series that appeared in 1964.

Each computer model has its own internal machine language. Originally, computers had to be programmed in their machine language. The first simplification permitted the use of mnemonics instead of numbers for instructions, for example, LOAD X for +3762000508. A big improvement came with the development of FOR-TRAN in 1957. This high-level language can be used on many different machines and is widely employed for scientific purposes. BASIC (Beginners All-Purpose Symbolic Code) is an almost conversational language, but one with rather limited applicability. On the other hand, PL/I is more flexible than FORTRAN and can be used for both business and scientific purposes. Other common languages are ALGOL (Algorithmic Language), the official international scientific language, and COBOL (Common Business-Oriented Language).

References

Books and articles concerning computers are so numerous and so widely available that we can only mention a few that we found particularly useful and interesting. The *Encyclopedia Britannica* article on computers is excellent. One of the best general articles on computers is "Behold the Computer Revolution" by P. T. White in the November 1970 issue of *National Geographic*, which has superb illustrations. There have been many articles in *Scientific American* over the past 25 years, and several of the most interesting are gathered together as part of the last section of *Mathematics in the Modern World*, edited by Morris Kline (San Francisco: W. H. Freeman, 1968).

For more information on the mathematics of finance, see either *Business Mathematics* by R. Robert Rosenberg and Harry Lewis (New York: McGraw-Hill, 1963) or *Mathematics of Finance* by Paul M. Hummel and Charles L. Seebeck (New York: McGraw-Hill, 1971), or any book listed in the library card catalog under mathematics of finance or business mathematics.

If you want to know more about programming, there are many texts available such as the following: Thomas Worth, *BASIC for Everyone* (Englewood Cliffs, N.J.: Prentice-Hall, 1976). J. R. Sturgul and M. J. Merchant, *Applied FORTRAN IV Programming*, 2d ed. (Belmont, Ca.: Wadsworth, 1976). R. Conway and D. Gries, *An Introduction to Programming*, 2d ed. (Cambridge, Mass.: Winthrop, 1975) (about PL/I).

For specific applications of computers, the best source is the *Readers Guide to Periodical Literature*, where articles are listed by specific application.

We highly recommend Jeremy Bernstein's *The Analytical Engine* (New York: Random House, 1964), which is a fascinating and very well written history of the computer through 1964, as well as an account of the author's own experiences with FORTRAN.

Answers

The answers to most of the exercises in the book are given below. To save space, answers in the form of one-column ($k \times 1$) matrices are written as row matrices with the superscript ^{T} to indicate that the matrix has been turned on its side. Numbers that should be read from top to bottom are printed from left to right. Thus, for example, the answer to 3-3(b),

$$[0.002 \quad 0.212 \quad 0 \quad 0.003]^T$$

means

$$\begin{bmatrix} 0.002 \\ 0.212 \\ 0 \\ 0.003 \end{bmatrix}$$

In general, the matrix answer

$$[a_1 \quad a_2 \cdots a_k]^T$$

should be read as

$$\begin{bmatrix} a_1 \\ a_2 \\ \vdots \\ a_k \end{bmatrix}$$

Answers are generally written in fraction rather than decimal form. However, if you are using a calculator, you may wish to convert numbers to decimal form in order to carry out some of the more complicated computations. If you do this, you must be conscious of the fact that replacing a fraction by the decimal obtained by dividing denominator into numerator usually introduces an error into the calculation. For example, 0.33333333 is not the same number as 1/3. Consequently, when arithmetic operations are performed using these decimal approximations, the errors are compounded and the eventual answer may differ somewhat from the correct answer—even though you went through the correct procedure. Therefore, if your answer is based on decimal calcula-tions, you can compare it to the answer given here by converting the fraction to decimal form, but you should not expect the numbers to be exactly the same.

Chapter 2

2-1 3/6
2-2 4/6
2-3 5/6
2-4 1/4
2-5 3721/5931
2-6 60/75
2-7 2/7
2-8 3/7
2-9 5/14
2-10 60/75
2-11 3/28
2-12 11/28
2-13 6/28
2-14 1
2-15 287/348
2-16 305/518
2-17 592/866
2-18 7/43
2-19 21/43
2-20 29/43
2-21 33/159
2-22 3/103
2-23 1/38
2-24 18/38
2-25 12/38
2-26 6/38
2-27 1000
2-28 1/1000
2-29 9/1000
2-30 990/1000
2-31 13/52

2-32 2/52
2-33 8/52
2-34 44/52
2-35 20/52
2-36 28/52
2-37 27
2-38 4
2-39 4/27
2-40 312
2-41 20
2-42 17,576
2-43 15,600
2-44 216
2-45 108
2-46 27
2-47 120
2-48 24
2-49 12
2-50 32
2-51 1024
2-52 1024
2-53 16
2-54 36
2-55 30,000
2-56 27
2-57 6
2-58 30
2-59 240
2-60 7000
2-61 24
2-62 33 cups
2-63 5 minutes 43 seconds
2-64 3 minutes 16 seconds
2-65 3 minutes 51 seconds

2-66 3 minutes 55 seconds

2-67 21; 20,160; 30,240; 84

2-68 3; 4; 5; n

2-69 336

2-70 24

2-71 10

2-72 560

2-73 336

2-74 210

2-75 5040

2-76 120

2-77 84

2-78 84/120

2-79 45

2-80 28

2-81 28/45

2-82 1/45

2-83 6

2-84 24

2-85 3024

2-86 22,100

2-87 1144

2-88 52

2-89 1144/22,100

2-90 52/22,100

2-91 three of a kind

2-92 343

2-93 210

2-94 210/343

2-95 273,000

2-96 27,885

2-97 111,540

2-98 35

2-99 15

2-100 22/52

2-101 28/52

2-102 .80

2-103 25

2-104 24/37

2-105 .155

2-106 77

2-107 .24

2-108 .85

2-109 .70

2-110 .06

2-111 .45

2-112 .15

2-113 29/36

2-114 9/36; 27/36

2-115 26/42

2-116 .20

2-117 .15

2-118 .97

2-119 .38

2-120 .16

2-121 .20

2-122 .95

2-123 7/8

2-124 11/16

2-125 0

2-126 120

2-127 20

2-128 20/120

2-129 100/120

2-130 .35

2-131 .60

2-132 .44

2-133 8/216

2-134 36

2-135 6; 3; 1

2-136 6/36; 3/36; 1/36

2-137 10/36

2-138 26/36

2-139 .20

2-140 1/11

2-141 3/11

2-142 3/7

2-143 .20

2-144 2/7

2-145 4/27

2-146 2/9

2-147 1/8

2-148 .25

2-149 .15

2-150 not independent

2-151 30/49

2-152 independent

2-153 5/7

2-154 not independent

2-155 independent

2-156 30/50

2-157 29/49

2-158 30/49

2-159 .36

2-160 not independent

2-161 15/55

2-162 C_3^{45}/C_5^{47}

2-163 23/36

2-164 5/6

2-165 not independent

2-166 C_{13}^{16}/C_{13}^{52}

2-167 C_8^{11}/C_8^{47}

2-168 1/36

2-169 1/9

2-170 1/12

2-171 1/6

2-172 5/18

2-173 .0375

2-174 3/156

2-175 10/156

2-176 75/156

2-177 1/1,679,616

2-178 .195

2-179 .755

2-180 .714

2-181 3/16

2-182 3/64

2-183 .0336

2-184 .3024

2-185 .0084

2-186 .3476

2-187 .9664

2-188 .14

2-189 .09135

2-190 1/16

2-191 22/208

2-192 1/4

2-193 24/169

2-194 1/20

2-195 .005

2-196 $10(.15)^2(.85)^3$

2-197 $35(.6)^4(.4)^3 + 21(.6)^5(.4)^2$
$+ 7(.6)^6(.4) + (.6)^7$

2-198 $6(.8)^5(.2) + (.8)^6$

2-199 5/16

2-200 $1 - [(.35)^6 + 6(.65)(.35)^5]$

2-201 $(12/13)^5$

2-202 $1 - [(12/13)^5$
$+ 5(1/13)(12/13)^4]$

2-203 56/1024

2-204 $45(.8)^8(.2)^2$
$+ 10(.8)^9(.2) + (.8)^{10}$

2-205 $45(1/4)^8(3/4)^2$
$+ 10(1/4)^9(3/4) + (1/4)^{10}$

2-206 $21(.15)^2(.85)^5$

2-207 $(.85)^7 + 7(.15)(.85)^6$
$+ 21(.15)^2(.85)^5$

2-208 57/64

2-209 $10(3/13)^3(10/13)^2$

2-210 $35(2/9)^4(7/9)^3$

2-211 11/32

2-212 466/512

2-213 $(.85)^5 + 5(.15)(.85)^4$

2-214 $(.95)^5$

2-215 $5(.95)^4(.05) + (.95)^5$

2-216 $1 - [(.7)^5 + 5(.3)(.7)^4]$

2-217 .9882

2-218 .38

2-219 $1 - (.9)^4$

2-220 $1 - (1/\sqrt{5})$

2-221 $(1 - p)^{12} = .15$

2-222 $1 - [(.76)^7 + 7(.24)(.76)^6$
$+ 21(.24)^2(.76)^5$
$+ 35(.24)^3(.76)^4]$

2-223 .135

2-224 .1825

2-225 .0078

2-226 .00111

2-227 66/1800

2-228 Republican

2-229 .0001

2-230 .0099

2-231 .0198

2-232 .9801

2-233 .00206

2-234 5/13

2-235 .60

2-236 $10,100

2-237 $-$2/37

2-238 $-$0.05

2-239 2.87 offspring

2-240 $45

2-241 $-$0.025

2-242 2 maidens

2-243 .8385; $-$0.3257

2-244 2756/2652 draws

2-245 153/221 cards

2-246 5/6

2-247 300/795

2-248 120/795

2-249 45/235

2-250 16/17

2-251 35/114

2-252 28/48

2-253 .25

2-254 2/5

2-255 60/86

2-256 100/108

2-257 21/41

2-258 225/475

2-259 200/395

2-260 392/410

2-261 1/3

2-262 1/3

2-263 approximately .89

2-264 approximately .107

2-265 approximately .14

2-266 approximately .86

2-267 approximately .01

2-268 approximately .01

2-269 approximately .25

2-270 approximately .95

2-271 statistics

2-272 215/216

2-273 .5725

2-274 350/715

2-275 $C_3^8 C_2^{40} / C_5^{48}$

2-276 2/5

2-277 40/59

2-278 .685

2-279 5325/5500

2-280 .06

2-281 $1 - [(.68)^{10} + 10(.68)^9(.32) + 45(.68)^8(.32)^2 + 120(.68)^7(.32)^3]$

2-282 55,272

2-283 60

2-284 $-$0.15

2-285 $1 - [(.65)^7 + 7(.35)(.65)^6]$

2-286 $(.25)^3(.15) + 4(.25)^3(.75)(.6) + 10(.25)^3(.75)^2(.25)$

2-287 1/2

2-288 .36

2-289 .76; independent

2-290 no

2-291 $(.7)^{17}(3/216) + 18(.7)^{17}(.3)(1/216)$

2-292 $1 - [(7/12)^4 + 4(5/12)(7/12)^3]$

2-293 3.975 points

2-294 63

2-295 .335

2-296 .0135

2-297 .05

2-298 240

2-299 11,760/19,445

2-300 not independent

2-301 51/99

2-302 3/8

2-303 200/1175

2-304 .6675

2-305 17,376,000

2-306 $35C_3^{65}/C_4^{100}$

2-307 $14.50/52

2-308 990

2-309 32/37

2-310 $32/37 \neq 37/94$

2-311 $(.69)^4$

Chapter 3

3-1(a) 0.293

(b) 58,600

(c) 0

(d) 2200

3-2(a) 0.014

(b) 0

(c) 1.065

(d) $48 million

(e) $480 million

(f) $1.288 billion

3-3(a) [0.298   0.002   0   0]

(b) $[0.002 \quad 0.212 \quad 0 \quad 0.003]^T$

(c) [0.029   0.003   0.004   0.030]

(d) there is none

(e) 0

(f) 0

(g) 4×1

(h) 119

(i) there is none

3-4(b) $[0 \quad 208 \quad 980 \quad 2025 \quad 3186]^T$

(c) [2025   1833   1038   0   1270]

(d) 1038

3-5(b) $[6 \quad 4 \quad 8]^T$

(c) 63

(d) there is none

3-6(a) $14

(b) $11.90

(c) $14.10

(e) 6×2

(f) [.28   0]

(g) 0

3-8 equal

3-9 equal

3-10 unequal; different sizes

3-11 equal

3-12 unequal; $0 \neq -3$

3-13 $x = 4; y = -1$

3-14 impossible

3-15 $x = -1; y = 1$

3-16 impossible

3-17 $x = -1; y = 5$

3-18 $\begin{bmatrix} -3 & -2 \\ 1 & -2 \end{bmatrix}$

3-19 $[1 \quad -1/2 \quad -2]$

3-20 $[-1 \quad -3 \quad -1]^T$

3-22 $\begin{bmatrix} -3 & 1 \\ -2 & 0 \\ 1 & 0 \end{bmatrix}$

3-23 $[-3 \quad -2]^T$

3-25 $[1 \quad -2 \quad -3]$

3-27 $[-4 \quad -3 \quad -4]^T$

3-28 $\begin{bmatrix} 2 & 1 \\ 3 & 2 \end{bmatrix}$

3-29 AB is 1×4; BA does not exist

3-30 neither exists

3-31 both are 2×2

3-32 AB does not exist; BA is 1×4

3-33 AB does not exist; BA is 1×4

3-34 AB is 2×2; BA is 4×4

3-35 neither exists

3-36 AB is 1×4; BA does not exist

3-37 [1]

3-38 [1]

3-39 $[-x + 2y + z]$

3-40 $[0 \quad -1 \quad 2 \quad -2]$

3-41 $[-9/2 \quad -17/2]$

3-42 $[2a + b - c \quad a + c \quad -c]$

3-43 $\begin{bmatrix} -1 & 3 & 1 \\ 1/2 & -1 & 2 \end{bmatrix}$

3-44 $\begin{bmatrix} 1 & 0 & 2 \\ 2 & 0 & 4 \\ -1 & 0 & -2 \\ -2 & 0 & -4 \end{bmatrix}$

3-45 $[-3/2 \quad 5/2]^{\mathrm{T}}$

3-47 $[-x+2y \quad 3x \quad -x-z]^{\mathrm{T}}$

3-48 $\begin{bmatrix} 1 & 2 & -1 \\ 0 & 1 & 1 \end{bmatrix}$

3-49 $\begin{bmatrix} 0 & -2 \\ 0 & 0 \end{bmatrix}\begin{bmatrix} 0 & 2 \\ 0 & 0 \end{bmatrix}$

3-50 $\begin{bmatrix} 8 & 4 \\ 3 & 9 \end{bmatrix}\begin{bmatrix} 8 & -3 \\ 4 & 9 \end{bmatrix}$

3-51 $\begin{bmatrix} -12 & -6 \\ 6 & 0 \end{bmatrix}\begin{bmatrix} -12 & -6 \\ 6 & 0 \end{bmatrix}$

3-52 $[3/2 \quad 2 \quad 5]$

3-53 $\begin{bmatrix} 2 & 1 \\ 0 & 1 \end{bmatrix}$

3-54 $[1 \quad -1]^{\mathrm{T}}$

3-55 $\begin{bmatrix} 4 & 4 & 4 \\ 1 & 0 & -4 \end{bmatrix}$

3-56 $[0]$

3-57 $[2 \quad -2]^{\mathrm{T}}$

3-58 $[3 \quad 1]^{\mathrm{T}}$

3-60(c) \$18,400

3-61(c) 11 man-hours

3-62(c) \$255,000; 120 shares

3-63(c) \$218

3-64(c) 230 ounces syrup; 680 ounces milk; 770 ounces ice cream

3-65(c) 7/5 loaves of bread; 5/2 quarts of milk; 1/5 pound of coffee; 9/16 pound of cheese

3-73 $\begin{bmatrix} 2 & -2 & 0 \\ 0 & 2 & 2 \end{bmatrix}$

3-74 $[-6 \quad -6]^{\mathrm{T}}$

3-75 $\begin{bmatrix} 0 & 0 & 0 \\ 0 & 0 & 0 \end{bmatrix}$

3-76 $[-4]$

3-77 $\begin{bmatrix} 1 & 6 \\ -2 & 9 \end{bmatrix}$

3-85 1/8

3-86 $-2/5$

3-87 $\begin{bmatrix} 1/2 & 1 \\ 5/2 & 3 \end{bmatrix}$

3-88 $\begin{bmatrix} 1 & -1 & 1 \\ -1/3 & 8/3 & -8/3 \end{bmatrix}$

3-89 $[0 \quad 2 \quad -1]^{\mathrm{T}}$

3-90 $\begin{bmatrix} 1/2 & -1/2 & -4/5 \\ -1/2 & 1/2 & 2/5 \end{bmatrix}$

3-91 $[0 \quad 1 \quad 5]^{\mathrm{T}}$

3-93 three chairs; two tables

3-94 100/15 units of type A; no type B; 100/15 units of type C

3-95 3/2 batches of love potion; 1 batch of cold remedy

3-96 twelve clerks; twenty secretaries; one receptionist

3-97 $\begin{bmatrix} 2 & 1/2 & 1 & 0 \\ 1 & -1 & 0 & 1 \end{bmatrix}$

3-104 $\begin{bmatrix} 1 & 2 & 1 & 0 \\ 0 & -3 & -2 & 1 \end{bmatrix}$

3-107 divide through the second row by -1

3-108 interchange the second and third rows

3-109 add 1/4 times the second row to the first row

3-110 add 1/2 times the second row to the first row

3-111 divide through the second row by $-17/2$

3-112 add 1/3 times the first row to the third row

3-113 add -3 times the first row to the second row

3-114 interchange the third and fourth rows

3-115 $\begin{bmatrix} 2 & 1 \\ -1/2 & 0 \end{bmatrix}$

3-116 $\begin{bmatrix} 2 & 3 \\ -1 & -2 \end{bmatrix}$

3-117 $\begin{bmatrix} 1/2 & 0 & 0 \\ 1 & -1 & 0 \\ 1 & -1 & 0 \end{bmatrix}$

3-118 $\begin{bmatrix} -1 & 1 & 2 \\ -3 & 2 & 4 \\ 1 & -1 & -1 \end{bmatrix}$

3-119 $\begin{bmatrix} 1/2 & 1/2 & 0 \\ 1/2 & 3/2 & -1 \\ -1/2 & 3/2 & -1 \end{bmatrix}$

3-120 $\begin{bmatrix} 5/8 & 1/4 & -1/8 \\ 1/4 & 1/2 & -1/4 \\ -3/4 & -1/2 & 3/4 \end{bmatrix}$

3-123 32 sodas; 128 milk shakes

3-124 4 units of type A; 5 units of type B

3-125 five vice-presidents; ten division managers; ten assistant managers

3-126 four mechanics; three attendants

3-127 $x = -2/3; y = -1/3$

3-128 $x = 4; y = 8$

3-129 $x = 3/2; y = 3/2; z = -1$

3-130 infinitely many solutions

3-131 no solutions

3-132 $x = -11/5; y = 3/5; z = 7/10$

3-133 infinitely many solutions

3-134 no solutions

3-135 infinitely many solutions

3-136 $w = -6; x = -3; y = 2; z = 1$

3-137 no solutions

3-138 no solutions

3-139 no solutions

3-140 infinitely many solutions

3-141 6 days in England; 4 days in France; 4 days in Spain

3-142 inconsistent

3-143 eight red cards; five black cards

3-144 inconsistent

3-145 insufficient information

3-146 300 shares of Eastern Airlines; 100 shares of Hilton Hotels

3-147(e) \$18,192,340,000 agricultural production; \$73,209,140,000 manufactured goods; \$66,790,490,000 labor

3-148(d) 195,433,182 pounds agricultural products; 25,969,790 pounds manufactured goods; 13,617,528 pounds energy

3-149(d) 16,191,566,000 pounds non-metal products; 9,731,306,000 pounds metal products; 2,504,381,000 pounds energy; 18,187,412,000 pounds services

Chapter 4

4-1 36%

4-2 40% fair; 30% cloudy; 30% rain

4-3 47%

4-4 .44

4-5 .2375

4-6 9% poor; 21.5% fair; 36.25% good; 33.25% excellent

4-7 39.25% As; 30% Bs; 19.5% Cs; 11.25% D/Fs

4-8 51%; 38.5%; 10.5%

4-9 0; .55; .45

4-10 22.25% arts; 37% humanities; 13.25% natural sciences; 27.5% social sciences

4-11 $[.11 \quad .89]^T$; $[.072 \quad .928]^T$; $[.0644 \quad .9356]^T$

4-12 $[.14 \quad .02 \quad .84]^T$; $[.506 \quad .028 \quad .466]^T$; $[.2824 \quad .1012 \quad .6164]^T$

4-13 $[0 \quad .4 \quad .6]^T$; $\quad$ $[.24 \quad .28 \quad .48]^T$; $[.24 \quad .256 \quad .504]^T$

4-14 $[.51 \quad .03 \quad .46]^T$; $[.531 \quad .003 \quad .466]^T$; $[.5331 \quad .0003 \quad .4666]^T$

4-15 .275; .2925; .30475

4-16 .45 fair, .25 cloudy, .30 rain; .46 fair, .31 cloudy, .23 rain; .4575 fair, .3195 cloudy, .223 rain

4-17 $[.33424 \quad .66576]^T$

4-18 $[.0735 \quad .7674 \quad .1591]^T$

4-19 39.25%

4-20 44.29% humanities; 17.01% social sciences; 28.7% natural sciences

4-26 $[.5 \quad .5]^T$

4-27 $[7/15 \quad 6/15 \quad 2/15]^T$

4-28 $[.59375 \quad .40625]^T$

4-29 $[.37 \quad .18 \quad .45]^T$

4-30 75/85

4-31 19/22 farmers; 5/44 craftsmen; 1/44 leaders

4-32 5/8

4-33 2/7 Jones; 1/8 Smith; 33/56 others

4-34 161/504 win; 224/504 draw; 119/504 lose

Chapter 5

5-1(a) no, not enough water
$\quad$ **(b)** no, too much potatoes
$\quad$ **(c)** yes
$\quad$ **(d)** yes

5-2(a) 1998; $22,490
$\quad$ **(b)** 1860; $9300
$\quad$ **(c)** 1995; $11,850
$\quad$ **(d)** 1950; $16,000

5-3 $x \geq 0; y \geq 0; x + y \leq 310;$ $x \leq 100; 15x + 6y \leq 2000$

5-4 $x \geq 0; y \geq 0; 2x + (7/4)y \leq 120;$ $(3/2)x + 4y \leq 200; x \leq 60;$ $y \leq 45$

5-5(a) yes
$\quad$ **(b)** no, too many tables
$\quad$ **(c)** yes
$\quad$ **(d)** no, not enough assembly man-hours

5-6(a) $4875
$\quad$ **(b)** $4500
$\quad$ **(c)** $4500
$\quad$ **(d)** $5875

5-7 yes; $a = 5, b = 2, c = 4$

5-8 yes; $a = 1, b = -1, c = 6$

5-9 no; x^2

5-10 yes; $a = -3, b = 1, c = 2$

5-11 yes; $a = -2, b = 4, c = -5$

5-12 yes; $a = 1, b = 0, c = 1$

5-13 no; x^3

5-14 yes; $a = 1, b = 1, c = 2$

5-15 yes; $a = -1, b = -1, c = -7$

5-16 $x \geq 0; y \geq 0; x + y \leq 7; x \geq 1;$ $y \geq 2x; \text{Calories} = 600x + 350y$

5-17 $x \geq 0; y \geq 0; x + y \geq 5;$ $x - y \geq 1; \text{Sugar} = (5/4)x + y$

5-18 $x \geq 0; y \geq 0; y \geq 1; 25x + 10y$ $\geq 100; 10x + 50y \geq 100; \text{Cost}$ $= 10x + 15y$

5-21 $A = -55; B = -43; C = -36.5;$ $D = -21; E = -13; F = 4;$ $G = 25; H = 39; I = 51; J = 59$

5-22 $A(-4.5, 0); B(-2, 2); C(-0.5, 3.5); D(3, 4); E(5.5, 0.5); F(2.2, 0); G(3.5, -0.5); H(2, -4); I(0, -3); J(-4, -3.5); K(-5, -0.5)$

5-27(a) yes
$\quad$ **(b)** no
$\quad$ **(c)** yes
$\quad$ **(d)** yes
$\quad$ **(e)** yes
$\quad$ **(f)** yes

5-49 (10, 0)

5-50 (1/2, 3/2)

5-51 (8/3, 1/3)

5-52 (5/9, 40/9)

5-53 small smelter 8 hours; large smelter 11 hours

5-54 20 cakes; 20 batches of cookies

5-55 none under diesel fuel; 10 hours under natural gas

5-56 4 hours under diesel fuel; 4 hours under natural gas

5-57 \$375 million in mortgages; \$125 million in business loans

5-58 neither

5-59 $A \geq B$

5-60 neither

5-61 $A \leq B$

5-62 $A \geq B$

5-63 $A \leq B$

5-64 $X = \begin{bmatrix} x \\ y \end{bmatrix}; \quad A = \begin{bmatrix} 1 & 1 \\ 6 & 1 \end{bmatrix};$

$B = \begin{bmatrix} 4 \\ 6 \end{bmatrix}; \quad C = [3 \quad 4]$

5-65 $Z = [x \quad y];$

$A = \begin{bmatrix} -1 & 0 & 6 & 1 \\ 0 & -1 & 5 & 6 \end{bmatrix};$

$B = \begin{bmatrix} 1 \\ 8 \end{bmatrix};$

$C = [-5 \quad -5 \quad 30 \quad 6]$

5-66 $X = \begin{bmatrix} x \\ y \end{bmatrix}; \quad A = \begin{bmatrix} -1 & 0 \\ 0 & -1 \\ 2 & 3 \\ 1 & 2 \\ 3 & 1 \end{bmatrix};$

$B = \begin{bmatrix} -1 \\ -2 \\ 10 \\ 7 \\ 12 \end{bmatrix}; \quad C = [5 \quad 9]$

5-67 $Z = [x \quad y]; \quad A = \begin{bmatrix} -1 & 3 & 5 \\ -1 & 2 & 6 \end{bmatrix};$

$B = \begin{bmatrix} 40 \\ 35 \end{bmatrix}; \quad C = [-12 \quad 8 \quad 15]$

5-69 $Z = [x \quad y \quad z]; A = \begin{bmatrix} 1 & 0 \\ 1 & 1 \\ 30 & 15 \end{bmatrix};$

$B = \begin{bmatrix} 50 \\ 150 \\ 3000 \end{bmatrix}; C = [207 \quad 200]$

5-70 $X = \begin{bmatrix} w \\ x \\ y \\ z \end{bmatrix};$

$A = \begin{bmatrix} 1 & 0 & 0 & 0 \\ 0 & 1 & -1 & 0 \\ 0 & 1 & 1 & 1 \end{bmatrix};$

$B = \begin{bmatrix} 12 \\ 0 \\ 15 \end{bmatrix}; \quad C = [2 \quad 3 \quad 1 \quad 1]$

5-71 $Z = [x \quad y \quad z];$

$A = \begin{bmatrix} 1/4 & 1/8 & 1/4 \\ 1/8 & 1/4 & 1/4 \\ 3/8 & 1/8 & 3/8 \end{bmatrix};$

$B = \begin{bmatrix} 1 \\ 2 \\ 3/2 \end{bmatrix}; C = [3 \quad 3 \quad 3]$

5-72 $X = \begin{bmatrix} x \\ y \\ z \end{bmatrix};$

$A = \begin{bmatrix} 1 & 1 & 1 \\ 0 & 0 & -1 \\ 1 & -0.5 & 0 \\ 1 & 0 & -1 \end{bmatrix};$

$B = \begin{bmatrix} 100 \\ -15 \\ 0 \\ 0 \end{bmatrix};$

$C = [0.09 \quad 0.075 \quad 0.05]$

5-75 $x = 3/5; y = 8/5$

5-76 $x = 1; y = 1; z = 1$

5-77 $x = 2; y = 1$

5-78 $x = 1/4; y = 1/4; z = 1$

5-79 $x = 5/3; y = 1$

5-80 $x = 4; y = 2; z = 6$

5-81 $x = 2; y = 1$

5-82 500 square feet of vegetables; 500 square feet of flowers

5-83 1/2 hour fast walking; 1/4 hour strolling; 1/4 hour talking

5-84 2.5 ounces of syrup; 1.5 ounces of cream; 4 ounces of soda water; 4 ounces of ice cream

5-85 $x = 0; y = 2$

5-86 $x = 0; y = 2$

5-87 $x = 1/2; y = 1$

5-88 $x = 1; y = 1; z = 1/2$

5-89 $x = 2; y = 8; z = 0$

5-90 $x = 2; y = 0$

5-91 $x = 1/2; y = 3/2$

5-92 $x = 1; y = 1; z = 0$

5-93 $x = 1; y = 0$

5-94 $x = 3/4; y = 1; z = 1$

5-95 $x = 1/2; y = 1$

5-96 $x = 0; y = 6; z = 4$

5-97 $w = 0; x = 10; y = 5; z = 0$ or $w = 0; x = 0; y = 5; z = 10$

5-98 $x = 0; y = 0; z = 1$

5-99 500 square feet of vegetables; 500 square feet of flowers

5-100 four rings; ten earrings; no pins; three necklaces

5-101 2 hours jogging; 4 hours bicycling; 2 hours swimming

5-102 3/2 hours jogging; 7/2 hours bicycling; 2 hours swimming

Chapter 6

6-1

		Children	
		Seashore	Mountains
Parents	Seashore	$(+1, +1)$	$(-1, -1)$
	Mountains	$(-1, -1)$	$(+1, +1)$

6-2

	Win	Lose
Even	1	-1
1–12	10	-5
17	36	-1

6-3

	Decline	Steady	Increase
Blue chip	$-\$2000$	$\$500$	$\$1500$
Growth	$-\$7000$	$-\$500$	$\$6000$

6-4

| | Blue | | |
	Mountain	Valley	Plain
Green Mountain	−30	20	20
Valley	15	−20	15
Plain	5	5	−10

6-5

	Highway	No Highway
Housing development	10%	40%
Shopping center	−40%	60%
Office complex	75%	−30%
Industrial	100%	−50%

6-6

	Rain	No rain
Picnic	−5	5
Home	2	−2

6-7

| | Company B | | |
	Beginning	Midyear	End
Company A Beginning	(0,0)	(3, 2)	(5, 5)
Midyear	(2, 3)	(0, 0)	(3, 2)
End	(5, 5)	(2, 3)	(0, 0)

6-8

| | Airline B | | | |
	Neither	Redesign	Snacks	Both
Airline A Neither	0	−5%	−5%	−10%
Redesign	5%	0	0	−5%
Snacks	5%	0	0	−5%
Both	10%	5%	5%	0

6-9

		Network *B*					
		SEC	SCE	ESC	ECS	CES	CSE
Network *A*	SEC	0	0	0	−2	0	2
	SCE	0	0	−2	0	2	0
	ESC	0	2	0	0	−2	0
	ECS	2	0	0	0	0	−2
	CES	0	−2	2	0	0	0
	CSE	−2	0	0	2	0	0

6-10 player *A*: row 3; player *B*: column 3

6-11 player *A*: row 1; player *B*: column 3

6-12 player *A*: row 3; player *B*: columns 1 or 3

6-13 player *A*: row 3; player *B*: column 4

6-14 player *A*: row 2; player *B*: column 2

6-15 player *A*: rows 1 or 4; player *B*: columns 1 or 3

6-16 player *A*: rows 1 or 3; player *B*: columns 3 or 4

6-17 player *A*: row 1; player *B*: column 1

6-18 player *A* choose −1 or 1; player *B* choose 1

6-19 blue chip stock

6-20 Mexican border

6-21 barley

6-22 stay home

6-23 subcompacts

6-24 guilty

6-25 0

6-26 −33/64

6-27 .54

6-28 −32/52

6-29 .116

6-30 −7/18

6-31 21/100

6-32 −37/36

6-33 $Q = [1 \quad 0]^T$

6-34 all the same

6-35 pure strategy, "row 3"

6-36 (*b*)

6-37 (*c*)

6-40 $P^* = [1/3 \quad 2/3]$
$Q_* = [1/3 \quad 2/3]^T; v = 5/3$

6-41 $P^* = [3/5 \quad 2/5];$
$Q_* = [3/5 \quad 2/5]^T; v = 4/5$

6-42 $P^* = [1/2 \quad 1/2];$
$Q_* = [1/2 \quad 1/2 \quad 0]^T; v = 3/2$

6-43 $P^* = [1/2 \quad 1/2];$
$Q_* = [1/2 \quad 0 \quad 1/2 \quad 0]^T;$
$v = -1/2$

6-44 $P^* = [1/5 \quad 4/5];$
$Q_* = [3/5 \quad 2/5]^T; v = -3/5$

6-45 $P^* = [1/4 \quad 3/4 \quad 0];$
$Q_* = [1/2 \quad 1/2]^T; v = 1/2$

6-46 $P^* = [0 \quad 1 \quad 0];$
$Q_* = [0 \quad 0 \quad 1 \quad 0]^T; v = 1$

6-47 $P^* = [3/4 \quad 1/4];$
$Q_* = [3/4 \quad 0 \quad 1/4]^T;$
$v = -1/4$

6-48 $P^* = [0 \quad 1 \quad 0];$
$Q_* = [1 \quad 0 \quad 0]^T; v = -1$

6-49 $P^* = [1/2 \quad 1/2];$
$Q_* = [1/2 \quad 1/2 \quad 0 \quad 0]^T;$
$v = 1/2$

6-50 $P^* = [2/3 \quad 1/3 \quad 0]$;
$Q_* = [1/3 \quad 2/3]^T; v = 1/6$

6-51 $P^* = [0 \quad 1]$; $Q_* = [0 \quad 0 \quad 1]^T$;
$v = 0$

6-52 $P^* = [1/2 \quad 1/2]$;
$Q_* = [1/2 \quad 1/2]^T; v = 0$

6-53 4/7 take jacket; 3/7 do not take jacket; $v = -1/7$

6-54 player A always plays the penny; player B always plays the dime; the value is a 1¢ gain for player B

6-55 invest all $10,000 in blue chip stocks

6-56 quarterback: 11/13 running, 2/13 pass; linebackers: 7/13 running, 6/13 pass; $v = 38/13$ (approximately 2.9 yards gained by the quarterback's team)

6-57 plant only sugar beets

6-58 both players make paper, rock, and scissors equally likely ($p = 1/3$); $v = 0$

6-59 green army: 3/19 mountain, 4/19 valley, 12/19 plain; blue army: 7/19 mountain, 3/19 valley, 9/19 plain; value is 5/19 in favor of the green army

Chapter 7

7-14 mean = 50 words per minute; variance = 25 words per minute

7-15 mean = $161; variance = $481

7-16 197,800 shares

7-17 .284

7-18 63%; .23

7-19 − .23

7-20 mean = 251.5 pounds; variance = 1020/7 pounds

7-21 proportion = .60; variance = .24

7-22 mean = 93°F; variance = 54°F

7-23 70

7-24 26%

7-25 93%

7-26 52%

7-27 27%

7-28 22%

7-29 1%

7-30 84%

7-31 76%

7-32 12%

7-33 1%

7-34 46%

7-35 approximately 90%

7-36 76%

7-37 46%

7-38 .0224; 38%

7-39 .0489; 96%

7-40 68%

7-41 98% +

7-42 .05

7-43 .02

7-44 .017

7-45 .98

7-46 .72

7-47 400; 712; 1600; 6400; 25,600

7-48 1057; 1878; 4225; 16,900; 67,600

7-49 690; 1225; 2757; 11,025; 44,100

7-50 169

7-51 256

7-52 361

7-53 441

7-54 676

7-55 1878

7-56 2845

7-57 4012

7-58 4900

7-59 7512

7-60 1842

7-61 9025

7-62 3440

7-63 2423

7-64 865

7-65 1600

Chapter 8

8-1 $45

8-2 $197.18

8-3 $10.62

8-4 $206.05

8-5 $313.70

8-6 $9764.28

8-7 $2406.15

8-8 $29,675.42

8-9 $185.52

8-10 $256.90

8-11 $155.70

8-12 $31.00

8-13 $579.75

8-14 $120.38

8-15 $8.70

8-16 9%

8-17 $8\frac{1}{2}\%$

8-18 11%

8-19 10%

8-20 bank A

Index
of
Symbols

Index